Engineering Geology for Underground Works

Paola Gattinoni • Enrico Maria Pizzarotti
Laura Scesi

Engineering Geology for Underground Works

Springer

Paola Gattinoni
DICA
Politecnico di Milano
Milan
Italy

Laura Scesi
DICA
Politecnico di Milano
Milan
Italy

Enrico Maria Pizzarotti
Pro Iter S.r.l.
Milan
Italy

ISBN 978-94-007-7849-8 ISBN 978-94-007-7850-4 (eBook)
DOI 10.1007/978-94-007-7850-4
Springer Dordrecht Heidelberg New York London

Library of Congress Control Number: 2014930757

Printed on acid-free paper

Springer is part of Springer Science+Business Media (www.springer.com).

Preface

The construction of tunnels involves the resolution of more or less complex technical problems depending on the geological and geological—environmental context in which the work fits.

Only a careful analysis of all the geological and geological—environmental issues and a correct reconstruction of their conceptual model, can lead to optimal design solutions from all points of view (including financial) and to ensure safety to the workers during construction, and to users, in the operation phase.

Therefore, the need to collect the synthesis of current knowledge about underground excavations in a volume is felt, especially with respect to: the geological and environmental issues related to the construction of underground works (Chaps. 1 and 2); the different methodologies used for the reconstruction of the conceptual model (Chap. 3); the underground excavation analysis (Chap. 4); the different risk typologies that it is possible to encounter or that can arise from the underground construction and the most important risk assessment, management and mitigation methodologies that are used in the underground work planning (Chaps. 5 and 6); the ground structure interaction (Chap. 7) and the characteristics and the equipment of the monitoring activity, which should be performed during an underground excavation (Chap. 8).

The authors are aware that the aim of this book is only to introduce the problems related to the construction of underground works rather than finding the solutions from them all and to provide readers useful concepts for a correct scientific approach to the subject.

Acknowledgments

The authors would like to thank Anna Agnieszka Surma for her contributions on specific topics, critical comments and corrections; Francesco Mungo, who helped to realize drawings and tables; Edvige Meardi for the not simple translation in English language of the text.

Contents

1 Geological Problems in Underground Works' Design and Construction 1
1.1 Introduction ... 1
1.2 Lithological and Structural Features ... 2
1.2.1 Lithological Features ... 3
1.2.2 Structural Features ... 5
1.3 Tectonic Setting ... 7
1.3.1 Faults ... 8
1.3.2 Folds ... 8
1.4 Scale Effect ... 10
1.5 In Situ Stress State ... 11
1.6 Morphological Conditions ... 14
1.6.1 Underground Works at Shallow Depth ... 14
1.6.2 Portals ... 15
1.7 Hydrogeological Setting ... 18
1.7.1 Aggressive Waters ... 19
1.8 Weathering and Swelling Phenomena ... 21
1.8.1 Weathering ... 21
1.8.2 Swelling ... 22
1.9 Geothermal Gradient ... 22
1.10 Seismic Aspects ... 23
1.11 Gas, Radioactivity and Hazardous Materials ... 25
1.11.1 Gas ... 25
1.11.2 Radon ... 27
1.11.3 Asbestos ... 28
References ... 29

2 Environmental-Geological Problems due to Underground Works ... 31
2.1 Introduction ... 31
2.2 Surface Settlements ... 32
2.3 Slope Instability ... 36
2.4 Interaction with Surface Water and Groundwater ... 38

2.5 Inert Waste ... 45
2.6 Noises and Vibrations During Excavation ... 47
References ... 50

3 Geological Conceptual Model for Underground Works Design ... 53
3.1 Introduction ... 53
3.2 Geological Studies and Investigations ... 54
3.2.1 Characterization of Shallow-Overburden Stretches ... 55
3.2.2 Characterization of Medium-High Overburden Stretches ... 55
3.2.3 Hydrogeological Surveys ... 56
3.3 Geological-Technical Characterization ... 58
3.4 Geomechanical Classifications ... 62
3.4.1 Bieniawski Classification (or of the RMR Index, Only Relevant for Rock Masses) ... 63
3.5 Rock Mass Excavability Index RME ... 67
3.5.1 Rock Mass index RMi ... 69
3.5.2 Surface Rock Classification SRC ... 70
3.5.3 Barton Q-System Classification ... 70
3.5.4 Q_{TBM} Classification System ... 81
3.6 Hoek-Brown Constitutive Model for Rock Mass ... 81
3.7 Strength of Discontinuities ... 91
3.7.1 Patton Criterion ... 91
3.7.2 Barton Equation ... 92
3.7.3 Ladanyi and Archanbault Criterion ... 94
References ... 96

4 Underground Excavation Analysis ... 97
4.1 Introduction ... 97
4.2 Discontinuous Medium and Equivalent Continuum ... 98
4.3 Convergence and Confinement ... 98
4.4 Underground Works at Shallow and Great Depth ... 104
4.5 Analysis Methods of the Excavation Behaviour ... 105
4.5.1 Block Theory ... 105
4.5.2 Characteristic Lines ... 106
4.5.3 Numerical Methods ... 109
4.5.3.1 Distinct Elements Method ... 110
4.5.3.2 Finite Elements or Finite Difference Methods ... 111
4.6 Squeezing and Time-Dependent Behaviour ... 111
4.6.1 Singh et al. (1992) Empirical Approach ... 112
4.6.2 Goel et al. (1995) Empirical Approach ... 113
4.6.3 Hoek and Marinos (2000) Semi-Empirical Method ... 114
4.6.4 Jehtwa et al. Method (1984) ... 115
4.6.5 Bhasin Method (1994) ... 116
4.6.6 Panet Method (1995) ... 117

4.7 Rock Burst ... 119
4.8 Face Stability Assessment ... 120
4.8.1 Shallow Overburden ... 121
4.8.1.1 Undrained Behaviour of Cohesive Soils ... 121
4.8.1.2 Grain Material with Drained Behaviour ... 123
4.8.1.3 Stability of the Excavation Face by Tamez (1985) ... 123
4.8.2 High Overburden ... 127
4.8.2.1 Face Stability as a Function of Characteristic Strength of Rock Mass ... 127
4.8.2.2 Face Stability with Convergence–Confinement Method ... 128
4.8.2.3 Face Stability as a Function of Shear Strength ... 128
4.8.2.4 Face Stability in Relationship to the Tensional Field and Mechanical Characteristics of Rock Masses ... 129
4.8.2.5 Face Stability with the Ground Reaction Curve Method ... 129
4.8.2.6 Face Stability Caquot Method ... 131
4.9 Ground Water Influence ... 132
4.9.1 Assessment of Tunnel Inflows ... 132
4.9.1.1 The Draining Process from an Advancing Tunnel ... 135
4.9.2 The Influence of Water on the Mass Behaviour ... 137
References ... 141

5 Geological Risk Management ... 143
5.1 Introduction ... 143
5.2 Definitions and General Concepts ... 147
5.3 Geological Risk Assessment for Underground Works ... 148
5.3.1 Qualitative Methods for Risk Analysis ... 149
5.3.2 Quantitative Methods for Risk Analysis: Safety Methods ... 149
5.3.3 Monte Carlo Method for Quantitative Risk Analysis ... 154
5.3.4 Risk Evaluation ... 155
5.4 Applicative Example: The Decision Aid in Tunnelling (DAT) ... 157
5.5 From Risk Assessment to Risk Mitigation ... 159
References ... 159

6 Risk Mitigation and Control ... 161
6.1 Introduction ... 161
6.2 Excavation Methods ... 161
6.2.1 Shielded and Pressurized TBM ... 164
6.3 Injections ... 170
6.3.1 Injections via Impregnation and Fracturing ... 170
6.3.2 Jet-Grouting ... 175

6.4 Freezing ... 178
6.5 Cutter Soil Mix (CSM) ... 180
6.6 Anchors ... 180
6.6.1 Nails ... 183
6.6.2 Bolts ... 190
6.6.3 Tiebacks ... 190
6.7 Drainage ... 191
6.8 Reinforced Protective Umbrella Methods (RPUM) ... 192
6.8.1 Forepoling ... 194
6.8.2 Jet-grouting Vaults ... 195
6.8.3 Precutting ... 196
6.8.4 Pretunnel ... 198
6.9 Linings ... 198
6.9.1 First Stage Linings ... 201
6.9.1.1 Shotcrete ... 201
6.9.1.2 Steel Ribs ... 204
6.9.2 Final Linings ... 206
6.9.2.1 In Situ Cast Concrete (Unreinforced and Reinforced) ... 206
6.9.2.2 Waterproofing and Water Management Systems ... 207
6.9.2.3 Prefabricated Linings ... 208
6.9.2.4 Single-Shell (Monocoque) Linings ... 213
References ... 214

7 Ground-Structure Interaction ... 215
7.1 Rabcewicz Theory ... 215
7.2 Method of Hyperstatic Reactions ... 216
7.3 Evaluation of the Loads Acting on the Linings ... 219
7.3.1 Vertical Loads ... 219
7.3.1.1 Soils: Caquot and Kerisel's (1956) and Terzaghi's (1946) Formulations ... 219
7.3.1.2 Rock masses: Terzaghi's (1946) Classification and Approaches Based on Bieniawski's Characterization ... 223
7.3.2 Horizontal Loads ... 225
7.3.3 Inclined Loads ... 226
7.3.4 Loads Assessment on the Lining in Case of Tunnel Under Groundwater Table ... 228
7.4 Nailing ... 231
7.4.1 Method of the Confinement Pressure ... 232
7.4.2 Homogenization Method ... 233
7.4.3 Modelling of the Cross Section with Continuum Discretization Methods ... 235
7.5 Spiling ... 238
7.6 Forepoling ... 241

7.7 Stabilization of the Excavation Face: Number and Length of the Forepoles 243
7.8 Characteristic Lines: Analysis of the Linings 245
7.9 Numerical Methods 250
7.10 Seismic Aspects 253
7.11 Final Considerations 261
References 262

8 Monitoring 265
8.1 Introduction 265
8.2 Geomechanical Surveys 266
8.3 Measurements of Convergence 267
8.4 Measures of Rock Deformations 272
8.4.1 Face Extrusion 272
8.4.2 Radial Deformations 273
8.5 Measures on Linings 274
8.5.1 Assessment of the Strain with 'Strain Gauges' 274
8.5.2 Assessment of the Stress 277
8.6 Measurements of Pressure and Flow Rate 278
8.6.1 Piezometers 279
8.7 Measures of Acoustic Emissions 281
8.8 Monitoring in Excavation by TBM 282
8.8.1 Measure of the Machine Parameters 282
8.8.2 Geophysical Seismic Surveys 283
8.8.3 Geoelectic Surveys of the Cutting Head (Shielded TBM) . . . 285
8.9 Surface Settlements and Surrounding Infrastructures Monitoring . . . 287
8.9.1 Settlement Gauges and Multibase Extensometers 289
8.9.2 Inclinometers 289
8.9.3 Other Instruments for Buildings and Facilities Monitoring 295
8.9.4 Settlements Monitoring 297

Index 301

Chapter 1
Geological Problems in Underground Works Design and Construction

1.1 Introduction

Underground excavations consist of progressive removal—by different methods, timings and techniques—of natural ground (rock mass or soil) in order to obtain a cavity of chosen shape and size. Before the excavation, the ground is generally in an equilibrium condition in its original state of stress. Therefore, no deformations or displacements occur. The excavation progressively modifies the state of stress in the ground by generating a stress deviation around the cavity, with particular stress concentration close to its boundary surfaces. As a consequence, the ground is forced to reach a new equilibrium state through deformations and, in case of fractured rocks, relative displacements of rock blocks. The magnitude of such deformation phenomena and the related kinematics depends on:

- The shape and the dimension of the cavity
- The method, timing and technique of excavation
- The nature and the original stress state of the ground

In particular, a stable condition can be expected at the cavity opening only for those materials which are defined as self-supporting. This type of condition is possible only due to their good geomechanical features. Materials having self-supporting characteristics are generally massive or slightly fractured rock masses or fractured rock masses in which the release of blocks is prevented (i.e. characterised by high shear joints strength or by favourable joints orientation).

The behaviour of the mass being excavated essentially depends on three main aspects: first of all, on the lithological nature, which determines the mechanical characteristics of the matrix; then, on the structural features (stratification, schistosity, fracturing etc.) which determine the mechanical properties of the mass itself; and lastly, on the state of stress existing before the excavation. In particular, the variation of the above-described factors can induce a broad spectrum of instability and deformation phenomena, from the already mentioned kinematics of rock blocks (Fig. 1.1) to major cavity wall movements both in brittle (rock burst, Fig. 1.2) or ductile (squeezing, Fig. 1.3) conditions.

P. Gattinoni et al., *Engineering Geology for Underground Works*,
DOI 10.1007/978-94-007-7850-4_1,

Fig. 1.1 Collapse due to block instability. (By Pizzarotti)

Fig. 1.2 Collapse under severe rock-burst conditions. (Hoek and Brown 1980)

Furthermore, soils and rocks are multiphase media; consequently, another factor affecting their behaviour during the excavation is related to the groundwater presence and flowing, which depend on the hydrogeological characteristics of the medium.

Last but not the least, other aspects also can be very relevant for good underground construction performance. These can be the location of the excavation in relation to the topographic surface, risk of natural gas finding, presence of aggressive water, weathering and swelling minerals, increase in temperature with depth (geothermal gradient), seismicity, radioactivity and the presence of hazardous minerals.

1.2 Lithological and Structural Features

From an engineering point of view, the geomechanical quality of a rock mass is the set of properties that affectsits behaviour, for example, when an underground excavation is opened. In Chaps. 3 and 4, the main and the most used methods to

Fig. 1.3 Heavy deformation due to intense squeezing. (Agostinelli et al. 1995)

assess the geomechanical quality of a rock mass will be described, as well as some methods that allow a rapid and preliminary evaluation of the excavation behaviour. In general, the lower the geomechanical quality of the rock mass, the more the problems during the excavation within. It is obvious that the most favourable conditions, from the static point of view for the excavation of an underground cavity, exist in the presence of massive rocks (i.e. not significantly disjoined, fractured or laminated) that have high mechanical strength. On the contrary, if the cavity has to be excavated in soft or highly fractured rocks or, in an extreme condition, in soils, precarious stability conditions always occur.

As stated before, the rock quality, and thus the rock mass behaviour, is influenced both by lithological nature that affects the strength of the rock matrix and by structural features.

1.2.1 Lithological Features

The geomechanical behaviour of the rock mass depends primarily on its lithological features, e.g. its mineralogical-petrographic composition and on the type of process which generated the lithology itself.

The magmatic rocks (with the exception of pumice and obsidian) and the metamorphic non-schistose rocks are generally of lithological types with the best strength characteristics; considering the same fracturing and weathering conditions, massive sedimentary rocks rank second, followed by metamorphic schistose ones, highly stratified sedimentary rocks and, at last, soils.

Table 1.1 Range of uniaxial compressive strength (UCS) for some common rock materials

Term for uniaxial compressive strength	Symbol	UCS Strength (MPa)	Range for some common rock materials				
			Granite, basalt, quarzite, marble	Schist, sandstone	Limestone, marl	Claystone slate	Soil
Extremely weak	EW	0.25–1				X	X
Very weak	VW	1–5		X	X	X	X
Weak	W	5–25		X	X	X	
Medium strong	MS	25–50	X	X	X	X	
Strong	S	50–100	X				
Very strong	VS	100–250	X				
Extremely strong	ES	> 250	X				

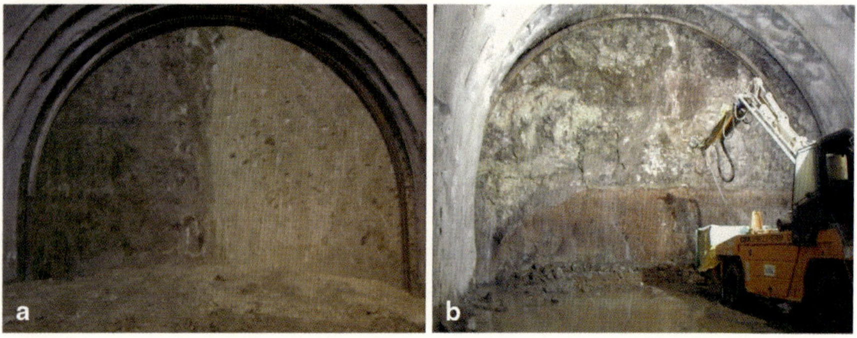

Fig. 1.4 Examples of mixed lithology sections (by Pizzarotti): **a** on the face of the tunnel excavation, a tectonic contact is clearly visible between alternation of basalts and vulcanoclastiti (left) and calcarenites with a high degree of cementation (right), **b** on the face of the tunnel excavation, contact between Plio-Pleistocene basic vulcanites (low) and an alteration layer (paleo soil, high) can be observed

According to the Basic Geotechnical Description given by ISRM (1980), the parameters used to define the limits between soils, weak rocks and hard rocks are uniaxial compressive strength and cohesion (Table 1.1).

The materials having cohesion lower than 0.3 MPa and uniaxial compressive strength less than 2 MPa are classified as soils; the materials with compressive strength between 2 and 20 MPa are defined as "weak rocks", while the materials with uniaxial compressive strength higher than 20 MPa are considered "hard rocks".

From a purely lithological point of view, a rock is "weak" because of the weak links among its components (for example shales, siltstones, marls, chalks, phyllites etc.)

The technical behaviour of a rock mass can be also be affected by the simultaneous presence of different lithology in the same cavity stretch (Fig. 1.4). This can be a factor causing instability or major difficulties during the advancement.

1.2.2 Structural Features

A further very important factor that affects the behaviour of the rock mass is undoubtedly its structural setting. It depends on:

- The processes that led to the formation of the different types of rock; they generate primary structural weaknesses, such as layering, schistosity or cooling joints.
- The tectonic phenomena to which rocks were subjected during their geological history; in this case, secondary structural weaknesses develop in different ways depending on the brittle or ductile response and on the stress acting on the rock mass.

It is evident that the type of response depends on the lithology, on the conditions of temperature and pressure and on the duration of deformation events. It is therefore essential to collect all data related to the following structural characteristics: geometry (inventory of all brittle or ductile structures), kinematics (examination of the displacements and movements that led to the change of position, orientation, size and/or shape of the rock bodies) and dynamics (reconstruction of the nature and orientation of the stresses that produced the deformation).

In presence of bedded and/or fractured rock masses, the following parameters should be carefully evaluated:

- The layer thickness and/or the fracturing degree, i.e. number of fractures per linear meter, or rather the inverse of the distance between the discontinuities (strata or fractures)
- The joint characteristics (persistence, roughness, aperture, filling, alteration etc.)
- The joint orientation relative to the walls of the underground cavity

Taking as a reference, by way of example, a family of discontinuities (i.e. the bedding) the following cases can be schematically analyzed:

- Horizontal layers (Fig. 1.5): The issues are becoming more pronounced with the thinning of the layer thickness. In particular, if the layers are constituted by banks of high thickness, a behaviour similar to that of massive rock masses can be expected (especially if the more resistant banks are located at the ceiling and along the sides); if the layers are thin, or even worse if they have reduced strength, instability at ceiling will be frequent, caused by flexural break of the layers.
- Sub-vertical layers (Fig. 1.6): If a generic cross-section of a cavity of undefined length (tunnel) is considered, conditions are much more favourable in case of interception of layers whose direction is perpendicular to the axis: in each crossed layer, the stresses can be laterally deviated with respect to the ceiling (arch effect), as in an intact rock; as the angle between the tunnel axis and the layer direction decreases, conditions gradually become more unfavourable with the development of failure phenomena of the layers (especially in presence of thin layers with low shear strength of the joints) caused by load concentration on the sides. Obviously, similar conditions are present at the face in case of a tunnel developing perpendicularly to the direction of the layers.

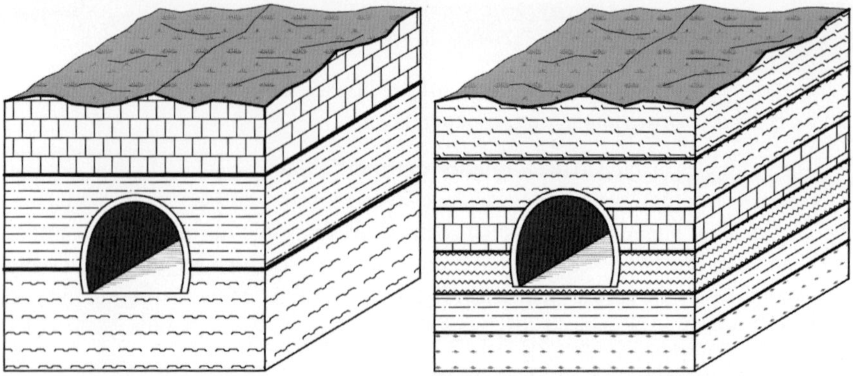

a b

Fig. 1.5 Tunnels excavated in horizontally stratified rock masses: **a** high thickness of the layers, **b** thin layers

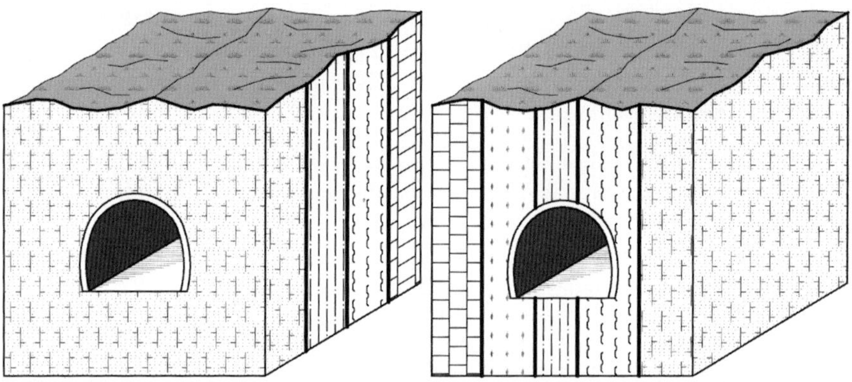

a b

Fig. 1.6 Tunnels excavated in vertically stratified rock masses: **a** tunnel axis perpendicular to the layers' direction, **b** tunnel axis parallel to the layers' direction

- Inclined layers (Fig. 1.7): Equilibrium conditions vary considerably depending on the direction of the tunnel axis with respect to the layers orientation. If the cavity is parallel to the direction of the layers ("tunnel in direction"), lateral dissymmetrical and almost continuous deformations or instability phenomena can develop longitudinally. If the tunnel axis is perpendicular to the direction of the layers, these phenomena are distributed symmetrically, whereas it is possible to have a strength change in the longitudinal direction depending on the nature and thickness of the crossed layers. In case of "obliquely" inclined layers, an intermediate situation between the two above-described cases occurs, even in case of prevailing dissymmetrical kinematics and deformations. Moreover, it is evident that in the presence of a low or high dip angle of the layers,

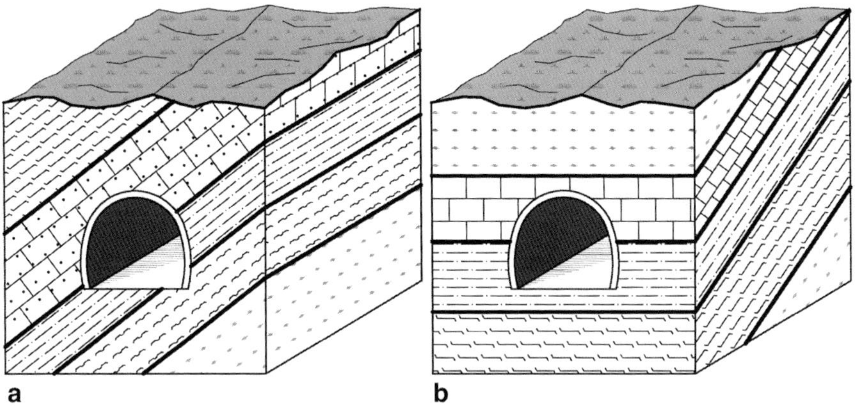

Fig. 1.7 Underground works in rocks with inclined bedding planes: **a** the underground work (in direction) always develops in the same strata: possible kinematics due to bending on the left side and sliding blocks along the layers on the right side; **b** the underground work crosses obliquely the layers for length greater than the layers' thicknesses: possible kinematics due to bending of the layers at the ceiling and sliding along the face

the situation will not be exactly the same as the ones previously described, as the rock mass tends to show a behaviour similar to the already described cases of horizontal or vertical layers.

Finally, it should be noted that all the features described above for bedded rock mass are totally transferable to other situations in which the presence of a systematic disjunctive element confers a layered attitude (cleavage, schistosity, lava plans, cooling layers, etc.) to the rock mass.

The above-outlined concepts also apply in presence of two or more discontinuity systems. In this case, potential mechanisms of sliding and/or falling wedges and, less frequently, toppling must be considered.

1.3 Tectonic Setting

It is well known that the lithosphere is continuously modified by internal forces that tend to deform it. Therefore, the lithosphere is divided into plates that may converge, diverge or scroll side by side. As a consequence, much of the geological hazards (volcanism, earthquakes, continental drift, expansion of the oceans, orogenesis etc.) are results of this interaction between plates.

It is therefore clear that an underground work carried out in a tectonically active area (recurrently the margin of the plates) will meet a stress state that depends, in terms of orientation and intensity, on the prevailing movement between the plates.

In case of divergent or transform tectonic movements, brittle tectonic structures as faults will be generated. If, on the contrary, the movements are convergent, folds and thrusts will frequently develop.

1.3.1 Faults

It is very well known that the presence of faults along the layout of an underground opening can cause significant problems.

If the shear stress along the discontinuity was particularly high, the rock mass became so fractured that can behave like a soil. Such deformations can interest more or less wide bands of rock mass. Particular attention is given to these fracture zones within underground works, since they are usually affected by the toughest structural-geological and hydrogeological problems. Such materials at the opening of the cavity often have limited, if any, self-supporting features.

Moreover, fracture zones frequently form preferential paths for groundwater: Therefore, water inflows, also of significant extent, are quite common in those situations. Similarly, the presence of major discontinuities may allow harmful gases to channel inside them and reach the excavation. Materials originated in correspondence of friction zones are defined as "fault rocks" and distinguished according to the classification in Table 1.2 (from Sibson 1977, modified).

Due to the above-mentioned problems, during the preliminary geological survey, it is important to accurately define the presence of tectonised zones in the area involved by future underground works. If a cataclastic band is intercepted, this should be crossed as orthogonally as possible in order to minimize its interference with the cavity.

The presence of overthrusts may cause similar problems. In this case, the low dip angle of the tectonic element implies the retrieval of poor material during the excavation of particularly long stretches.

1.3.2 Folds

The interception of a fold structure by underground works causes some particular consequences from the structural point of view, such as dissymmetry of the defromation and lithological inhomogeneity.

Folds can also contain residual stresses; there are, in particular, compressive stresses in correspondence of the core and tensile at its hinge.

If the folds are located at great depth, the residual stresses can be particularly high due to the difficulty of geological units to stress release because of the presence of heavy lithostatic confinement.

Therefore, it is extremely important to know not only where underground works intercept a fold, but also the fold type (Fig. 1.8): for example, the crossing of a syncline along its axial plane involves strong lateral stresses and important water inflows, while crossing an anticline in its hinge can facilitate releases and collapses at the ceiling and sides deformations.

Table 1.2 Texture classification and deformation type of fault rocks

"Random fabric" rocks				Foliated rocks		
Brittle deformation				Ductile deformation		
Not cohesive	Fault breccia (if the rock fragments are more than 30 %)		Cataclasis	Foliated gouge	Cataclasis	
	Fault gouge (if the rock fragments are less than 30 %)		Cataclasis			
Cohesive	Pseudotachilite (vitreous rock)		Frictional melting	Tectonites	Dissolution reprecipitation	
	Breccia	0–10 % matrix	Cataclasis	Mylonites		
	Protocataclasites	10–50 % matrix	Cataclasis	Protomylonites	10–0 % matrix	Intracristallin plasticity
	Cataclasites	50–90 % matrix	Cataclasis	Mylonites	50–0 % matrix	Intracristallin plasticity
	Ultracataclasites	90–100 % matrix	Cataclasis	Ultramylonites	90–100 % matrix	Intracristallin plasticity

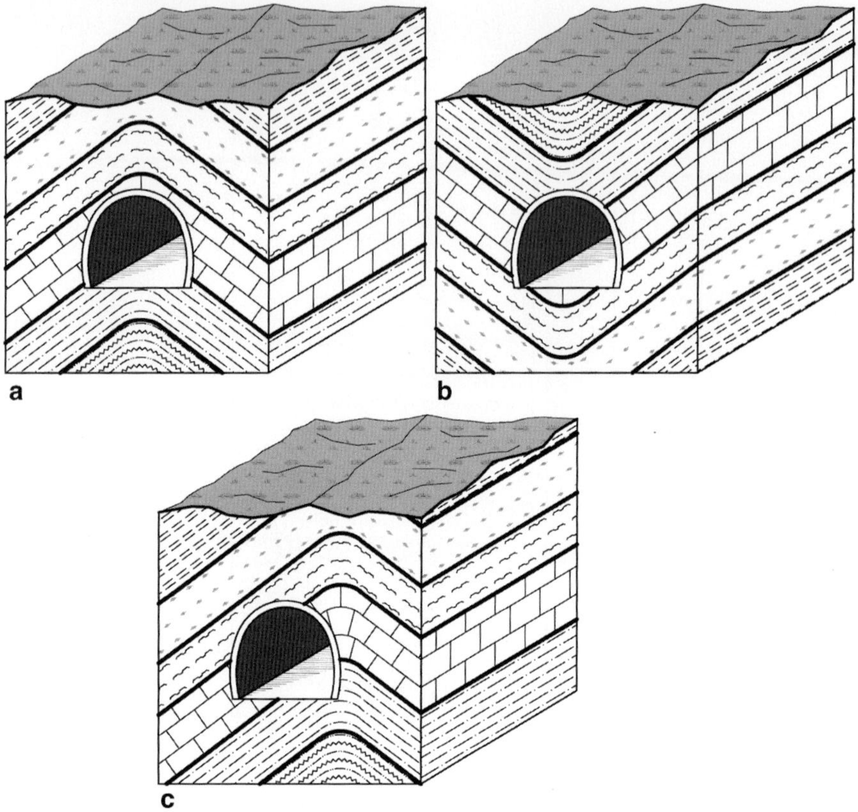

Fig. 1.8 Relation between underground works and folds: **a** tunnel at the anticline core; **b** tunnel at the syncline core; and **c** tunnel at the syncline hinge

1.4 Scale Effect

Strength features of rock masses are highly dependent on the scale of analysis. If the underground cavity size is small with respect to the joint spacing, the number of intercepted discontinuities is reduced. Then, the intact rock behaviour assumes great importance. On the contrary, if the tunnel diameter increases with respect to the joint spacing, the role of the discontinuities becomes more and more important in defining the rock mass behaviour. In this case, the strength of a joint rock mass depends on the properties of the intact rock blocks and also on the freedom of these blocks to slide and rotate under different stress conditions.

Of course, when defining the scale effect, the degree of fracturing of the rock and the size of the cavity have to be considered.

In general, it is reasonable to suggest that, when dealing with large-scale rock masses, the strength will reach a constant value when the size of individual rock

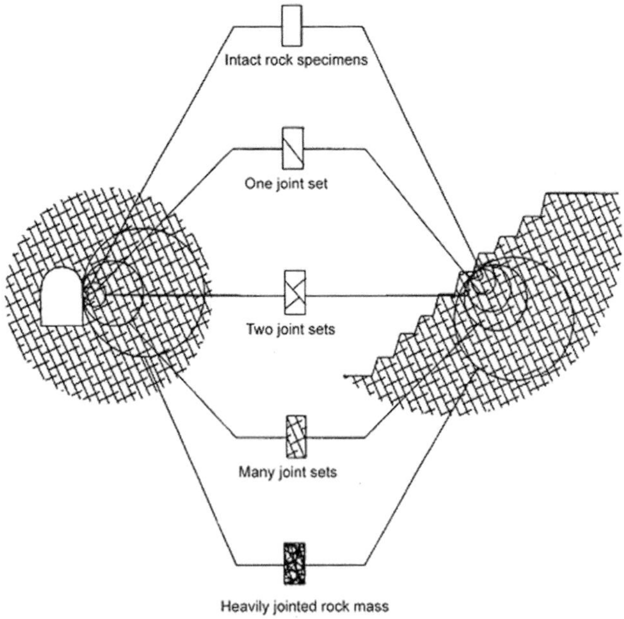

Fig. 1.9 Idealised diagram showing the transition from intact to a heavily jointed rock mass with increasing sample size. (From Hoek 2013, modified)

blocks is sufficiently small in relation to the overall size of the cavity being considered. This suggestion is embodied in Figure 1.9, which shows the transition from an isotropic intact rock specimen, through a highly anisotropic rock mass in which failure is controlled by one or two discontinuities, to an isotropic heavily jointed rock mass.

1.5 In Situ Stress State

Rock at depth is subject to stresses resulting from the weight of the overlying strata and from locked-in stresses of tectonic origin.

The weight of the vertical column of rockresting on a rock element is the product of the depth and the unit weight of the overlying rock mass (Fig. 1.10).

The horizontal stresses acting on an element of rock at a depth z below the surface are much more difficult to estimate than the vertical stresses. Measurements of horizontal stresses at civil and mining sites around the world show that the ratio of the average horizontal stress to the vertical stress tends to be high at shallow depth and that it decreases at depth (Hoek and Brown 1980; Herget 1988). Sheorey (1994) provided simplified equation which can be used for estimating the horizontal to

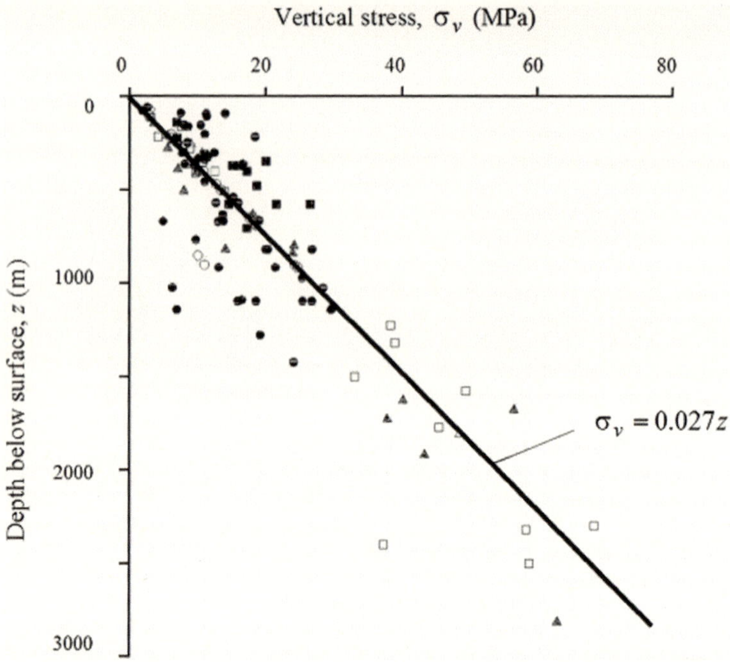

Fig. 1.10 Vertical stress measurements from mining and civil engineering projects around the world. (Modified from Hoek and Brown 1980)

vertical stress ratio k:

$$k = 0.25 + 7E_h(0.001 + 1/z)$$

where z (m) is the depth below surface and E_h (GPa) is the average deformation modulus of the upper part of the earth crust measured in a horizontal direction.

The Sheorey's theory does not explain the occurrence of measured vertical stresses that are higher than the calculated overburden pressure, the presence of very high horizontal stresses at some locations or why the two horizontal stresses are seldom equal. These differences are probably due to local topographic and geological features, strictly related to the tectonic setting (see Sect. 1.3). In this regard, the World Stress Map will give a good first indication of the possible complexity of the regional stress field and possible directions for the maximum horizontal compressive stress. A map showing the orientation of the maximum horizontal compressive stress for the Mediterranean is reproduced in Fig. 1.11. Afterwards, the results of in situ stress measurements can be used to refine the analysis. Where regional tectonic features such as major faults are likely to be encountered, the in situ stresses in the vicinity of the feature may be rotated with respect to the regional stress field and the stresses may be significantly different in magnitude from the values estimated from the general trends.

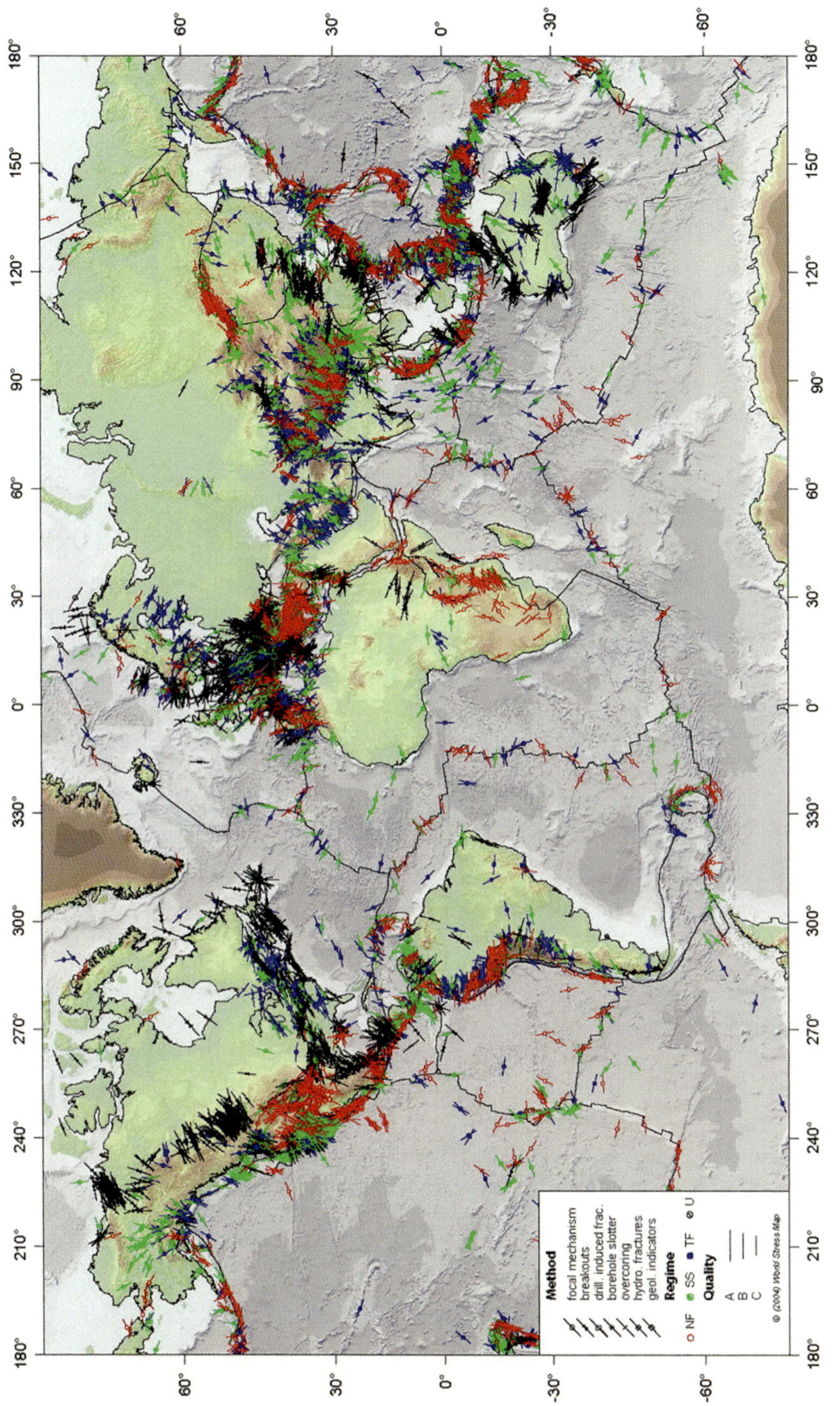

Fig. 1.11 World stress map giving orientations of the maximum horizontal compressive stress. (From www.world-stress-map.org)

Other relevant modifications of the lithostatic stress state also at great depth can be linked to the surface morphology, that is to the position of the underground work with respect to the side or to the valley ridge and to the morphodynamic evolution of the site, e.g. the presence of glaciers in the past.

When an opening is excavated in this rock, the stress field is locally disrupted and a new set of stresses are induced in the rock surrounding the opening. Knowledge of the magnitudes and directions of these in situ and induced stresses is an essential component of underground excavation design since, in many cases, the strength of the rock is exceeded and the resulting instability can have serious consequences on the behaviour of the excavations.

1.6 Morphological Conditions

Morphological conditions also play an important role during the execution of underground works. For this reason, it is important to distinguish among shallow, deep underground works and tunnels close to the side of the slope, and to analyze the different geomorphological problems characterizing each type. In addition to these aspects, specific problems present in portal areas must be taken into consideration.

1.6.1 Underground Works at Shallow Depth

Underground works can be referred to as "shallow" when the disturbed area around the tunnel interferes with the ground surface. This situation may lead to instability also involving surface materials, with serious effects on general environmental equilibrium.

As an indication, these situations can take place when the overburden thickness is less than four times the excavation diameter.

Shallow underground works are also strongly affected by meteoric events; therefore, they are often subject to significant water inflows, also depending on material permeability. Water inflows may also result in alterations that are responsible for a weathering of the mechanical properties of rocks and soils (Figs. 1.12 and 1.13).

Due to these reasons, the construction of shallow underground works is often preceded by the implementation of systematic consolidation measures which, in the more critical cases, allow the improvement of soil mechanical properties before the excavation.

Shallow underground works are also strongly affected by topography and surface loads. An example is that of tunnels which extend within a valley side with a pattern transversal to the valley itself and maintain particularly low overburden conditions in correspondence to the downstream side.

These tunnels are affected by the same problems described for shallow underground works, but in addition, they are also affected by dissymmetrical stress

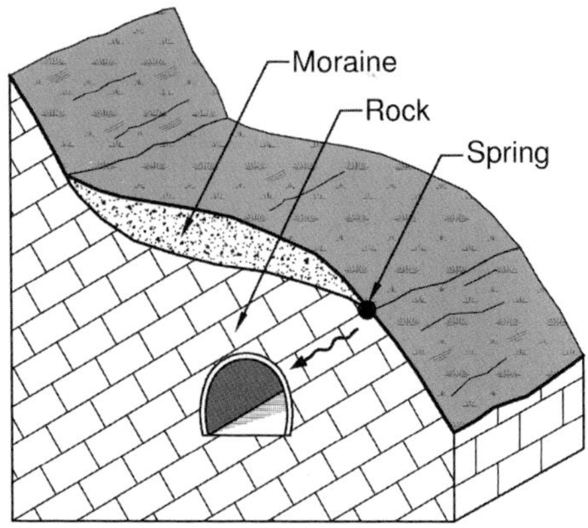

Fig. 1.12 Debris and glacial deposits allow the groundwater flow through the fractures towards the tunnel

distributions and by consequent deformation phenomena, which lead to design and construction difficulties depending on layers' arrangement and fracturing degree (Figs. 1.14–1.16).

For deep underground works (overburden approximately four times greater than the tunnel diameter), geomorphological conditions progressively lose their importance, unless the work is located in very steep slopes or on the boundary of glacial valleys, where the influence of surface morphology and morphodynamics can have effects even at a great depth. Another exception is constituted by areas characterized by deep-seated landslides or important karst phenomena.

1.6.2 Portals

Among geomorphological issues, tunnel portal areas also must be examined in detail, since they are characterized by specific problems independent from the excavation.

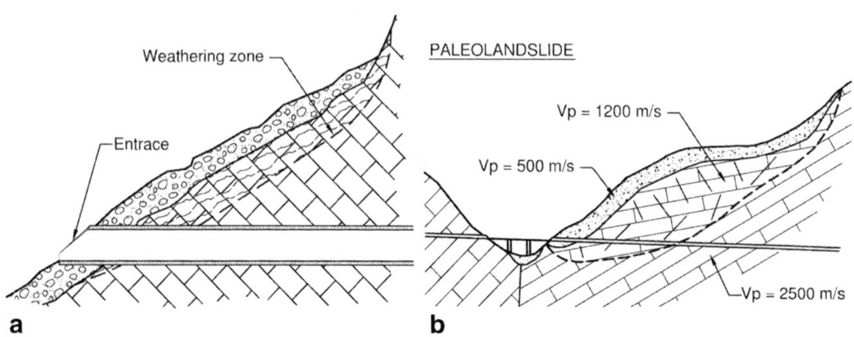

Fig. 1.13 Debris instability condition (**a**) and a paleo landslide (**b**) at the tunnel portal

Fig. 1.14 An example of a shallow tunnel along the side of a valley. (By Pizzarotti)

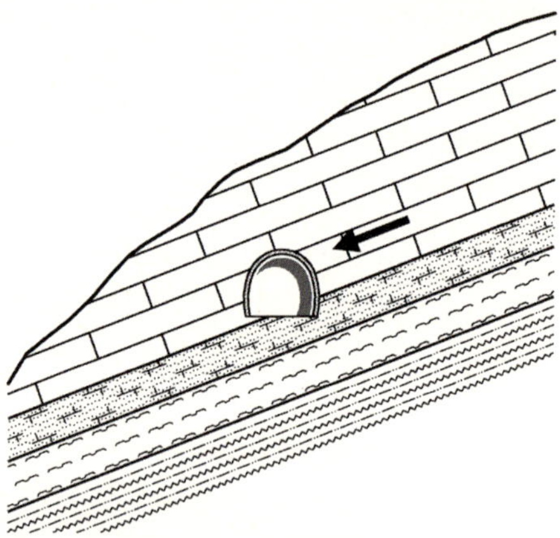

Fig. 1.15 Dissymmetric stresses affecting a shallow tunnel along slope

The underground excavation disturbs the pre-existing equilibrium condition of a natural slope as a notch is realized to host the portal (Fig. 1.17).

In presence of rock masses, problems at tunnel portals can be ascribed to decompression, alteration or fracturing phenomena undergone by the rocks. On the contrary, if the portal is excavated in loose materials (slope debris, glacial deposits, etc.), issues are to be found in the poor geotechnical properties of these soils (e.g. their low cohesion) and in slope steepness. These factors strongly influence slope stability and groundwater flow, which may require peculiar works to stabilize excavation in portal areas. These works can range from punctual excavation support (for example the nailing of unstable rock blocks) to actual support of the side (Fig. 1.18), stretches of artificial tunnels, etc. (Fig. 1.19).

In any case, portal excavation should be performed frontally to the slope, or at least keeping the highest possible angle of incidence, since this condition greatly facilitates the achievement of a new equilibrium.

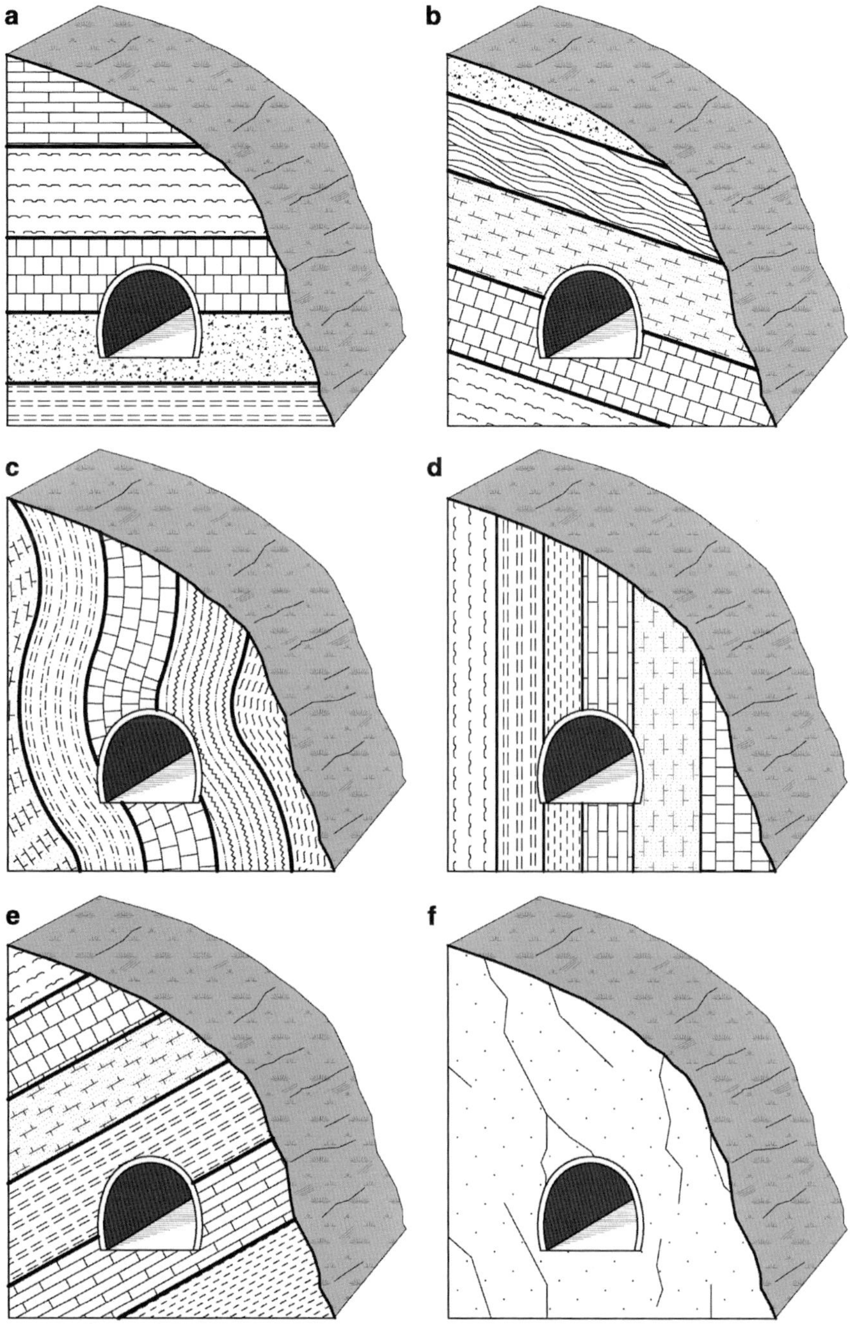

Fig. 1.16 Stability conditions for shallow tunnel along slope in relation to joint orientation: **a**, **d** and **e** very stable; **c** quite stable; **b** and **f** instable

Fig. 1.17 An example of a tunnel portal with nailing of the slope. (By Pizzarotti)

1.7 Hydrogeological Setting

As far as hydrogeological conditions are concerned, groundwater inflow during underground excavations is a very common occurrence: Therefore, it is important to envisage the hydrogeological situations which could lead to significant water inflow.

Factors favouring water inflow within the excavation are related to (Fig. 1.20):

- High permeability materials (granular soils, rocks permeable for porosity or fracturing etc.)
- Sudden changes in permeability
- Tectonic structures (faults, overthrust etc.) having a great water supply
- Karst phenomena
- Syncline folds
- Buried river beds

Fig. 1.18 Stabilization of a tunnel portal by retaining walls with anchors **a** during construction, **b** final configuration. (By Pizzarotti)

Fig. 1.19 An example of a portal stabilized by means of retaining walls with anchors **a** during construction, **b** final configuration. (By Pizzarotti)

The depth of the underground work with respect to the groundwater table has to be considered, as well as the characteristics of the aquifer itself. If the tunnel is located above the water table, problems due to water inflow are small and basically connected with water reaching the excavationby infiltration or percolation. Only in karstified rock masses a large, although temporary, inflow rateis possible even above the piezometric surface. On the contrary, if the excavation develops below the water table, water inflow can become very important and make excavation difficult.

Hence, during the design phase, the identification of tunnel stretches that can be subject to severe hydraulic problems leads to the adoption, both in design and execution phases, of peculiar techniquesaimed at draining, conveying and pumping out the water from the tunnel (Fig. 1.21).

1.7.1 Aggressive Waters

During underground excavation, it is possible to intercept water that can chemically attack the concrete. Its identification during the design phase is of primary importance, because this water could lead to a complete breakdown of the final lining, with very significant economic losses.

This risk is directly related to the lithological features of the rock formations intercepted by the underground work, since aggressive substances are released into groundwater by the geological materials in which the water flows. Aggressive water may originate from grounds other than those directly intercepted by the cavity:

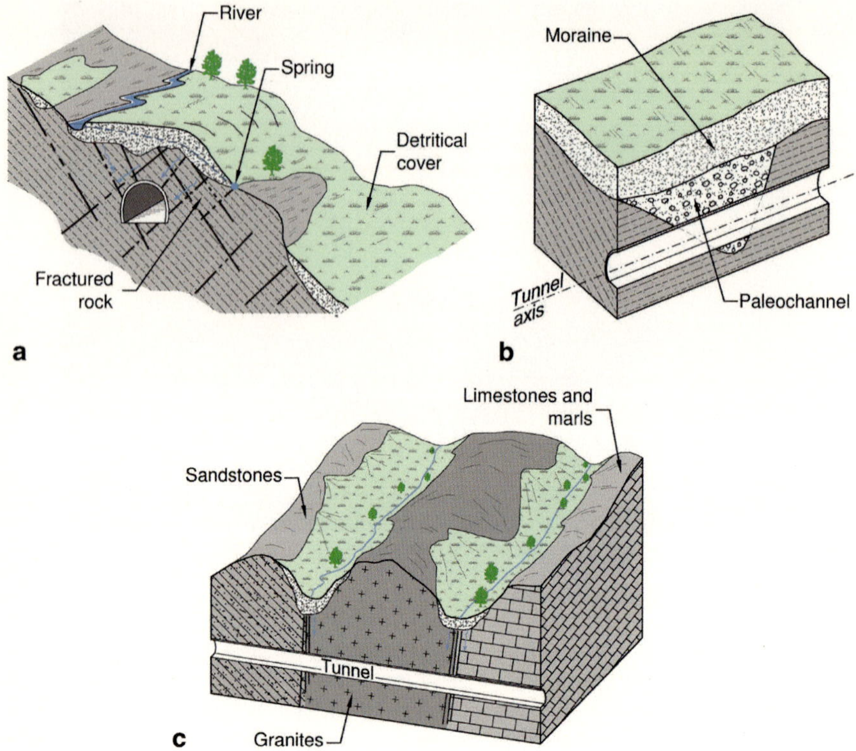

Fig. 1.20 Examples of geological structures which can lead to great groundwater inflow during tunneling: **a** fractured rocks, **b** paleo-river, and **c** faults

therefore, as in the case of gases, the forecasting study should cover a stratigraphic succession reasonably larger than the one directly affected by the work.

The most frequent aggressive water types are schematically listed below, together with the geological deposits generally responsible for their presence:

Selenitic water, that is water rich in calcium sulphate, is by far the most aggressive: it is released by anhyidritic or gypseouslithotypes.

Water rich in sulphuric acid can be encountered in peaty soils or in clayey soils containing pyrite as well as chalk masses.

Water rich in free carbon dioxide can be found in peaty deposits, surface deposits covered with forests and ground hosting mineralized water sources related with magmatic phenomena.

Water rich in chloride and magnesium sulphate (always associated with sodium chloride) can be exceptionally met in grounds belonging to the evaporitic series.

Water with pH lower than 6.5 are to be considered aggressive, too.

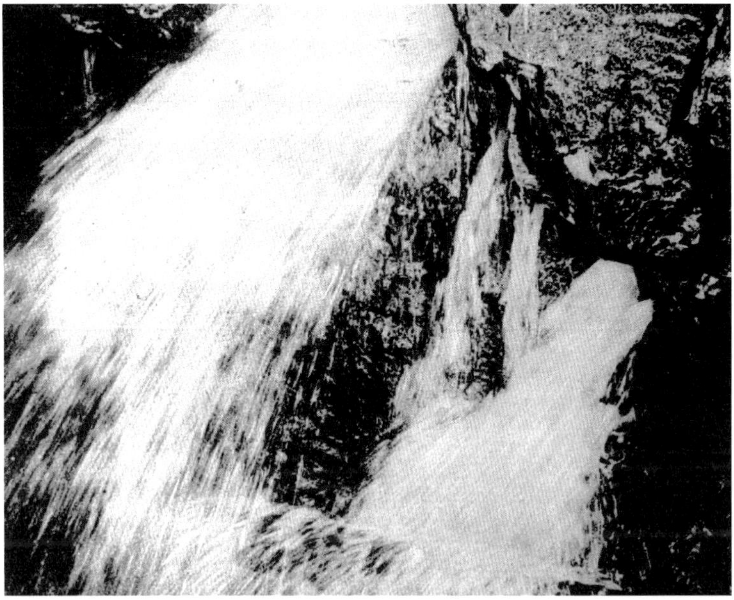

Fig. 1.21 An example of water inflow in the Simplon tunnel (historical photo)

1.8 Weathering and Swelling Phenomena

Water circulation induced by excavation or just by exposure of ground to hygrometric conditions different from the original ones, due to cavity opening, may cause a deterioration of the mechanical properties of soils and rocks, with a consequent reduction of strength characteristics, an increase in deformability or in volume.

The phenomenon of strength loss and deformability increase (softening) due to the change of water content in clayey materials is already well known. Weathering and swelling are further processes that negatively affect excavation stability.

1.8.1 Weathering

In a broad sense, weathering involves two processes (chemical and mechanical) that often act together to decompose rocks:

Chemical weathering involves a chemical change in at least some of the minerals within a rock.

Mechanical weathering involves physically breaking rocks into fragments without changing the chemical make-up of the minerals within it.

Weathering is usually a surface or near-surface process. Nevertheless, it can extend in ground mass if unprotected altered surfaces gradually tend to release, and it can heavily affect stability if it involves the material filling rock mass joints, since it can drastically alter its shear strength. Furthermore, in case of water circulation along joints, it is often associated to transport phenomena of de-structured material within the rock mass, together with volume loss and associated deformation.

1.8.2 Swelling

Some materials expand or contract in response to changes in environmental factors (wet and dry conditions, temperature etc.).

Adsorption or absorption of water due to differences in concentration, unsaturated or partially saturated bonds and differences in potential is frequently associated with a time-dependent volume increase that can lead to swelling phenomena.

In general, swelling phenomena are closely connected to the lithological features of rocks and soils affected by excavation. Most common swelling materials are those containing minerals of the smectite group and, to a lesser extent, of illite group or even kaolinite group.

Swelling phenomena can occur even when anhydrite turns into gypsum due to water imbibition.

These materials, once they are deprived of their natural confinement due to the cavity opening, tend to increase significantly in volume, sometimes even in a quite impressive way. A method to see whether the materials are susceptible to swelling phenomena or not is to analyze their mineralogical composition and to use the graph shown in Figure 1.22 (Bonini et al. 2009).

1.9 Geothermal Gradient

It is well known that temperature and pressure increase with depth. The average pressure increase is about 27 MPa each 1,000 m of depth, whereas the temperature increase (geothermal gradient) is highly variable and ranges from 1.5° C/100 m to 5.0°C/100 m (average 3° C/100 m). This variability is caused, for example, by cold water infiltration due to melting glaciers, by permafrost or cooling magmatic masses at shallow depth, as well as by the local geodynamic evolution.Geothermal gradient is inversely proportional to the thermal conductivity of the involved material: geothermal gradient is approximately 0.05/k (°C/m) where k is the thermal conductivity. The values of temperature and geothermal gradient measured in some famous tunnels are listed in Table 1.3 (see also Fig. 1.23).

The main problem caused by temperature increase is related to working conditions; workers actually operate in optimal conditions at temperatures below 25° C, whereas temperatures above 30° C become unbearable. Thus, the installation of particularly effective ventilation systems becomes essential in such circumstances.

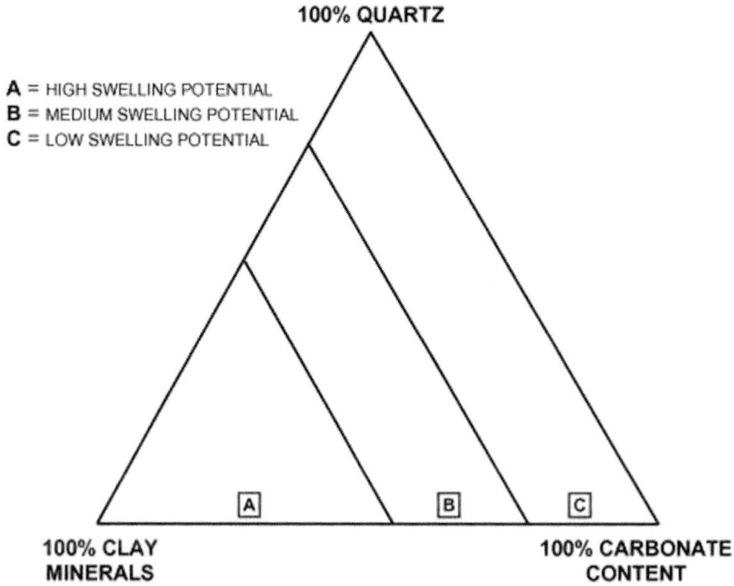

Fig. 1.22 Swelling potential chart

Table 1.3 Examples of geothermal gradient observed during the excavation of some famous tunnels

Tunnel	Geothermal gradient (°C/100 m)	Temperature at different depths
Lötschberg (Switzerland)	~2	34° C at 1500 m
Sempione (Italy–Switzerland)	~2.5	56° C at 2000 m
Gottardo (Switzerland)	~2	30° C at 1500 m
Monte Bianco (Italy–France)	~1.5	30° C at 2000 m

1.10 Seismic Aspects

Studies carried out during earthquakes demonstrated that underground structures have a much lower seismic vulnerability than surface infrastructures (roads, railways, bridges, etc.). Underground works are actually flexible enough to withstand the strains imposed by the surrounding soil without reaching their breaking point. A proper design of lining and section geometry can grant these structures a good seismic behaviour. In any case, even if the seismic behaviour of underground works is usually good, violent seismic events may become hazardous, especially if the surrounding ground is affected by liquefaction phenomena or by displacements along faults. The vulnerability of underground structures to earthquakes depends on the following factors:

- Geological conditions: soils with different stiffness
- Tectonic setting

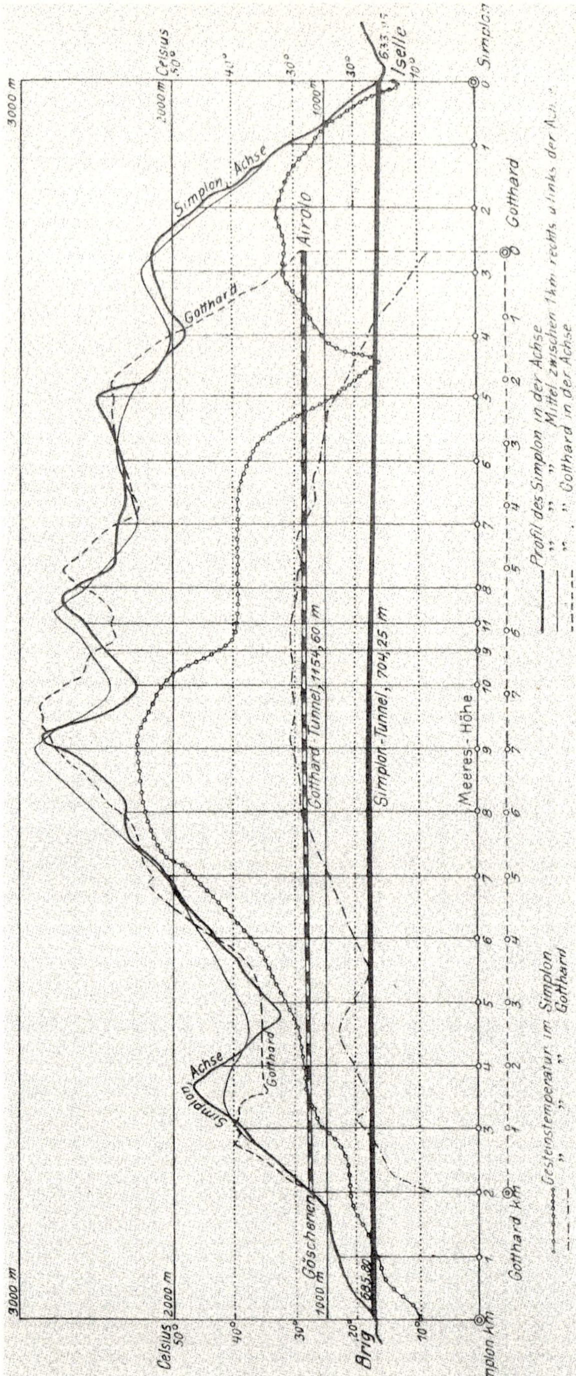

Fig. 1.23 Geothermal gradient–Simplon and Gottard tunnels. (Kovari and Fechtig 1996)

- Depth: observed damage extent decreases with depth
- Location of underground work with respect to the valley side (side of the slope, surface, depth, etc.)
- Size (the larger the section, the greater the seismic vulnerability)
- Section type (with or without invert)

Generally, the main strain types affecting an underground work due to seismic waves are (Fig. 1.24):

- Longitudinal deflection and axial strain: These deformations are observed along underground works having horizontal or predominantly horizontal axis, when seismic waves propagate in parallel or obliquely to the longitudinal axis.
- Cross section ovalization or distortion: They occur when seismic waves propagate in a direction which is perpendicular (or nearly perpendicular) to the longitudinal axis of the underground work.

In any case, tunnel mass is generally small if compared to the mass of the surrounding ground, and the complete confinement of an underground work by the soil allows a considerable damping of the seismic phenomenon.

1.11 Gas, Radioactivity and Hazardous Materials

Materials or minerals can be found inside rock masses that, although of natural origin, may be dangerous both for workers during construction, and for users in the operating phase. They include, in particular, some gases, radon and asbestos.

1.11.1 Gas

Gas retrieval during excavation can cause particularly risky situations for the workers safety, especially if the gas is under pressure. The presence of gas is related to the lithological nature of intercepted formations and/or to the existence of open fractures which may constitute preferential paths for the conveyance of such gases in areas (reservoir rocks) different from source rocks. For this reason, in forecasting studies on the presence of gas, it is important to consider all the stratigraphic and structural elements of the soil volume affected by the work, both directly and indirectly.

The list below contains the gaseous substances that can be found in underground works, together with the rock types most commonly responsible for their presence:

- Methane is an odourless and colourless gas that can easily explode when mixed with air in proportions ranging from 5 to 14 % (mixture called "firedamp" or "grisou"). It is generally contained in carbonaceous rocks, in marshy deposits, as well as in flysches formations rich in clay or belonging to the "scaly clays" rock type.

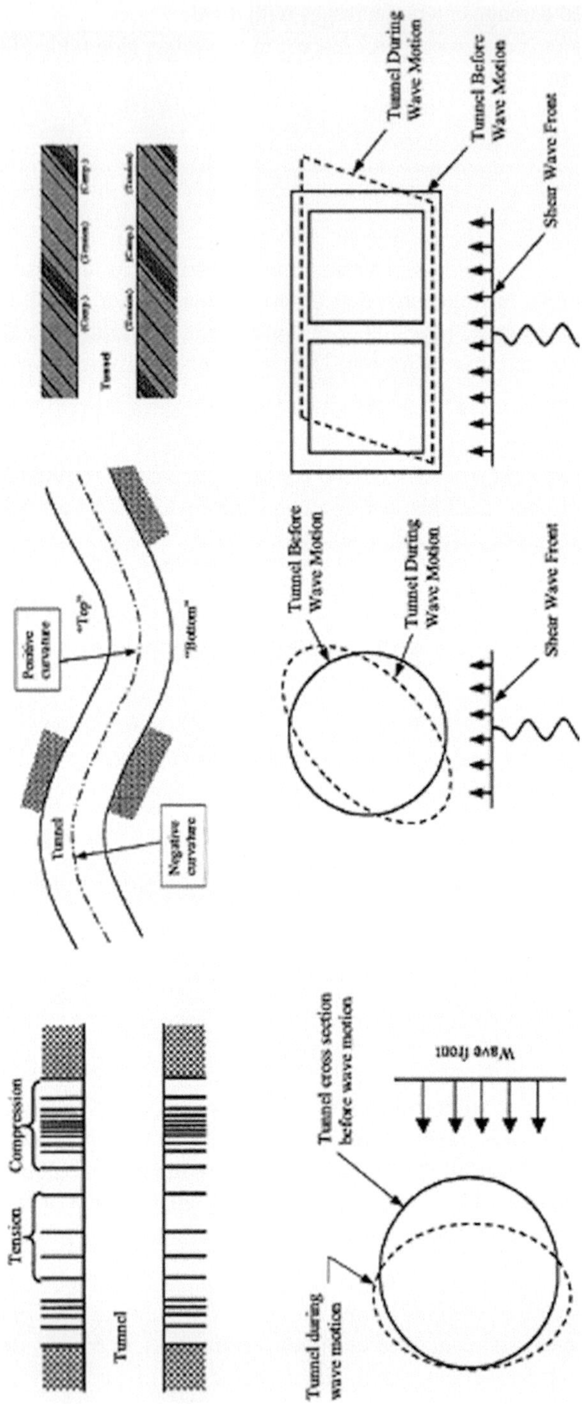

Fig. 1.24 Main types of tunnel deformation as the consequence of an earthquake. (Owen and Scholl 1981)

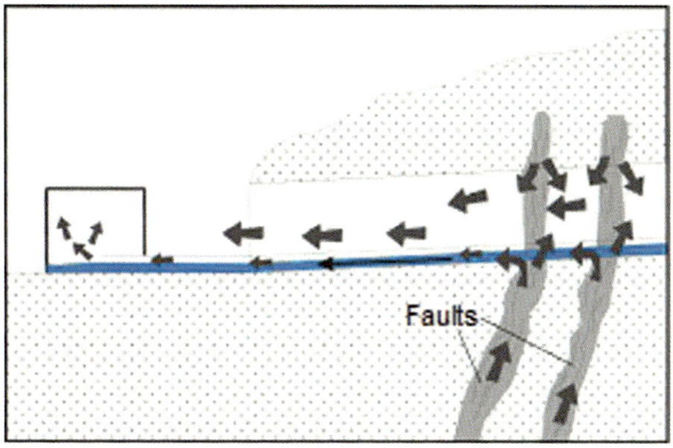

Fig. 1.25 Schematic section of gas and water inflow into a tunnel. The release mechanism of CO_2 through the fractures and faults (in *light grey*) linked to a fault is shown (*black grey arrows*). The water flow (in *blue*) at the tunnel bottom brings parts of the CO_2 towards the cistern, where it gathers

- Carbon dioxide, frequently associated with methane, is an odourless and colourless gas. It is usually found in carbonaceous, organic clayey or volcanic soils. In addition to its poisonousness to humans, it is very aggressive towards concrete.
- Carbon monoxide, colourless and odourless and very poisonous, is mostly present in carbonaceous rocks.
- Nitrogen oxides, similar to carbon dioxide, are usually found in carbonaceous rocks or in rocks containing decaying organic substances, or even within volcanic soils; nitrogen is not toxic but, being lighter than air, it can accumulate in large amounts in the ceiling and cause death by asphyxiation.
- Hydrogen sulphide (H_2S) is a toxic gas, combustible and explosive if mixed with air, characterized by a very unpleasant smell. It is typically connected to volcanic exhalations, but it can also be produced by bacterial reduction from the decomposition of sulphates or sulphur, or released from water containing putrefying organic substances.
- Sulphur dioxide (SO_2) is found, as a rule, in igneous rocks; in addition to its high toxicity, this gas, like carbon dioxide, is aggressive towards concrete.

In presence of gas or in anticipation of its retrieval, underground works must proceed with extreme caution. It is essential to carry out a continuous and thorough monitoring of the air quality in order to allow prompt evacuation in case of danger, to use shielded flame-proof machines and, most important of all, to implement an effective ventilation system. In any case, to overcome these zones, the choice of appropriate operation methods should be taken, considering these situations case by case Fig. 1.25.

1.11.2 Radon

Certain areas have a high natural radioactivity, a phenomenon resulting from high concentrations of radioactive minerals such as, for example, uranium, thorium and all those elements which originate from their radioactive decay.

Radon is quite common among these elements, in particular, its Rn-222 isotope (decay time 3.82 days) belonging to the decay chain of uranium U-238.

Radon is produced by some igneous rocks (e.g. lavas, tuff, pozzolans etc.), by some granites, marbles, marls and flysches that contain uraniferous minerals or radium. The amount of radon released depends on permeability, density and grain size, as well as on soil conditions (dry, wet, frozen, snow covered) and weather conditions (ground and air temperature, atmospheric pressure, wind speed and direction).

At standard temperature and pressure radon is colourless, odourless and water-soluble, whereas its concentration in atmosphere is typically extremely low, since it disperses quickly. Nevertheless, this gas can be conveyed very far from its release point when it is dissolved in a fluid.

The processes governing radon circulation are essentially three: diffusion, convection and transport by a fluid (mostly water).

Diffusion and convection only allow radon migration on a centimetre–meter scale. On the contrary, transport by a fluid can spread the gas over a wider area.

Radon migration in water is obviously affected by permeability and moisture and by a number of geological features, e.g. karstification, fracturing degree and lithology.

The measure unit for radon is Bq/m^3 (Becquerel per cubic metre), which indicates the number of nuclear disintegrations that takes place in a second in a cubic metre of air. If high concentrations of radon in gaseous state are measured in underground works, the problem can be solved by enhancing ventilation processes, whereas if radon is found in collected water, specific water treatments have to be implemented.

1.11.3 Asbestos

Asbestos is a set of naturally occurring silicate minerals (e.g. chrysotile, anthophyllite, tremolite etc.) belonging to the mineralogical groups of serpentine and amphibole. Asbestos can be found either in veins or in small dispersed fibres in ultrabasic rocks (e.g. peridotite) or in metamorphic rocks (serpentines and amphiboles).

It is now widely known that dust containing asbestos fibres is very dangerous for human health. For this reason, when geological survey and investigation indicate the possible presence of asbestos-containing rocks, specific procedures have to be followed in order to:

- Retain dust at the excavation face and during loading and transport of excavated materials (muck). This result is usually achieved by means of systematic watering (eventually using water with surfactant additives). After this process, water must be treated in specific plants. Moreover, once watering is performed, the excavation face must be immediately covered with shotcrete in order to isolate the excavation from the origin of a potential asbestos release.
- Use means of transport with air-conditioned cabs, dust filters and closed load area.

- Monitor air quality next to the excavation face and the grinder. Extractor fans with dust filters must be adopted.
- Apply a detailed work plan for the staff.
- Arrange adequate storage sites for excavated materials.

References

Agostinelli G, Comin C, Pedemonte S, Mair V, Veligogna A (1995) Aspetti geologici applicativi conseguenti allo scavo della galleria Fleres—Terme di Brennero. Geologia Tecnica & Ambientale 3/95

Bonini M, Debernardin D, Barla M, Barla G (2009) The mechanical behaviour of Clay Shales and implications on the design of tunnels. Rock Mech and Rock Eng 42(2):361–388

Herget G (1988) Stresses in rock (p. 175). Rotterdam. Balkema

Hoek E (2013) Rock mass properties. In Hoek's corner, Rocscienceinc (http://www.rocscience.com/education/hoeks_corner)

Hoek E, Brown ET (1980) "Underground excavations in Rock" Institution of Mining and Metallurgy. Chapman & Hall (UK)

International Society for Rock Mechanics (1980) Commission on classification of Rocks and Rock Masses (p 26) Lisbon, Portugal. The Society

Kovári K, Fechtig R (1996) Trafori alpini storici in Svizzera. S. Gottardo Sempione Lötschberg, Società per l'arte delle costruzioni. Zurigo, Stäubli AG

Owen GN, Scholl RE (1981) Earthquake engineering of large underground structures. Repot no. FHWA/RD-80/195.Federal Highway Administration and National Science Foundation rock.

Sheory PR (1994) A theory for in situ stresses in isotropic and transversely isotropic Int J Rock Mech Min Sci Geomech Abstr 31(1):23–34

Sibson RH (1977) Fault rocks and fault mechanisms. J Geolo Soc London 133: 191–213

Chapter 2
Environmental-Geological Problems due to Underground Works

2.1 Introduction

Considering the environmental impact, the realization of underground works usually implies some advantages, such as the availability of free space on the surface, lower visual and noise impact and little effects on ecosystems (Barla and Barla 2002). But sometimes there are disadvantages as well, such as higher construction and maintenance costs, the alteration of complex natural balances, a local increase in noise and vibrations in the construction and operating phases. Actually, underground works are usually more expensive, but the cost and benefit balance is positive nevertheless if one considers the social-economical value of the area used, in particular, in urban areas or areas with beautiful or protected landscapes and ecosystems. Obviously, the negative aspects of the infrastructure are not cancelled, still the more evident ones are removed or reduced and others are produced on different elements, subjects and temporal scales (Table 2.1).

The main environmental problems linked to the construction of underground works are listed below:

- Triggering of surface settlements, structures collapses and slope instabilities
- Drying up of springs and groundwater alterations
- Storage and use of excavated materials
- Noise
- Vibrations
- Pollution of groundwater, mainly after the realization of stabilization works by injections
- Emission of dust and air pollutants

As far as air pollution is concerned, apart from the obvious increase of pollutants in the construction phase, in theory a road tunnel is built to reduce traffic, noise and pollutant emissions, in particular in urban areas. Actually, the cost–advantage balance also has to take into account a local increase in those pollutants (in particular, carbon monoxide, nitrogen oxides, suspended solids, dusts and benzene) in the area of a few hundred metres surrounding the tunnel portal. Those aspects, as well as

P. Gattinoni et al., *Engineering Geology for Underground Works*,
DOI 10.1007/978-94-007-7850-4_2,

Table 2.1 Explanatory confrontation among the effects of road and railway tunnels compared to corresponding infrastructures on the surface

		Construction	Operating phase	Maintenance
Environmental aspects	Air	+	+	
	Geomorphology	+		
	Groundwater	−		
	Surface waters	+		
	Ground	+		
	External noise	+	+	+
	Inside noise	−	−	−
	Vibrations	−	−	−
	Vegetation	+		+
	Fauna	+	+	+
	Landscape	+	+	+
	Surface ground heritage	+		
	Underground heritage	−		
	Soil use	+	+	+
	Infrastructures/facilities	+	+	
	Safety	−	+	
	Efficiency		+	
	Human psychology		−	
Economic aspects	Costs	−	−	−

+ environmental aspects determining the advantages of an underground work compared to the same work on the surface

− environmental aspects determining the disadvantages of an underground work compared to the same work on the surface

groundwater pollution, are not discussed in this book; this book is focused on the analysis of the more typical geological and hydrogeological problems.

2.2 Surface Settlements

The opening of underground works causes a deformation of the soils and rocks around the excavation area. Such deformations may trigger sudden collapses, subsidence and sinking that can damage both the work under construction and pre-existing nearby structures, in particular, if the work is being constructed in developed areas (Fig. 2.1).

The consequences and the damages depend on the intensity of the phenomenon and on the vulnerability of the elements on the surface (buildings, rivers, industrial settlements, facilities etc.). Generally, the response to the opening of the underground work and, as a consequence, the extent of settlements depends on the following elements:

- Excavation technique
- Dimension and geometry of the excavation
- Type of excavated material

Fig. 2.1 Examples of the formation of crown holes and surface settlements induced by tunnelling: **a** Crater of Saint-Laurent Place (Metro Lausanne-Ouchy; Seidenfuß 2006). **b** Kaohsiung collapses (from T&T International 2006). **c** Shanghai Metro (P.R. of China), 2003. (Wannick 2006)

The sudden collapse with cavity filling (Fig. 2.2) usually happens in case of limited overburden that can reach, in some particular cases, up to ten times the cavity diameter.

In case of subsidence phenomena, the deformation can be delayed with respect to the excavation phase and can involve wider areas. In this case, the typical parameters

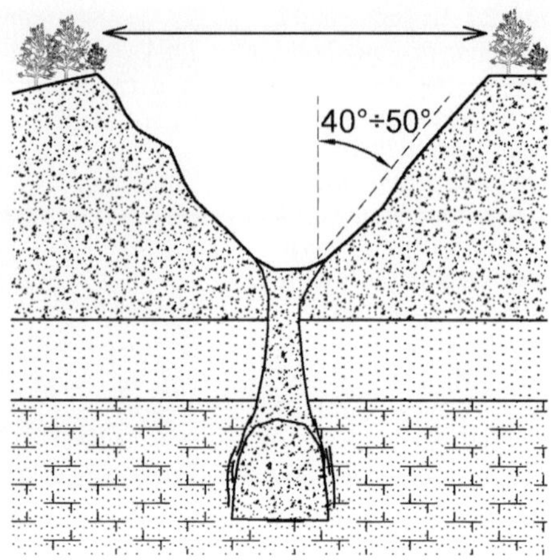

Fig. 2.2 Mechanism of sudden collapse with cavity filling

are the shifts, mainly vertical and variable on the surface from point to point, and the width of the area interested by subsidence. Usually, this is included in a dihedral angle whose opening varies between 15° and 45° according to the depth of the excavation and the features of the soil.

A generalized and uniform sinking does not cause big damages, whereas discrete settlements may seriously damage overlying structures and facilities.

Usually, the forecast of surface movement associated with the excavation, when no structures are present on the ground level, is carried out by means of empiric relations based on the following hypothesis: In a section transverse to the tunnel axis and at an adequate distance from the excavation face, the outline of the collapses can be approximated to an inverted Gaussian distribution (Fig. 2.3). In order to characterize the width of the area interested by deformations, Peck (1969) proposed to interpolate the deformed area of the topographic surface with a Gaussian curve and to take the horizontal measures between the tunnel axis and the inflection point of the curve as characteristic dimension.

Data on the behaviour of models in 1:1 scale show that the relation $i/2R$ (R being the tunnel radius) increases when $a/2R$ increases (a being the tunnel depth from land surface) according to different laws in different soils.

Nevertheless, it is well known that the presence of a building or a structure may change the subsidence profile. Numerical model can be used if the possible impact of an underground work on pre-existing structures must be evaluated. Possibly, the modelling should be carried out in a three-dimensional field to be able to correctly simulate the different relative tunnel-structure positions, which means the study should not be limited to the case of the tunnel axes perpendicular to the main level of the structure. Obviously, the reliability of the results of the numerical modelling depends on the quality of input data of the model, in particular those concerning the deformation parameters of the ground material and its heterogeneity (Gattinoni et al. 2012).

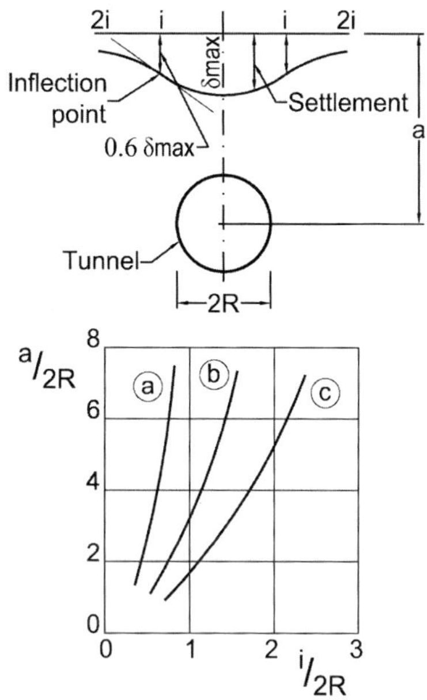

Fig. 2.3 Surface settlements caused by the construction of a tunnel with radius R situated at shallow depth, in the presence of: **a** rocks, hard clays, sands over the groundwater table, **b** soft to hard clays and **c** sand under the groundwater table

In case the load change is due to a lowering of the groundwater table, the subsidence phenomenon is ruled by the following equations:

$$S_\infty = \Delta q \cdot \frac{H}{E} \quad \text{with:} \ \ \Delta q = \sigma'_\infty - \sigma'_{in}$$

where:

s_∞ final settlement
Δq change of the applied load
H thickness of the soil interested by the load change
E deformation modulus of the material
σ'_{in} initial effective stress
σ_∞ final effective stress

According to a first estimate, the effective stress at the depth of the underground work, before its construction, can be calculated as follows:

$$\sigma'_{\text{in}} = \sigma_{\text{in}} - u_{\text{in}}$$
$$\sigma_{\text{in}} = \gamma_d \cdot h_t - \gamma_d \cdot h_w + \gamma_{\text{sat}} \cdot h_w = \gamma_d (h_t - h_w) + \gamma_{\text{sat}} \cdot h_w$$
$$u_{\text{in}} = \gamma_w \cdot h_w$$

where:

γ_s specific weight of the grains
γ_d specific weight of dry material
γ_{sat} specific weight of saturated material

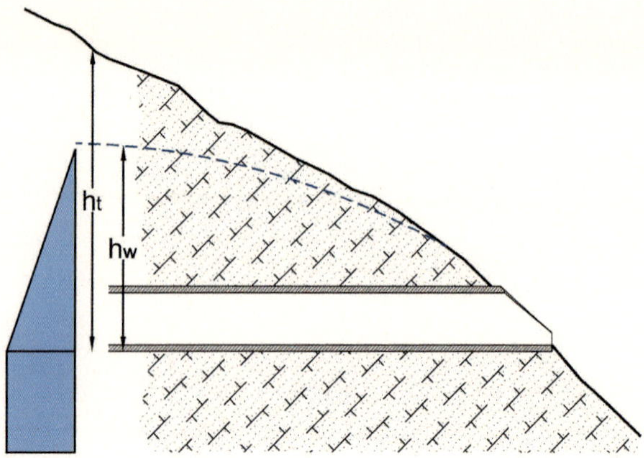

Fig. 2.4 Load change induced by the groundwater table drawdown caused by the opening of an underground work

γ_w specific weight of water
m porosity of the material
h_t excavation depth with regards to land surface
h_w piezometric height above the excavation level

If a total drainage of the soil up to the excavation level is considered, the effective stress after excavation becomes:

$$\sigma_\infty = \gamma_s \cdot (1 - m) \cdot h_t = \gamma_d \cdot h_t$$
$$u_\infty = 0$$
$$\sigma'_\infty = \sigma_\infty - u_\infty = \sigma_\infty$$

It is therefore possible to calculate the maximum load change at the level of the underground work and to consider a decreasing trend toward the surface (Fig. 2.4).

2.3 Slope Instability

The interaction with slope stability is a typical problem of the portal stretches and underground works carried out close to the side of a slope (Figs. 2.5–2.7). The excavation of an underground work implies the annihilation of the stress state in correspondence of its boundaries, the redistribution of stresses with local increases of the deviatoric stresses. In terms of stress state, the global effect depends on the following elements:

- Characteristics of the excavation (site, shape and dimension)
- Excavation technique
- Material constitutive laws
- Initial stress state (including the water neutral pressures)

Figure 2.8 shows an example of change in the slope stability conditions after the construction of a tunnel close to the side of the slope, according to its depth. In

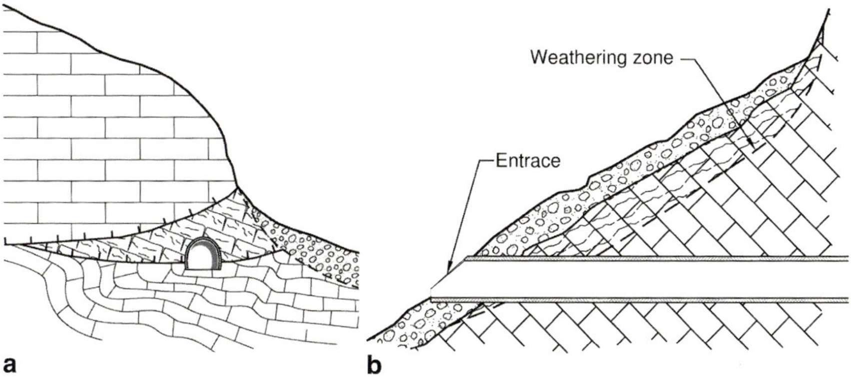

Fig. 2.5 Examples of a geological structure potentially interested by excavation interactions with the slope stability: **a** Overthrust zone, **b** portal stretches

Fig. 2.6 Tunnel portal affecting the slope obliquely: stabilization with tie rods. (By Pizzarotti)

that case, the analysis was carried out with limit equilibrium techniques in a two-dimensional field, and it refers to dry conditions. The problem is complex and should be studied in a three-dimensional field considering the tunnel progress. Furthermore, when dealing with materials having small grain size, the stability analysis should be referred to three different phases:

- Initial (short term)
- Transient
- Final (long term)

In practical terms, the assessment of slope stability conditions in relations to tunnelling can be assimilated to that of any slope. Therefore, specific books can be consulted for its detailed dissertation.

Fig. 2.7 Tunnel portal affecting the slope obliquely: in this case, stabilization structures were: **a** placement of tie rods on a reinforced concrete head-beam and a steel ribs-reinforced portal, **b** completion of the reinforced concrete structure. (By Pizzarotti)

As far as slope stability is concerned, another quite typical problem in tunnelling is related to the presence of deep-seated gravitational deformation (DSGD), which is a common and widespread type of large slope instability in the Alps (Agliardi et al. 2013). The presence of DSGD assumes a great importance both during the tunnel construction and during its lifetime, as it is often correlated with important cataclastic systems. Actually, DSGDs can affect the tunnel-boring machine (TBM) parameters and then the excavation speed, just like Pont Ventoux hydroelectric power plan tunnel (Venturini et al. 2001). They also have a very important impact on groundwater flow and tunnel inflow (Vincenzi et al. 2010). Moreover, the progressive deformation of DSGD can lead to important deformative phenomena within a tunnel if its lining is not properly designed, by taking into account the stresses related to DSGD. This is the case of the Mt. Piazzo tunnel (Northern Italy), which cuts a metamorphic mica schist formation ("Scisti dei Laghi" Auct.): The tunnel suffers from the M. Legnoncino DSGD, which causes deformations in the order of several millimetres in the cladding, requiring constant maintenance; hence, the tunnel was recently closed as a consequence of the increasing deformations.

The geognostic investigations useful to detect and characterize DSGDs consist of a detailed morpho-structural and morpho-dynamic analysis (Cadoppi et al. 2007), remote sensing (Strozzi et al. 2013) and geophysical system (Supper et al. 2013).

2.4 Interaction with Surface Water and Groundwater

The interaction between tunnelling and groundwater is a very relevant problem not only due to the need to safeguard water resources from impoverishment and pollution risk, but also to guarantee the safety of workers and the effectiveness of the draining works. One of the most emblematic examples (Table 2.2) in Europe concerns the construction of the Gran Sasso (Italy) motorway tunnels, that were interested by water inflows exceeding 2,000 l/s, and the railway tunnel for the Bologna–Firenze (Italy) high-speed stretch (Rossi et al. 2001), with drained flows reaching 650 l/s.

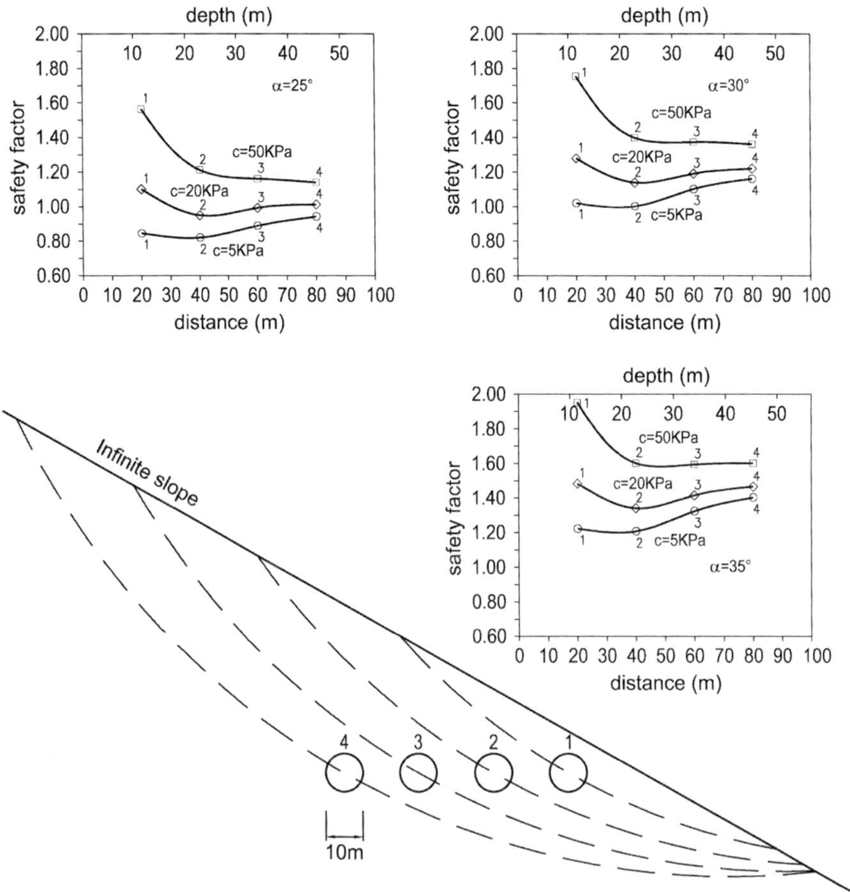

Fig. 2.8 Example of change in a slope stability after the construction of a tunnel near the side of the slope, according to its depth. (Picarelli et al. 2002)

On the contrary, when the tunnels in San Pellegrino Terme (Bergamo–Italy) were constructed, the interdisciplinary approach of the study and the attention to the environment guaranteed the safeguard of the hydrothermal springs of the area (Barla 2000).

It is well known that the excavation of tunnels has a relevant draining effect leading to a more or less generalized drawdown of the groundwater table (Figs. 2.9 and 2.10), whose effects may be undesirable, such as: the drying up of springs (Figs. 2.10 and 2.11) and/or wells, qualitative changes of the groundwater, changes in the vegetations, changes in the slope stability, changes in the flow and in the quality of thermal waters, changes of the hydrogeological balance at the basin scale. These phenomena can persist even after the tunnel construction if the final alignment is not completely waterproof.

Table 2.2 Examples of tunnel inflow during excavation. (Modified from Civita et al. 2002)

Tunnel	Type	*L.* (km)	Q_{max} (m^3/s)	Q_{min} (m^3/s)	Aquifer
Sempione (ITA–CH)	Railway	19.8	1.700	0.864	Limestone
Vaglia (BO–FI)	Railway	18.6	0.080	–	Limestone, calcarenite, sandstone
Direttissima (BO–FI)	Railway	18.5	1.200	0.060	Sandstone
Pavoncelli bis (AV)	Hydraulic	15.5	0.800	0.070	Limestone, clay
Firenzuola (BO–FI)	Railway	15.1	0.277	0.070	Sandstone and marl
Santomarco (Paola–CS)	Railway	15.3	0.100	0.038	Metamorphites
Frejus (T4)	Highway	12.9	0.007	0.001	Several
M. Bianco (Ti)	Highway	11.6	0.800	0.440	Granite
Raticosa (BO–FI)	Railway	10.4	0.037	–	Sandstone, marl and clay
Gran Sasso (A24)	Highway	10.2	3.000	0.600	Limestone
S. Lucia (NA–SA)	Railway	10.2	1.000	0.250	Limestone
Putifigari (SS)	Road	9.8	0.070	0.050	Vulcanites
Zuc del Bor (UD–AUT)	Railway	9.3	0.700	0.650	Limestone
S. Stefano (GE–F)	Railway	7.9	–	high	Marly limestone, sandstone
M. Olimpino 2 (MI–CO)	Railway	7.2	very high	–	Limestone, sands
Serena (PR–SP)	Railway	6.9	medium	–	Calcarenites, breccia, flysch
M. La Mula	Hydraulic	6.3	0.200	0.800	Limestone, dolomite
Turchino (GE–AT)	Railway	6.4	0.110	0.075	Calceschysts
Satriano (1°salto)	Hydraulic	6.4	very high	–	Milonitic granite
Gran S. Bernardo (T2)	Highway	5.9	low	very low	Gneiss, schist
S. Leopoldo (UD–AUT)	Railway	5.7	3.600	high	Limestone
Gravere (TO–FRA)	Railway	5.6	very high	0.013	Calceschysts
Vado Ligure (ITA–FRA)	Railway	4.9	0.200	0.050	Dolomite
Colle Croce (ITA–FRA)	Road	4.1	low	very low	Calceschysts
Col di Tenda (ITA–FRA)	Railway	3.2	0.600	0.200	Limestone
Bypass Spriana	Hydraulic	3.2	0.300	0.040	Gneiss, limestone, dolomite
Villeneuve (A5)	Highway	3.2	0.200	0.001	Calceschysts
Prè Saint Didier (A5)	Highway	2.8	0.100	0.080	Calceschysts, sandstone
Moro (AN–BA)	Railway	1.9	0.080	–	Gravels and sands
Colle della Scala	Railway	–	very high	high	Limestone
Croccetta (Paola–CS)	Road	1.5	0.022	0.028	Tectonised schists

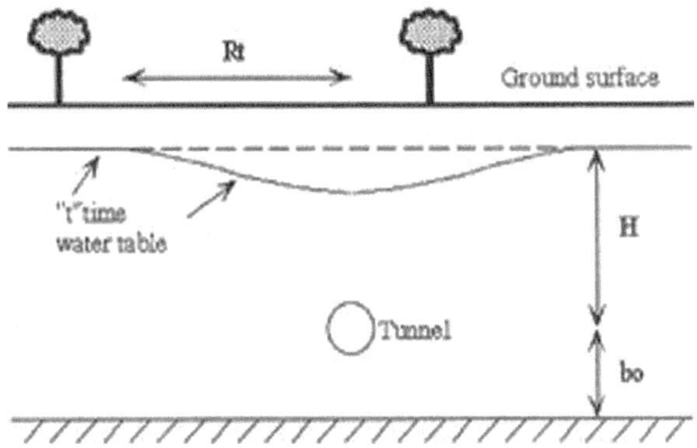

Fig. 2.9 Water table drawdown induced by the opening of a tunnel. *H* is the initial piezometric level above the tunnel, b_o the depth of the substratum with respect to the tunnel level, R_t the radius of influence of the tunnel. (Lunardi and Focaracci 2001)

From the point of view of the underground work, the importance of forecasting the interaction between the excavation and groundwater is linked to the fact that water flows (Fig. 2.12) could involve serious problems for the stability of the underground work and the safety of the workers. Actually, seepage forces acting on discontinuities may affect the cavity stability, both in soils and in rocks.

Generally speaking, it can be said that if the underground work is above the water table, the problems are limited. Actually, above the water table, important water flows can only occur in presence of karstified rock masses. If, on the contrary, the underground work is drilled under the water table, water flows can be very important, in particular in presence of highly permeable materials (granular soil, permeable rocks due to porosity of fracturing), sudden changes in permeability, tectonic dislocations having a great water supply, rock masses subject to karst phenomena, synclines, paleo-rivers, faults, overthrusts, etc. (see Fig. 1.20 Chap. 1).

The effect of a tunnel on the hydrogeological setting depends on the feeding conditions and on the aquifer permeability, as well as on the system used to excavate the tunnel. In the last few years, many studies have been carried out that allowed the definition of the contribution that hydrogeology can provide to the different stages of tunnel projecting, in particular, in relation to the tunnel inflows assessment, and

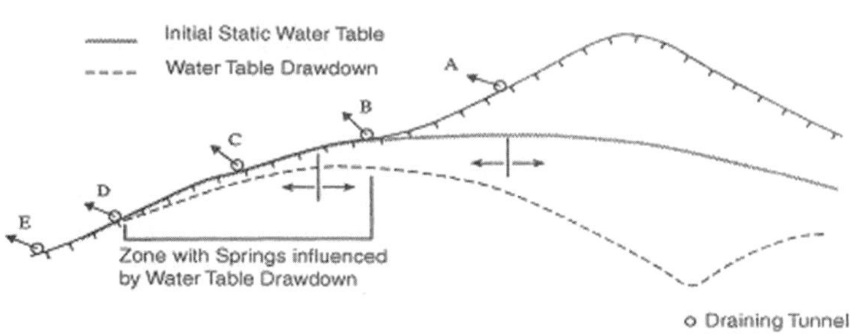

Fig. 2.10 Example of spring drying up after the water table drawdown due to the opening of a tunnel. (Loew 2002)

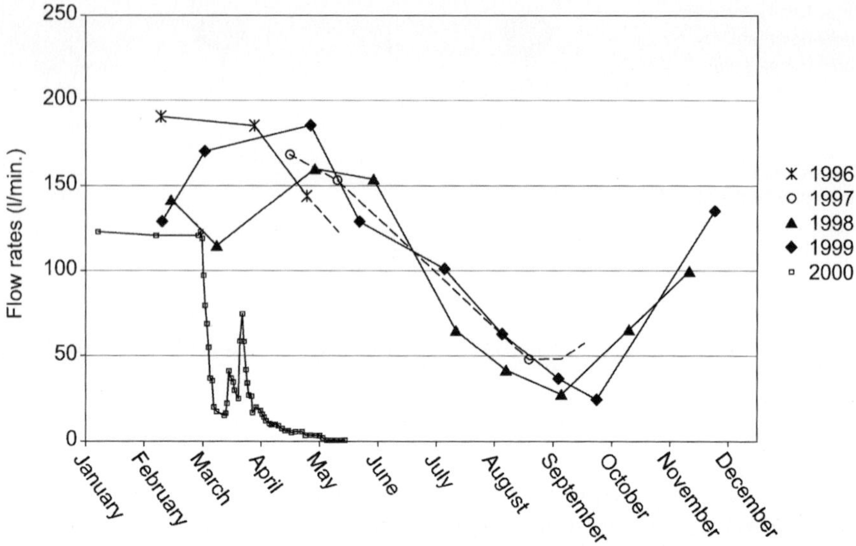

Fig. 2.11 Example of spring drying up after the opening of a tunnel. (Gisotti and Pazzagli 2001)

Fig. 2.12 Example of tunnel inflow during excavation works. (Wilhelm and Rybach 2003)

to the impact on the hydrogeological conditions of the surrounding environment, in particular, on the regime of springs, groundwater and superficial waters (Dematteis et al. 2001).

From the environmental point of view, the impoverishment, the drying up, or generally the change of regime of springs are only some of the most difficult risks to forecast and to quantify during the tunnel design, as these elements are ruled by very complex and often unpredictable phenomena. Moreover, considering that the hydraulic characteristics of the rock mass are neither homogeneous nor isotropic, the water flow is ruled by the orientation and by the hydraulic characteristics of the joints, as well as by the fracturing degree. In shear zones, for example, permeability increases (of some orders of magnitude) can occur along strikes predetermined by the orientation of the same shear zones; that flow orientation strongly influences the form and the extension of the area that is potentially interested by the draining process (Scesi and Gattinoni 2009).

The study of the hydrogeological impact of a tunnel is generally articulated in two main phases (Gattinoni et al. 2005):

- The definition of the perimeter of the area affected by geological risk: this operation is carried out by reconstructing the groundwater flow in different conditions, assessing the tunnel inflow and the radii of influence.
- The statistic quantification of the hydrogeological risk both for the tunnel and the environment, through the calculation of the probability that the tunnel inflow or the piezometric drawdown due to the excavation exceeds the acceptable values.

In order to bind the potentially interested area by tunnel construction, it appears extremely important to integrate the geological, geological-structural, geomechanical and traditional hydrogeological studies with more in-depth analyses about the individuation and the characterization of the "shear zones" and of karst circuits. The interpretation of the results obtained by those analyses allows a quite detailed reconstruction of the conceptual model of the underground hydraulic flow. In particular, following steps are fundamental:

- The identification of the tunnel sections situated over the piezometric surface, where only the water percolating inside the discontinuities directly intercepted by the excavation are drained, and those sections fully situated in the groundwater, where a general drawdown of the piezometric level is expected (Fig. 2.13).
- Identification of possible intercommunication areas among different aquifers.

Actually, the presence of shear zones where rock fractures are mainly vertical determines a local interconnection between the deep aquifer intercepted by the tunnel and superficial waters (Fig. 2.14); as a consequence, the superficial catchment area must be defined according to the hydraulic features of the previously identified shear zones.

The estimate of tunnel inflows can be obtained using geomechanical classifications, analytical formulations or numerical models (see Sect. 4.9). Some heuristic approaches (Dematteis et al. 2001; Gattinoni et al. 2001) have been developed starting from the Rock Engineering System (Hudson 1992), which help to evaluate tunnelling impact on groundwater flow. With this aim, the system aquifer-tunnel is described by means of different variables together with their interactions in order to detect the critical sectors and the areas not at risk.

Afterwards, to delineate the tunnel influence zone in the critical sectors, also the typical anisotropy of rock masses has to be taken into account (Gattinoni et al. 2005). Therefore, it is better to refer not to the radius of influence, but to the ellipse of influence, whose shape can be easily obtained on the basis of the permeability tensor. Actually, knowing the permeability tensors on structurally and geologically homogeneous areas, it was possible to reconstruct the shape of the ellipse of influence on the horizontal plane, with the relevant values and direction of maximum (K_{max})

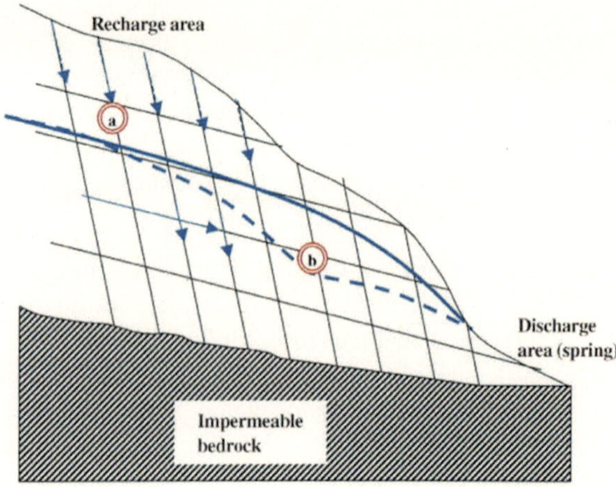

Fig. 2.13 Schematic representation of the flow condition typical of a rock slope, with indication of the water-flow direction within the discontinuity network. The water path is conditioned by the hydraulic gradient and the discontinuity distribution and orientation. The tunnel in position (**a**) only drains water flowing within the discontinuities directly intercepted; the tunnel in position (**b**), i.e. underneath the groundwater, causes a generalized drawdown of the piezometric level, whose shape and extension are ruled by the permeability tensor

Fig. 2.14 3D scheme of shear zones that cause the interconnection between superficial and deep waters, with the associated water-flow scheme

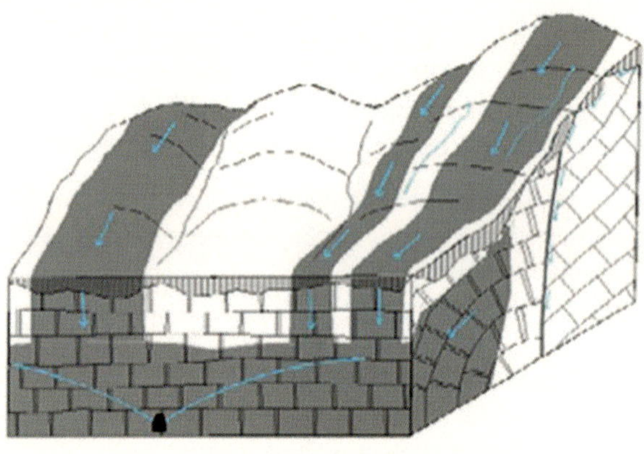

and minimum ($K_{\min}$) permeability. Starting from the average radius of influence R, estimated for the isotropic medium, the semiaxes of the ellipse of influence can be obtained (Fig. 2.15) as a function of the anisotropy ratio k:

$$a = R\sqrt{\frac{K_{\max}}{K_{\mathrm{eq}}}} \rightarrow \text{major semiaxes}$$

$$b = R\sqrt{\frac{K_{\min}}{K_{\mathrm{eq}}}} \rightarrow \text{minor semiaxes}$$

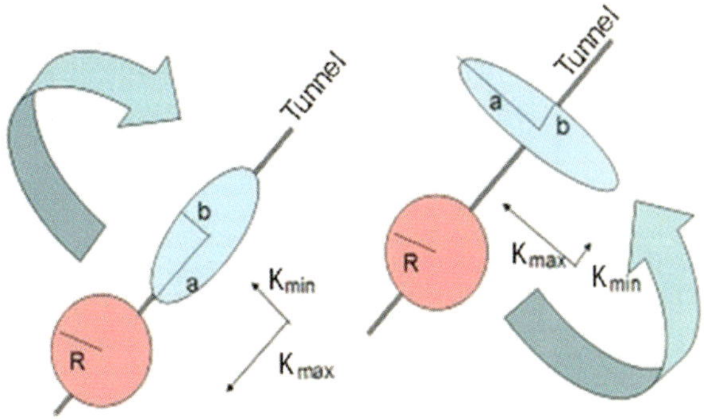

Fig. 2.15 Anisotropy effect in the medium rock on the radius of influence of the tunnel: **a** K_{max} is parallel to the tunnel axis, **b** K_{max} is orthogonal to the tunnel axis

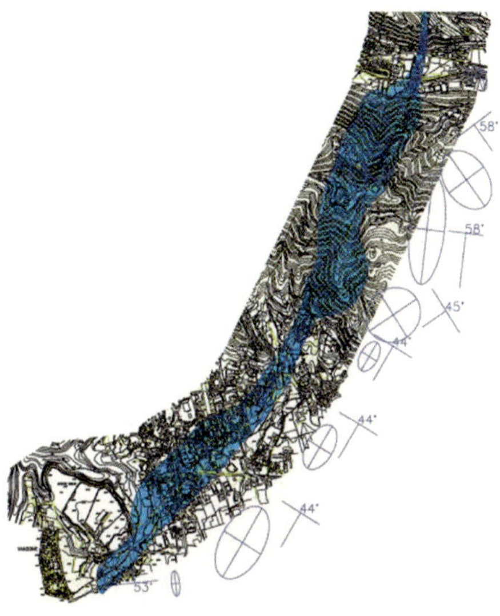

Fig. 2.16 Example of delimitation of the influence zone of a tunnel (*light blue*), according to the rock mass anisotropy (*navy blue ellipses*)

The orientation of the main discontinuities determines a marked orientation of the hydraulic flow, bringing about a deformation of the radius of influence. Higher the anisotropy ratio, higher the deformation. In that way, it was possible to bind the area of the aquifer that was potentially influenced by the draining of the tunnel along its whole length (Fig. 2.16).

2.5 Inert Waste

The storage of excavated material often brings along serious environmental problems. Therefore, during the discussion of the project, the relevant authorities should encourage the use of those materials for useful purposes, mainly, if possible, as construction materials (limestone and marl for concrete), gneiss, asbestos-free serpentine for railway beds and river banks or road embankment (Fig. 2.17).

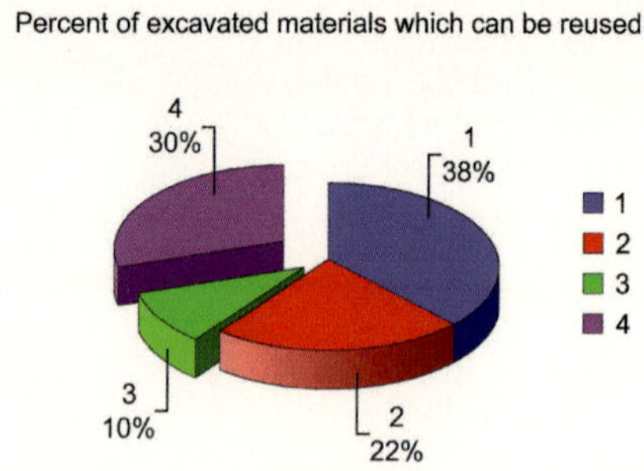

Fig. 2.17 Estimate of the percentage of excavated materials that can be potentially reused as inert in the construction of a tunnel: *1* massive green rocks, *2* gneiss, *3* carbonate and schistose rocks difficult to reutilize, *4* serpentinite and serpentine-schists that cannot be reused. (Modified from Bottino 2002)

With this aim, natural substances or minerals can be detected inside rock masses that may be dangerous both for the workers, during the construction phase, and for the users, during the operation phase. Among those substances, there are radioactive minerals and asbestos, already discussed in Chap. 1.

Therefore, in the projecting stage of an underground work, in particular when in environmentally protected areas, it is necessary to foresee and plan the management and reutilization of excavated materials in order to minimize the extension of the areas destined to the inert storage as well as the volumes of material to be removed (Fig. 2.18), thus lowering the impact on local traffic. Actually, people and material handling during the construction phase implies a number of undesired effects, some of which are:

- Traffic jams, with an increase in the number of accidents and road wear
- Noise
- Air pollution
- Dust production

Some of the most common solutions to limit this environmental impact include:

- The in situ reutilization of the excavated materials
- The remodelling and reutilization of abandoned quarries
- The optimization and correction of the road network and of transportation methods

For example, when the Kleinensiel/Dedesdorf tunnel (Germany, in a protected area of about 4.1 km^2) was constructed, excavated inert materials for a total volume of about 1,800,000 m^3 were separated, converted and reintroduced in the construction cycle as:

- Infill for the road construction (silt content $< 5\,\%$)
- Inert in walls to decrease noise
- Inert in foundation works

This kind of approach allowed the reutilization of over 70 % of the produced inerts.

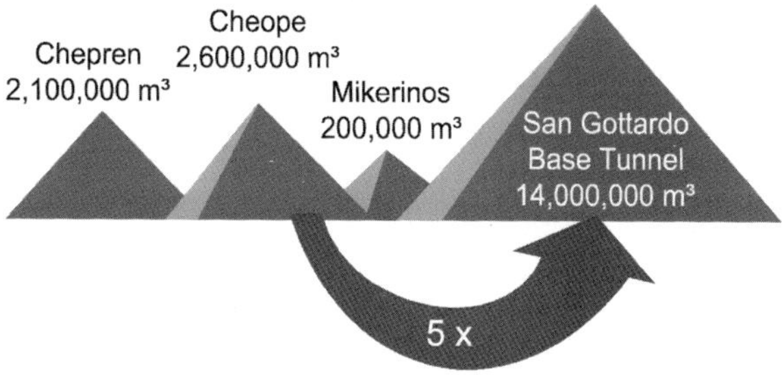

Fig. 2.18 Volume of inerts extracted during the construction of the St. Gotthard Base Tunnel. (Lanfranchi 2002)

2.6 Noises and Vibrations During Excavation

Noise is a typical problem linked to underground works in urban areas that can be effectively solved using deafening elements and soundproof materials. Of course, such a problem is particularly evident during construction, but it can be limited by shielding the working area (e.g. the portal zone).

Ground vibrations are mainly linked to excavation by blasting, but they can also be present in mechanical excavations.

In order to be able to predict the ground vibration, it is important to have undertaken testing or have a full understanding of the excavation system (blasting or mechanical) and the ground types for a given study area. The parameter which defines the vibration danger level is the peak particle velocity. For instance, for mechanical excavation, the peak particle velocity (PPV) can be expressed as:

$$\mathrm{PPV} = \frac{K}{d} e^{-\alpha d,}$$

where d is the distance in metre from the source, K and α , respectively, depend on the site characteristics and on the excavation means (Speakman and Lyons 2009).

Geophones are used to measure in situ excavation-induced vibrations, as they can evaluate the frequencies and the displacement speed at varying distances from the excavation face. Obviously, the kind of vibration propagation depends on the lithotype being excavated (Fig. 2.19).

If the vibration is measured at a number of locations at varying distance from the source, then the site-specific operators can be determined as:

$$\alpha = -\frac{\ln (V_2 d_2 / V_1 d_1)}{(d_2 - d_1)}$$

$$K = \frac{V_1 d_1}{e^{-\alpha d_1}}$$

where V_1 and V_2 are the measured PPV at distance d_1 and $d_{2,}$ respectively.

Figure 2.20 shows a typical vibration propagation curve for TBM vibration. The site-specific operators must be measured at a number of representative locations throughout the study area in order to account for varying ground types.

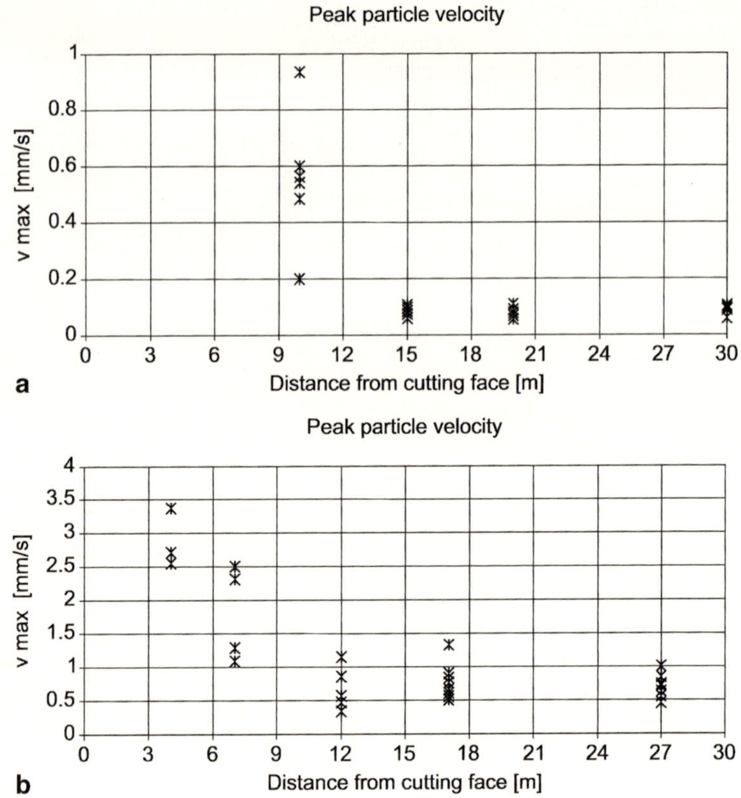

Fig. 2.19 Peak speed measured by geophones at varying distance from the excavation face: **a** in porphyry and **b** in dolomite. (Speakman and Lyons 2009)

To evaluate the effect produced by blasting, it can be considered that the particle peak velocity PPV (expressed in mm/s) is proportional to the distance R (in m) from the blasting point as well as to the explosive Q (in kg) (site law; Langefors and Kihlström 1963):

$$PPV = k\left(\frac{R}{\sqrt{Q}}\right)^{\alpha}.$$

where, also in this case, the coefficients k and α must be determined for each single case on the base of instrumental measures (Figs. 2.19 and 2.20).

As far as vibrations affecting the buildings are concerned, the acceptable limit varies according to the laws in force in the different countries (Fig. 2.21). German guidelines (DIN 4150) are the most commonly used also in Italy.

Generally speaking, in case of vibrations affecting rocks, it can be said that:

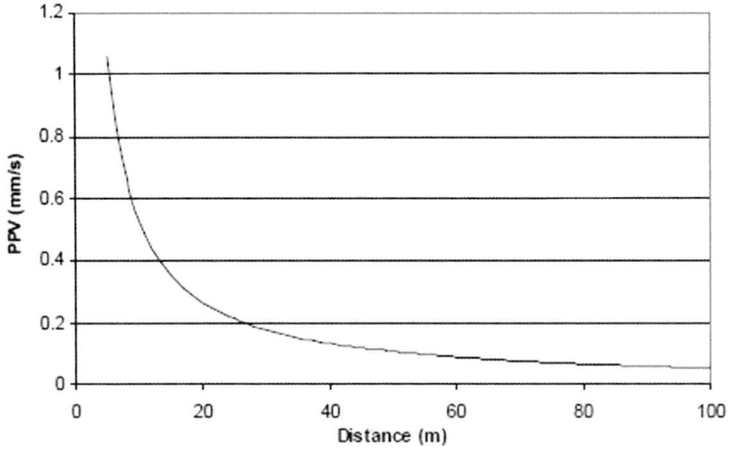

Fig. 2.20 Typical TBM vibration propagation. (Speakman and Lyons 2009)

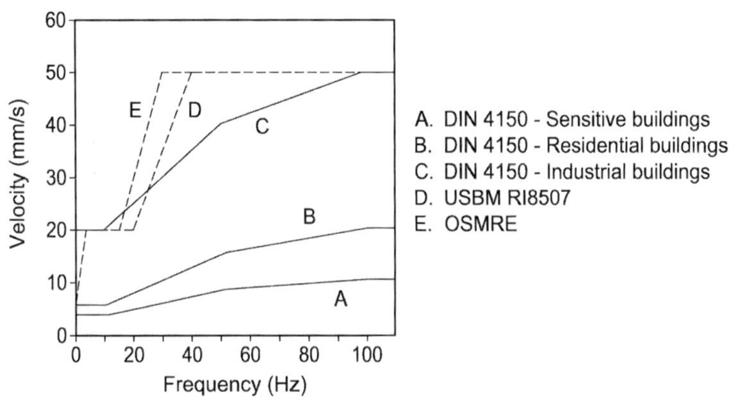

Fig. 2.21 Examples of definitions in the *V–f* plane of the intensity limits considered acceptable by law in the different countries. *DIN 4150* = Germany; *USBM RI 8507* and *OSMRE* = the USA (in Italy the most commonly used are *A–B–C*). (Piovano and Sorlini 1994).

- With PPV > 600 mm/s new fractures in rock are generated
- With 300 < PPV < 600 mm/s existing fractures propagate
- With PPV < 300 mm/s rock blocks may detach

A relevant rock detach can occur when:

- PPV > 200–600 mm/s for low quality rocks
- PPV > 600–2,000 mm/s for good quality rocks

In any case, the study has to be carried out for each single case.

References

Agliardi F, Crosta GB, Frattini P, Malusà MG (2013) Giant non-catastrophic landslides and the long-term exhumation of the European Alps. Earth Planet Sci Lett 365(1 March):263–274

Barla G (2000) "Lessons learnt from the excavation of a large diameter TBM tunnel in complex hydrogeological conditions"-GeoEng2000. International conference on geotechnical and geological engineering, pp 938–995

Barla G, Barla M (2002) Le opere in sotterraneo e il rapporto con l'ambiente, in "Le opere in sotterraneo e il rapporto con l'ambiente, IX ciclo di Conf di Meccanica e Ingegneria delle Rocce MIR 2002. Patron Editore, pp 9–43

Bottino G (2002) Problematiche geologiche e morfologiche, IX ciclo di Conferenze di Meccanica e Ingegneria delle Rocce "Le opere in sotterraneo e il rapporto con l'ambiente", Torino 26–27 novembre 2002 (Patron Ed.), pp 45–71

Cadoppi P, Giardino M, Perrone G, Tallone S (2007) Litho-structural control, morphotectonics, and deep-seated gravitational deformations in the evolution of Alpine relief: a case study in the lower Susa Valley (Italian Western Alps). Quaternary Int 171–172:143–159

Civita M, De Maio M, Fiorucci A, Pizzo S, Vigna B (2002) "Le opere in sotterraneo e il rapporto con l'ambiente: problematiche idrogeologiche". Meccanica e Ingegneria delle rocce: MIR, Torino, pp 73–106

Dematteis A, Kalamaras G, Eusebio A (2001) "A systems approach for evaluating springs drawdown due to tunnelling"-World Tunnel Congress AITES-ITA 1:257–264

Gattinoni P, Papini M, Scesi L (2001) "Geological Risk in Underground Excavations" – World Tunnel Congress AITES-ITA 2001, Milano 10-13 giugno 2001, Patron Editore, Vol 1, pp 309–318

Gattinoni P, Pizzarotti E, Rizzella A, Scesi L (2012) "Geological risk in tunneling: the example of Teheran underground" in Proceedings of the World Tunnel Conference 2012, Bangkok 18–23 May 2012, p 1–8

Gattinoni P, Scesi L, Francani V (2005) "Previsione del rischio idrogeologico derivante dalla costruzione di gallerie in roccia a media copertura", Atti del 2° Congresso Nazionale AIGA, 15–17 febbraio 2006 pubblicati sul Giornale di Geologia Applicata Vol 2, anno 2005, pp 13–19

Gattinoni P, Scesi L, Francani V (2005) "Tensore di permeabilità e direzione di flusso preferenziale in un ammasso roccioso fratturato", Quaderni di Geologia Applicata, n. 12-1(2005), Pitagora Editrice (Bologna), pp 79–98

Gisotti G, Pazzagli G (2002), L'interazione tra opere in sotterraneo e falde idriche. Un recente caso di studio, World Tunnel Congress AITES-ITA 2001, Vol 1, pp 327–334

Hudson JA (1992) Rock engineering systems: Theory and practice (p. 185). Ellis Horwood, New York.

Lanfranchi P (2002), Gestione del materiale della galleria di base del San Gottardo, IX ciclo di Conferenze di Meccanica e Ingegneria delle Rocce "Le opere in sotterraneo e il rapporto con l'ambiente", Torino 26–27 novembre 2002 (Patron Ed.), pp 357–372

Langefors U, Kihlström B (1963) The modern technique of rock blasting. Wiley, New York, pp 405

Loew S (2002) "Groundwater hydraulics and environmental impacts of tunnels in crystalline rocks". Meccanica e Ingegneria delle rocce: MIR. Torino, pp 201–217

Lunardi P, Focaracci A (2001) "Action to reduce the hydrogeological impact produced by underground works"-World Tunnel Congress AITES-ITA. 1:509–515

Peck RB. (1969); "Deep Excavations and Tunnels in Soft Ground". Proceedings of the 7th International Conference on Soil Mechanics and Foundation Engineering, Mexico City, State of the Art Volume, pp 225–290

Picarelli L, Petrazzuoli SM, Warren CD (2002) Interazione tra gallerie e versanti, in "Le opere in sotterraneo e il rapporto con l'ambiente, IX ciclo di Conf di Meccanica e Ingegneria delle Rocce MIR 2002, pp 219–248, Patron Editore

Piovano G, Sorlini A (1994) Normative e raccomandazioni sulle vibrazioni: criteri di salvaguardia, tollerabilità degli edifici e delle persone, scelta dei parametri valutativi. Explosives and blasting, No. 1

Rossi S, Ranfagni L, Biancalani P, Calzolai L (2001) "Geological and hydrogeological analysis in large scale tunnelling and impact forecasting on groundwater resources: Bologna-Florence High Speed Railway (Italy)"-World Tunnel Congress AITES-ITA 1:649–656

Scesi L, Gattinoni P (2009) Water circulation in rocks. Springer, p 172, ISBN: 9048124166

Seidenfuß T (2006) Collapses in Tunnelling. Master Degree Foundation Engineering and Tunnelling. Stuttgart, Germani, p 194

Speakman C, Lyons S (2009) Tunnelling indiced ground-borne noise modelling. Proc of ACOUSTICS, pp 1–5

Strozzi T, Ambrosi C, Raetzo H (2013) Interpretation of Aerial Photographs and Satellite SAR Interferometry for the Inventory of Landslides. Remote Sens 5(5):2554–2570

Supper R, Baron I, Ottowitz D, Motschka K, Gruber S, Winkler E, Joichum B, Romer A (2013) Airborne geophysical mapping as an innovative methodology for landslide investigation: evaluation of results from the Gschliefgraben landslide. Austria Nat Hazards Earth Syst Sci Discuss 1:2281–2318

Tunnels &T (2006) Kaohsiung collapses twince in one week. World News, January

Venturini G, Damiano A, et al. 2001- Productivity parameters from TBM excavations of "Pont Ventoux" hydroelectric power plants tunnels (AEM Torino S.p.A), ITA-AITES World Tunnel Congress, Milano

Vincenzi V, Piccinini L, Gargini A, Sapigni M (2010) Parametric and numerical modelling tools to forecast hydrogeological impacts of a tunnel. AQUA mundi AM02017:135–154

Wannick HP. (2006) The code of practice for risk management of tunnel works future tunnelling insurance from the insurers' point of view, ITA Conference Seoul, April 25

Wilhelm J, Rybach L (2003) The geothermal potential of Swiss Alpine tunnels. Geothermics 32 (4–6):557–568

Chapter 3
Geological Conceptual Model for Underground Works Design

3.1 Introduction

The main geological and environmental problems that can be faced during the realization of underground works have been defined in the previous chapters. It is clear that the recognition and the following analysis of those problems are based on the knowledge of a reference model of the work to be constructed.

To this aim, three types of conceptual models have to be defined (Fig. 3.1):

- The geological model, where materials and processes are identified, obtaining a representation of the spatial distribution of materials, tectonic structures, geomorphological and hydrogeological data concerning the area of influence of the underground work
- The geological-techical model, representing the lithotechnical and hydrogeological characterization of materials, together with their geomechanical characterization. This model also takes into account the risk factors that may influence the mechanical behaviour of materials and the safety of the underground work
- The behaviour model, representing the ground behaviour during and after construction

The geological, hydrogeological, geothecnical and geomechanical characterization of rocks and soils involved in the excavation of an underground work is very important not only to choose the best layout, but also to identify the best location and most suitable type of portal the excavation techniques, the features of the lining and of stabilization works, the methodologies to be used for rock and soil consolidation etc. This characterization takes place through geognostic surveys aimed at a detailed knowledge of the geological-technical and hydrogeological behaviour of the grounds being excavated.

In particular, the geological model synthesizes the results of geological studies and surveys carried out on a large scale, providing a first indication about the geological, geotechnical and hydrogeological conditions of the layout and identifying the critical or the most complex sites from the geological point of view. In that sense, the geological model represents the first fundamental step for the following technical

P. Gattinoni et al., *Engineering Geology for Underground Works*,
DOI 10.1007/978-94-007-7850-4_3,

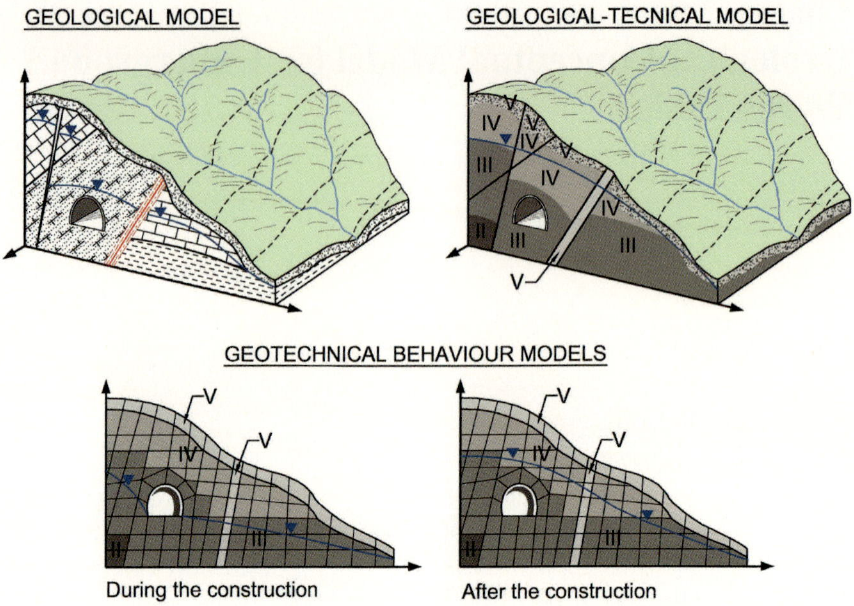

Fig. 3.1 Examples of reference models in tunnel design. (From Gonzalez de Vallejo et al. 2006, modified)

characterization of the materials. That characterization is carried out by means of detailed geognostic surveys and geothecnical tests, resulting in the geological-technical model. The latter allows to carry out the modelling of the stress-strain behaviour of the material, and therefore the forecast of its behaviour during the excavation, using methodologies based on the application of geomechanical classifications and the development of analytical or numerical models.

3.2 Geological Studies and Investigations

Studies and geological surveys are basically aimed at defining the geological reference model, an essential element for the correct design and realization of an underground work.

Initially, a geological survey at adequate scale (1:10,000–1:25,000) must be carried out based on the existing geological cartography. This will highlight the lithological orientation and stratigraphic aspects of the different geological units, together with the main tectonic features of the area. Then, detailed geomechanical surveys must be carried out on rock outcroppings, aimed at knowing the fracturing degree of rock masses and, therefore, assessing their quality.

The geomorphological study of the area follows. It is aimed at identifying possible instabilities phenomena (with particular attention to the portal and stretches with limited overburden), karst phenomena and Paleo Rivers. A detailed hydrogeological analysis is carried out as well, whose procedures and aims are described in the paragraphs that follow.

The last step is the assessment of the seismic and climate aspects of the area examined.

On the basis of those studies, technical outputs as, for example, the geological, geomorphological, hydrogeological map etc. are produced, as well as the preliminary profiles, in which critical areas are highlighted where detailed geognostic surveys are required.

According to the overburden or the stretch considered (portal or central part of the tunnel), the necessary surveys will be different in order to characterize the terrains interested by the underground work in the best way.

3.2.1 *Characterization of Shallow-Overburden Stretches*

The leading survey methods to characterize shallow-overburden stretches consist in:

- Geological, geomechanical and geomorphological surface survey (the latter should be accompanied by the analysis of aerial photographs)
- Mechanical continuous core drilling testing
- Seismic refraction geophysical survey, required to assess the depth of surface deposits or of weathered and fractured rock, as well as to determine the elastic characteristics of the materials to be excavated (they are useful for the following geological-technical characterization)

The above-described analyses are usually enough to produce a reliable forecast of the underground geological setting. The installation of instruments may be required to record possible instability phenomena affecting the slopes (inclinometers, topographer rods) and the features of groundwater that will interfere with the excavation (piezometers).

3.2.2 *Characterization of Medium-High Overburden Stretches*

In these tunnel stretches, the above-mentioned surveys tend to lose much of their reliability, if they are possible at all.

In presence of relevant overburden, the data obtained by a survey carried out on the surface cannot be easily extrapolated at the excavation level. Drillings, for example, whose costs and time required to increase by far with the increase of the desired depth, provide punctual data that cannot be easily horizontally correlated (still, they are essential to obtain the samples for lab tests).

Moreover, also the data obtained by seismic refraction cannot always be interpreted immediately. Therefore, geognostic surveys referred to these stretches make use of other indirect methodologies. In this case, seismic reflection and other techniques such as seismic tomography can be very useful.

Also geognostic surveys aimed at the qualitative and quantitative study of stress fields within the rock mass are very important for these stretches, due to the heavy lithostatic load.

Sometimes, the importance of the work and the geological complexity are such that they require the drilling of an exploratory tunnel to carry out a number of tests and surveys useful for the following enlarging phases.

3.2.3 Hydrogeological Surveys

When talking about geognostic surveys aimed at the realization of underground works, the hydrogeological aspect requires to be considered by itself, as the presence of groundwater has a big influence on the excavation and drilling technique. Moreover, it may cause many serious problems.

Water is actually fundamental in redefining the equilibrium state of the cavity that has to be opened. For example, its presence in soil may cause a drastic reduction of the stability whereas, in rocks, open joints are preferential paths for underground water that can also lead to important tunnel inflows.

Furthermore, there are geological and structural situations (already discussed in Chaps. 1 and 2) that can determine very dangerous hydrogeological conditions for the construction of underground works, as they are associated with violent water flows or settlements. Yet, these situations cannot always be easily recognized on the surface.

As a consequence, before the designing phase of the excavation, specific surveys are required to identify the possible presence of water within the rock mass to be excavated. In particular, following hydrogeological aspects must be defined (Fig. 3.2):

- Geometry and reconstruction of the groundwater flow path (Fig. 3.3)
- Permeability determination of the intercepted materials and estimation of tunnel inflow (with forecasts about their location and entity)
- Forecast concerning the interference with existing springs and/or wells

This approach requires the execution of geognostic drilling tests, with in borehole permeability tests and piezometric level measures. Then, an articulated piezometric network (e.g. concerning the number of points and the depth) must be implemented on the basis of the geostructural and permeability features of the rock mass. This allows the monitoring of the aquifer, whose variation will be influenced by rainfalls, the extension of the supply area, etc.

To this purpose, it is important to gather the data about precipitations in the area, according to which the potentiality of the underground aquifer supply area can be assessed, possibly after a careful geomorphological study of the area.

In presence of different aquifers, it is necessary to identify the possible interconnection areas by means of focused surveys, carried out in the critical tunnel sections.

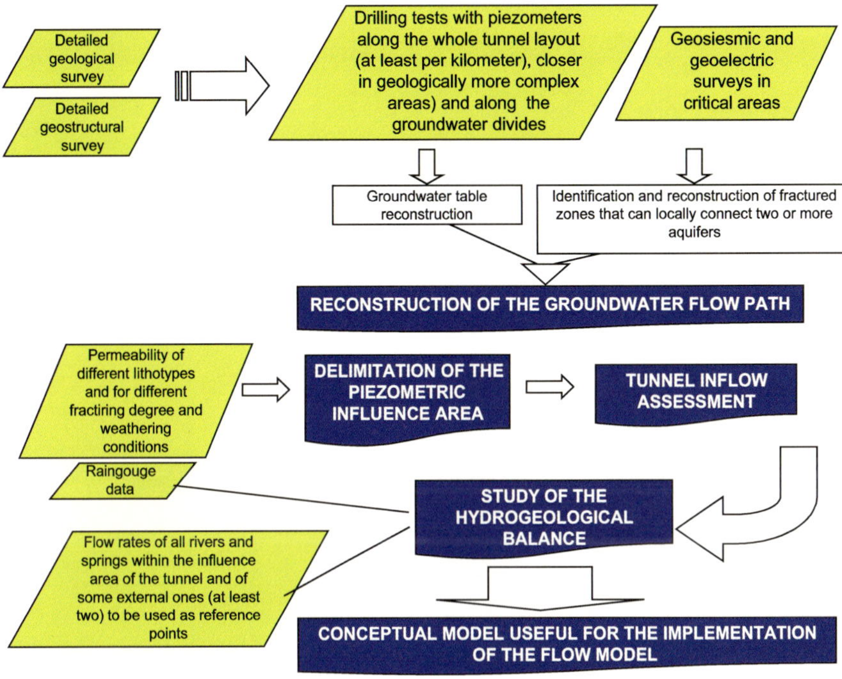

Fig. 3.2 Scheme of useful surveys to forecast the hydrogeological risk connected to the construction of a tunnel

Fig. 3.3 Example of the reconstruction of a deep groundwater flow path interested by the construction of a tunnel

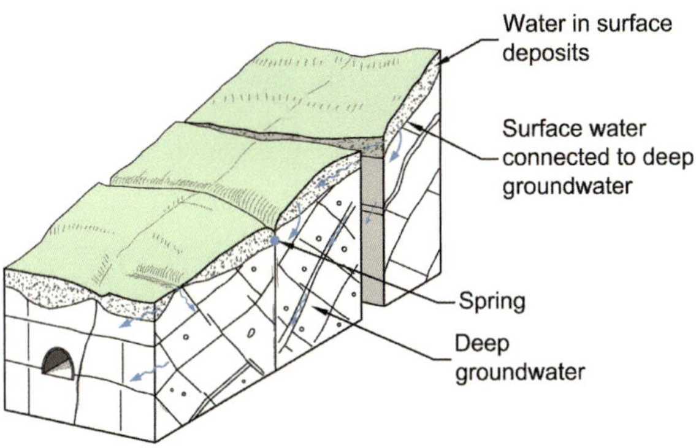

For examples, high-definition geoseismic reflection surveys allow to point out the presence and the trend of very fractured rock zones that constitute the supply areas of deep hydraulic circuits. Figure 3.4 provides an example of the interpretation of the geoseisimic survey results that highlight the presence of some anomalies in the propagation speed of seismic waves.

Those anomalies have to be related to the presence of very fractured rock zones, that probably constitute the supply areas of deep hydraulic circuits, that can locally

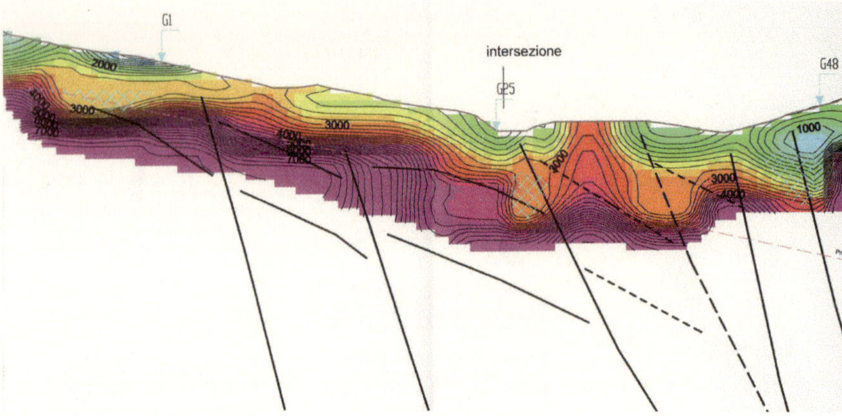

Fig. 3.4 Example of high definition reflection seismic profile. Speed is expressed in m/s. The blue dotted formation shows the shear zones, the red dotted line shows the tunnel layout

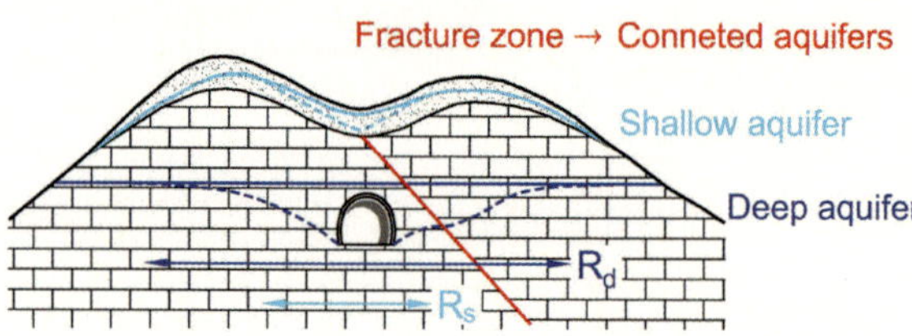

Fig. 3.5 Exemplification of the draining effects of the tunnel on the two aquifer levels (superficial level in light blue and deep level in blue navy) at a shear zone that causes the percolation in the unsaturated medium along the direction identified by the red lines. The lines represent the undisturbed water table, whereas the dotted lines represent the new water table following the tunnel draining effect on the two water levels

connect two or more aquifers (Fig. 3.5). In this case, the draining process induced by the tunnel excavation might lead to consequences on the deeper aquifer directly interested by the excavation as well as on more superficial waters, having potential negative effects on the regime of springs and rivers in the area.

3.3 Geological-Technical Characterization

The geological-technical characterization of the material interested by the excavation is carried out mainly by means of mechanical drilling tests (with collection of samples for lab tests and in borehole tests, Table 3.1), in situ tests and geophysical surveys (geo-seismic reflection or refraction and geo-electrical). In rock masses, geological-structural and geomechanical surveys on the surface are also required.

Table 3.1 In situ and lab geognostic tests

In situ geognostic surveys	
Tests and surveys to determine the stress-strain behaviour of the materials	Tests with flat jack[a]
	Plate bearing test
	Dilatometric tests[a], pressiometric tests[b], scissometric tests[b]
	Geophysical tests (sonic wave speed, seismic reflection and refraction, geoelectrical tests)
	Penetrometer tests[b]
	Joint Wall Compression Stress tests, sclerometer tests[a]
	Shear strength tests
Tests to determine the strike and the intensity of stresses	Doorstopper method[a]
	USBM method (two-dimensional stresses reconstruction)[a]
	CSIR triaxial strain cell (three-dimensional stresses reconstruction)[a]
	Flat jack[a]
	Hydraulic fracturing tests[a]
Tests to determine the hydrogeological behaviour of materials	Permeability tests, Lugeon tests[a], Lefranc tests[b]
	Pressiometric tests[b]
	Geophysical tests (seismic and electric tomographies)
	Tracer tests
Lab geognostic tests	
Tests to determine the material physical properties	Physical indexes of materials, sonic wave speed[a], grain size distribution[b], Atterberg limits[b],
Tests to determine the material mechanical properties	Uniaxial and triaxial tests, direct shear tests, tensile tests, oedometer tests[b], sonic wave propagation test[a], point load tests[a], Pocket penetrometer[b], swelling tests
Tests to determine permeability	Permeameters[b]

[a]Test carried out only on rocks
[b]Test carried out only on soils

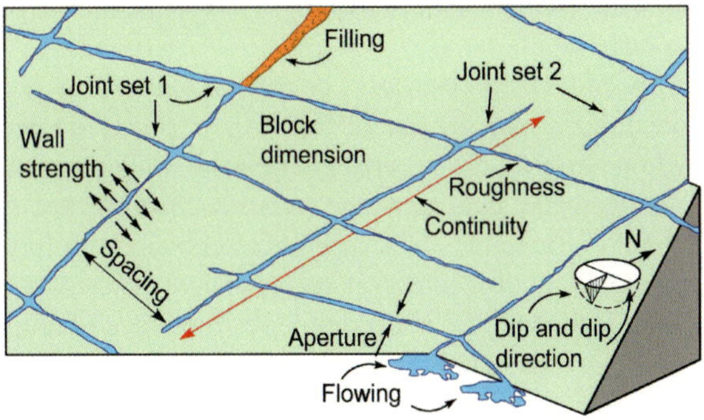

Fig. 3.6 Schematic representation of the joint features useful for rock mass characterization. (From Gonzalez de Vallejo et al. 2006, modified)

Those surveys allow to gather all data referring to the geometric, kinematics and dynamic characteristics of (brittle and/or ductile) deformative structures present inside rock masses, and their weathering condition (Table 3.2). In particular, great

Table 3.2 Weathering conditions of discontinuities (WD)

Term	Description	Symbol
WD		
Not weathered	There are no evidences of weathering on the discontinuity surface	WD1
Discoloured	The original colour of the rock is partially or completely changed	WD2
Weathered	The weathering interests the discontinuity surface as well as the rock for a thickness of about 1 mm The original texture is still evident	WD3
Very weathered	The weathering interests the in depth rock for several mm with the complete transformation of the rock into soil	WD4

attention is given to discontinuities (Figs. 3.6, 3.7a, b), constituting weak structural surfaces and determining the overall behaviour of the rock mass (Figs. 3.7c, d).

It's obvious that the type and number of geognostic surveys depend on different elements, such as:

- The complexity of the geological-structural context where the work is located, which is defined in the geological reference model.
- The length of the tunnel.
- The overburden (shallow, medium, high).

Obviously, the quantity of necessary data to formulate quite reliable geomechanical forecast is different for the excavation of a few hundred meters in a single geological formation or the excavation of some kilometres inside a highly tectonized mountain with thousand-metre high overburden.

In particular, the geological-technical characterization of the material interested by the excavation can be done both by means of in situ surveys, usually carried out in the boreholes or lab tests, using samples collected inside the same boreholes (more rarely from outcroppings or excavations). These geomechanical tests provide reliable parameters about physical, strength, deformation and permeability characteristics of the materials.

Specific in situ tests must be carried out both to determine existing stresses (e.g. doorstopper tests, hydraulic fracturing tests etc. on rock masses) and to define the hydrogeological characteristics (Lefranc, Lugeon tests etc.).

All gathered data allow the reconstruction of a geological-technical profile longitudinal to the tunnel axis, that has to contain both forecast information concerning geological, hydrogeological and technical aspects of the formations present at the tunnel level, and the main problems to be faced during the construction phase (instability of the excavation walls, water inflows, presence of gas, high temperature, etc.) (Fig. 3.8).

a

STATION: RGM 1 LYTHOLOGY: Dolomia Principale

SITE: S.P. 9 Valvestino (BS – Italy))

	STRATIFICATION	DISCONTINUITIES	
	S	K1	K2
Orientation	270°/32°	58/64	204/43
Spacing [cm]	13.2	5.1	25
Aperture [mm]	closed	closed	closed
Persistence (P1)	<50	<50	<50
Shape	undulating	undulating	undulating
Alteration	WD3	WD3	WD3
Filling	absent	absent	absent
Water	dry	dry	dry
Vertical intercept [cm]		19	
Horizontal intercept [cm]		16.5	
URV [cm^3]		1502	
VRU shape		Irregular	
Roughness (JRC)	1	9	5
γ [kN/m^3]		23	
σ_c app [MPa]	20.8	30.7	26.9

b

Parameter	Value	Score	
Compressive strength	58.34 MPa	15	
RQD	44.81%	8	
Joint spacing	5.16 cm	5	
Joint conditions	closed	30	
Water conditions	dry	15	
		RMR = 73	Class II

c

Parameter	Score	
RQD	44.81	
Jn	9	
Jr	2	
Ja	2	
Jw	1	
SRF	7.5	
	Q = 0.664	Very poor

d

Fig. 3.7 Example of **a** photo of the rock mass, **b** the sheet for geostructural and geomechanical survey data, **c** Bieniawski classification, **d** Barton classification

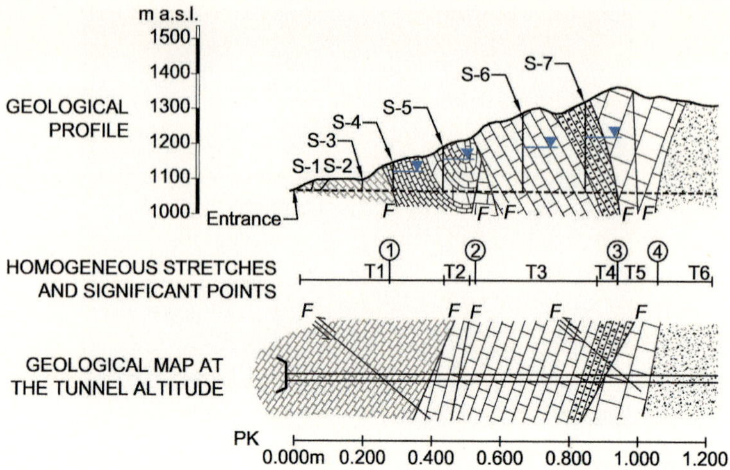

STRETCH	*T1*	*T2* *T3*	*T4* *T5* *T6*
SIGNIFICANT POINT	①	②	③ ④
LENGHT (m)	420	100	340
CHARACTERISTIC HITHOLOGY	marly limestone	dolomites	dolomites
WATER INFLOW	low with decreasing flow rates	variable with inrush points	sudden inflow with high flow rates
RMR CLASSIFICATION	45 - III - medium	72 - II - good	61 - II - good
σ_{ci} (MPa)	43	105	105
E_M (GPa)	3 - 6	5 - 13	6 - 14
EXCAVABILITY	rippable	blasting	blasting
SUPPORTS	Type III	Type I	Type I
STEP LEGHT (m)	from 1.5 to 2	4	from 2 to 3
SPECIAL TREATMENT	dowels and drains	no one	jet grout

Fig. 3.8 Example of geomechanical zoning in a tunnel. (From Gonzalez de Vallejo et al. 2006, modified)

3.4 Geomechanical Classifications

The design and realization of an underground work require not only geological and geological-technical data synthesised respectively in the geological and geological-technical model, but also strength and deformation data that are essential to define a reliable behaviour model of the materials as well as for the following design phase of the lining, of stabilization works, soil improvement works and also to choose the correct excavation method. If soils are present, those parameters can be obtained directly from in situ and lab tests carried out on the materials.

On the contrary, if a work develops through a rock mass, the correct determination of strength and deformability of that mass (considered as the combination of intact rock and discontinuities) by means of in situ and lab tests is difficult. The mechanical properties of the rock mass are preferably determined starting from the matrix properties, using empirical correlations and models based on the definition of a geomechanical "quality".

To that aim, during the years, different authors developed a number of criteria for the attribution of geomechanical qualities, known as Geomechanical Classifications. Some of them are essentially qualitative, others are based more on quantitative data.

Geomechanical classifications are very useful tools as they allow the categorization, within certain limits, of the behaviour of the material according to its geomechanical characteristics. Therefore, they represent a universal technical base, a common language for technicians of this sector. Nevertheless, it is always important to be extremely careful and to take into account their historical genesis and geographical origin. Moreover, it must be remembered that their origin is empirical, the result of their application depends on the single case and that the most updated versions has to be used.

Here is a brief description of the main classifications:

- Rock Mass Rating RMR
- Rock Mass Excavability RME
- Rock Mass index RMi
- Surface Rock Classification SRC
- Quality System Q
- Q_{TBM}

Some of these (RME and Q_{TBM}) have been created expressively for mechanical excavations and are aimed at determining TBM working conditions. Others can be applied to rock masses in general. In particular, Bieniawski classification (RMR) represents the link between those parameters determined by in situ and in laboratory tests and the definition of the Hoek-Brown constitutive model, that is widely used in the modelling of rock-mass behaviour.

It has to be said that, originally, some classifications were conceived to directly determine the strength parameters of the rock mass, define its behaviour and the load on the lining, and choose the stabilization measures. At present, thanks to the developments of research in rock mechanics, these secondary aspects of classification cannot be sufficiently reliable designing tools and have to be considered only for their historical interest or, at the very least, be used for preliminary considerations.

3.4.1 Bieniawski Classification (or of the RMR Index, Only Relevant for Rock Masses)

Bieniawski Geomechanics Classification (1989) or Rock Mass Rating (RMR) is based on a rating to be given to the rock, according to five parameters:

1. Strength of intact rock material, obtained by uniaxial compression tests or Point Load Test
2. RQD, that represents the modified core-recovery percentage within a borehole, and it is the ratio between the sum of the core pieces with length over 10 cm and

Table 3.3 Parameters of Bieniawski classification and their numerical coefficients. (Bieniawski 1989)

Classification parameters and their ratings									
Parameter			Range of values						
1	Strength of intact rock material	Point-load strength index	> 10 MPa	4–10 MPa	2–4MPa	1–2 MPa	For this low range-uniaxial compressive test is preferred		
		Uniaxial comp. Strength	> 250 MPa	100–250 MPa	50–100 MPa	25–50 MPa	5–25 MPa	1–5 MPa	< 1 MPa
	Rating		15	12	7	4	2	1	0
2	Drill core *Quality RQD*		90%–100%	75%–90%	50%–75%	25%–50%	< 25 %		
	Rating		20	17	13	8	3		
3	Spacing of discontinuities		> 2 m	0.6–2 m	200–600 mm	60–200 mm	< 60 mm		
	Rating		20	15	10	8	5		
4	Condition of discontinuities		Very rough surfaces Not continuous No separation Unweathered wall rock	Slightly rough surfaces Separation < 1 mm Slightly weathered walls	Slightly rough surfaces Separation < 1 mm Highly weathered walls	Slickensided surfaces or Gouge < 5 mm thick or Separation 1–5 mm Continuous	Soft gouge > 5 mm thick or Separation < 5 mm Continuous		
	Rating		30	25	20	10	0		
5	Groundwater	Inflow per 10 m tunnel length (l/m)	None	< 10	10–25	25–125	> 125		
		(Joint water press)/(Major principal σ)	0	< 0.1	0.1–0.2	0.2–0.5	> 0.5		
		General conditions	Completely dry	Damp	Wet	Dripping	Flowing		
	Rating		15	10	7	4	0		

the total core length; when drilling tests are not available, the RQD value can be estimated using empirical relations, such as:

$$\text{RQD} = (115 - 3.3 \cdot J_v) \qquad (J_v = \text{number of discontinuities per volume unit})$$

$$\text{RQD} = 100(0.1f + 1)e^{-0.1f}(f = \text{number of discontinuities per meter or frequency})$$

3. Spacing of discontinuities
4. Condition of discontinuities, with particular attention to their aperture, roughness, weathering degree of the walls, and the filling.
5. Groundwater, expressed as the flow rate of tunnel inflows or as the ratio between joint water pressures and major principal stress in situ, or, in terms of general conditions

A rating (numerical value) is given to each parameter by means of a table (Table 3.3) or graphics (Fig. 3.9). The sum of the rating points of the five parameters is the "quality score" of the mass (RMR).

The higher this number, ranging from 0 to 100, the better the quality of the rock mass. The value is then corrected according to the orientation of the tunnel axis with respect to that of the joints (Table 3.4 and Fig. 3.10)

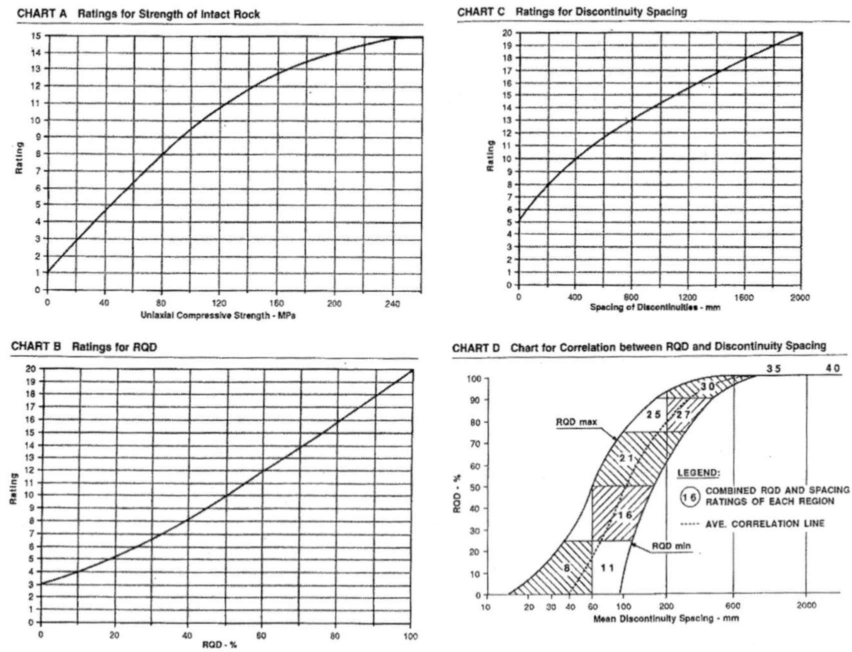

CHART E Guidelines for Classification of Discontinuity Conditions[a]

Parameter	Ratings				
Discontinuity length (persistence/continuity)	<1 m	1-3 m	3-10 m	10-20 m	>20 m
	6	4	2	1	0
Separation (aperture)	None	<0.1 mm	0.1-1.0 mm	1-5 mm	>5 mm
	6	5	4	1	0
Roughness	Very rough	Rough	Slightly rough	Smooth	Slickensided
	6	5	3	1	0
Infilling (gouge)		Hard filling		Soft filling	
	None	<5 mm	>5 mm	<5 mm	>5 mm
	6	4	2	2	0
Weathering	Unweathered	Slightly weathered	Moderately weathered	Highly weathered	Decompesed
	6	5	3	1	0

[a]Note: Some conditions are mutually exclusive. For example, if infilling is present, it is irrelevant what the roughness may be, since its effect will be overshadowed by the influence of the gouge.

Fig. 3.9 Charts for interpolating the ratings of the different parameters of the Bieniawski classification (1989)

With the value thus obtained it is possible to place the rock in one of the five classes. These classes also have a range of values concerning cohesion and shear strength angle (to be used only for very generalized evaluations). There is also a comment on

Table 3.4 Correction of the numerical coefficients according to the discontinuities orientation. (Bieniawski 1989)

	Strike and dip orientations	Very favourable	Favourable	Fair	Unfavourable	Very Unfavourable
Ratings	Tunnels and mines	0	−2	−5	−10	−12
	Foundations	0	−2	−7	−15	−25
	Slopes	0	−5	−25	−50	

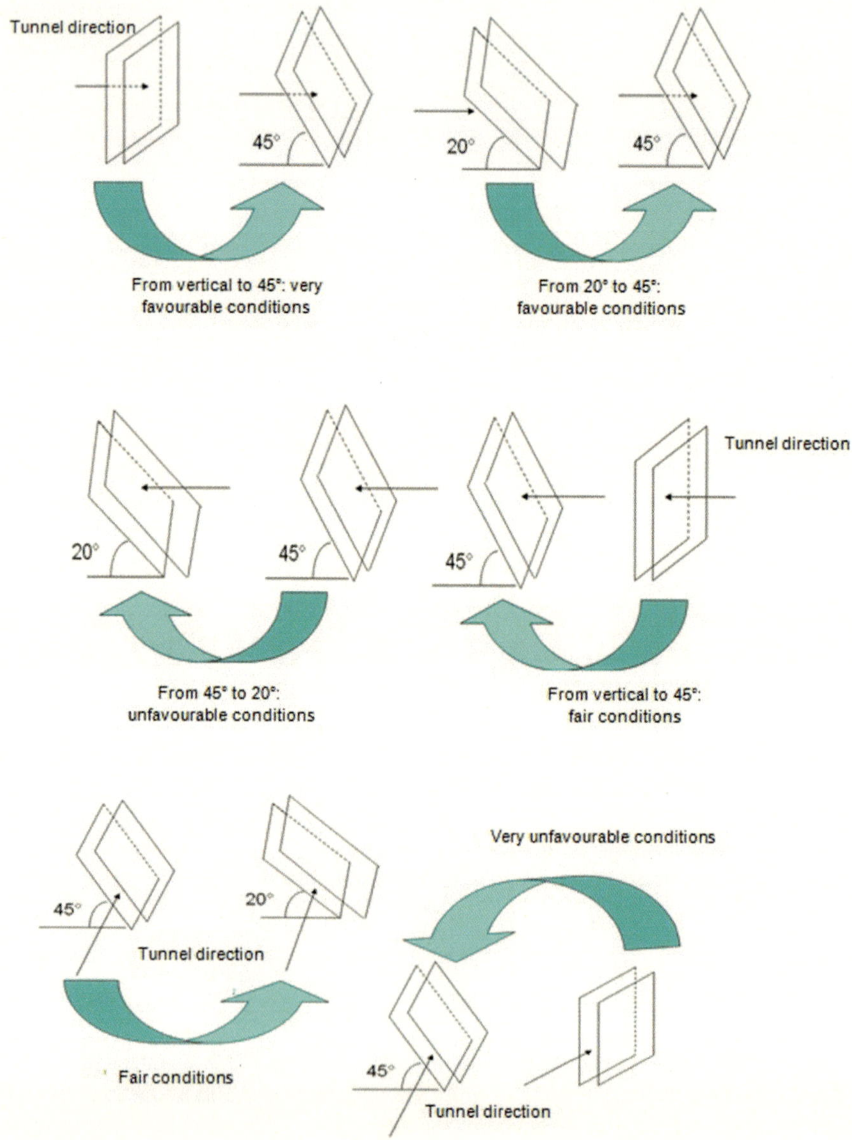

Fig. 3.10 Evaluation of the effects of discontinuities orientation with respect to the direction of the tunnel (for any angle between strike and tunnel axis, with dip between 0° and 20° = discrete conditions)

Table 3.5 Rock mass classes of the Bieniawski classification (1989) and their meaning

Rating	100 ← 81	80 ← 61	60 ← 41	40 ← 21	< 21
Class number	I	II	III	IV	V
Description	Very good rock	Good rock	Fair rock	Poor rock	Very poor rock
Average stand-up time	20 yrs for 15 m span	1 year for 10 m span	1 week for 5 m span	10 hrs for 2.5 m span	30 min for 1 m span
Cohesion of rock mass (kPa)	> 400	300–400	200–300	100–200	< 100
Friction angle of rock mass (°)	> 45	35–45	25–35	15–25	< 15

the excavation problems and the self-supporting qualities of the material (e.g. stand up time without any reinforcement) (Table 3.5).

Moreover, advancing modalities and the type of temporary support to be adopted are suggested: steel arches, bolts or shot-concrete (Table 3.6), also these indications can only be used for preliminary evaluations.

The RMR rating calculated without taking into account the Rating R6 (correction according to the discontinuities orientation), called Basic RMR or BRMR, can be correlated to elastic and deformation modulus by means of seldom used empiric relations:

$$E = RMR^{2.615} \times 0.451 \quad [MPa] \quad \text{(load perpendicular to discontinuities)}$$

$$E = RMR^{2.525} \times 0.913 \quad [MPa] \quad \text{(load parallel to discontinuities)}$$

$$E_{ed} = RMR^{2.977} \times 0.067 \quad [MPa] \quad \text{(load perpendicular to discontinuities)}$$

$$E_{ed} = RMR^{2.843} \times 0.183 \quad [MPa] \quad \text{(load parallel to discontinuities)}$$

3.5 Rock Mass Excavability Index RME

The RME (Bieniawski et al. 2006) evaluates rock mass excavability in terms of TBM performance and it serves as a tool for choosing the type of TBM most appropriate for tunnel construction in given rock mass conditions.

The RME index is calculated using five input parameters having these initial ratings (Table 3.7):

- Uniaxial compressive strength of intact rock material: 0–25 rating points
- Drilling rate index DRI: 0–15 points
- Number of discontinuities present at tunnel face, their orientation with respect to tunnel axis and homogeneity at tunnel face: 0–30 points
- Stand up time of the tunnel front: 0–25 points
- Water inflow at tunnel front: 0–5 points.

The sum of the rating of the above parameters varies between 0–100 rating points and it is expected that the higher the RME value, the easier and more productive

Table 3.6 Advancing methods and types of reinforcement according to the five classes of the RMR: indication for a 10 m-long excavation, u-shaped, vertical stress < 25 MPa and drill and blast construction. (Bieniawski 1989)

Rock mass class	Excavation	Rock bolts (20 mm diameter, fully grouted)	Shotcrete	Steel sets
I Very good rock RMR: 81–100	Full face, 3 m advance	Generally no support required except spot bolting		
II Good rock RMR: 61–80	Full face, 1–1.5 m advance. Complete support 20 m from face.	Locally, bolts in crown 3 m long, spaced 2.5 m with occasional wire mesh	50 mm in crown where required	None
III Fair rock RMR: 41–60	Top heading and bench 1.5–3 m advance in top heading. Commence support after each blast. Complete support 10 m from face	Systematic bolts 4 m long, spaced 1.5–2 m in crown and walls with wire mesh in crown	50–100 mm in crown and 30 mm in sides	None
IV Poor rock RMR: 21–40	Top heading and bench 1.0–1.5 m advance in top heading. Install support concurrently with excavation, 10 m from face	Systematic bolts 4–5 m long, spaced 1–1.5 m in crown and walls with wire mesh	100–150 mm in crown and 100 mm in sides	Light to medium ribs spaced 1.5 m where required
V Very poor rock RMR: < 20	Multiple drifts 0.5–1.5 m advance in top heading. Install support concurrently with excavation. Shotcrete as soon as possible after blasting	Systematic bolts 5–6 m long, spaced 1–1.5 m in crown and walls with wire mesh. Bolt invert	150–200 mm in crown, 150 mm in sides, and 50 mm on face	Medium to heavy ribs spaced 0.75 m with steel lagging and fore poling if required. Close invert

the excavation of the tunnel. Using the RME index, the theoretical average rate of advance (ARA_T) of TBM can be estimated:

$$ARA_T = 0.422RME - 11.61$$

Table 3.7 The ratings for RME input parameters

<table>
<tr><td colspan="11">Uniaxial compressive strength of intact rock [0 - 25 points]</td></tr>
<tr><td>σ_c [MPa]</td><td colspan="2"><5</td><td colspan="2">5-30</td><td colspan="2">30-90</td><td colspan="2">90-180</td><td colspan="2">>180</td></tr>
<tr><td>Rating</td><td colspan="2">4</td><td colspan="2">14</td><td colspan="2">25</td><td colspan="2">14</td><td colspan="2">0</td></tr>
<tr><td colspan="11">Drillability [0 - 15 points]</td></tr>
<tr><td>DRI</td><td colspan="2">>80</td><td colspan="2">80-65</td><td colspan="2">65-50</td><td colspan="2">50-40</td><td colspan="2"><40</td></tr>
<tr><td>Rating</td><td colspan="2">15</td><td colspan="2">10</td><td colspan="2">7</td><td colspan="2">3</td><td colspan="2">0</td></tr>
<tr><td colspan="11">Discontinuities in front of the tunnel face [0 - 30 points]</td></tr>
<tr><td colspan="3">Homogeneity</td><td colspan="5">Number of joints per meter</td><td colspan="3">Orientation with respect to tunnel axis</td></tr>
<tr><td colspan="2">Homogeneous</td><td>Mixed</td><td>0-4</td><td>4-8</td><td>8-15</td><td>15-30</td><td>>30</td><td>Perpendicular</td><td>Oblique</td><td>Parallel</td></tr>
<tr><td>Rating</td><td>10</td><td>0</td><td>2</td><td>7</td><td>15</td><td>10</td><td>0</td><td>5</td><td>3</td><td>0</td></tr>
<tr><td colspan="11">Stand up time [0 - 25 points]</td></tr>
<tr><td>Hours</td><td colspan="2"><5</td><td colspan="2">5-24</td><td colspan="2">24-96</td><td colspan="2">96-192</td><td colspan="2">>192</td></tr>
<tr><td>Rating</td><td colspan="2">0</td><td colspan="2">2</td><td colspan="2">10</td><td colspan="2">15</td><td colspan="2">26</td></tr>
<tr><td colspan="11">Groundwater inflow [0 - 5 points]</td></tr>
<tr><td>Liters/sec</td><td colspan="2">>100</td><td colspan="2">70-100</td><td colspan="2">30-70</td><td colspan="2">10-30</td><td colspan="2"><10</td></tr>
<tr><td>Rating</td><td colspan="2">0</td><td colspan="2">1</td><td colspan="2">2</td><td colspan="2">4</td><td colspan="2">5</td></tr>
</table>

Subsequently, to get the real average rate of advance (ARA_R) of TBM from ARA_T, some adjustment factors are needed, by considering the influence of the TBM crew, the excavated length, and the tunnel diameter.

3.5.1 Rock Mass index RMi

The Rock Mass index (RMi, Palmström 1996) combines numerical values of relevant parameters in the rock mass to express the RMi value. Most of these parameters, including the rock material and the joints intersecting it, can be found from common observations or measurements in the field.

The input data in the RMi is shown in Fig. 3.11. RMi is based on the principle that the joints intersecting a rock mass tend to reduce its strength. It is therefore expressed as: $\text{RMi} = \sigma_c JP$, where, σ_c is the uniaxial compressive strength of intact rock, measured on 50 mm samples, and JP is the jointing parameter, expressing the reduction in strength of the intact rock caused by the joints (Fig. 3.11). The JP can be calculated as:

$$JP = 0.2\sqrt{jC} \times Vb^D$$

Where, jC and Vb are respectively the joint condition factor and the block volume (Table 3.8), and D is:

$$D = 0.37 jC^{-0.2}$$

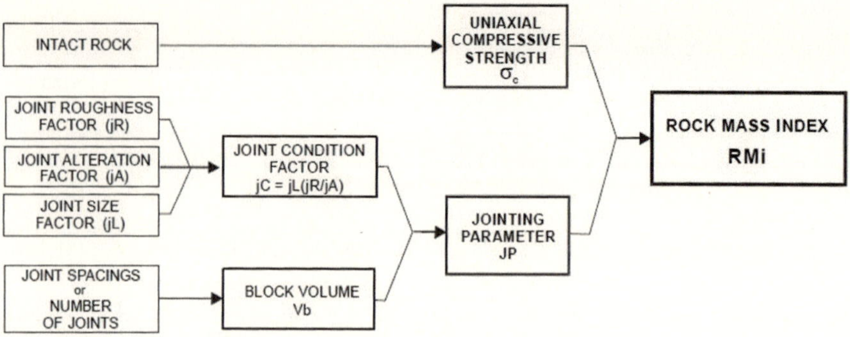

Fig. 3.11 RMi calculation scheme. (Palmstrom 1996)

The RMi-value is an approximate measure of the uniaxial strength of the rock mass. It can be used in several calculation methods in rock engineering and rock mechanics, such as for rock support estimates in underground excavations, input parameters to the Hoek-Brown failure criterion for rock masses, and for estimating penetration rate of TBMs (tunnel boring machines). In addition, RMi can be useful in estimation of some input data used in numerical modelling.

3.5.2 Surface Rock Classification SRC

The Surface Rock Classification (SRC) system (Gonzalez de Vallejo 2003) was developed from RMR index to take into account in situ stress, data from outcrops and tunnel construction conditions. The SRC index is calculated from the parameters shown in Table 3.9a, to which the correction factors shown in Table 3.9b are applied. The scores obtained, and the corresponding geomechanical rock classes, classify the rock mass in conditions prior to excavation and represent the SRC basic. To account for effects due to constructions, the correction factors shown in Table 3.9c are applied to give the SRC-corrected. To characterize the properties of the rock mass, the criteria used in RMR classification are directly applied to the value obtained for the SRC.

3.5.3 Barton Q-System Classification

This method is based on the definition on the quality index Q (Rock Mass Quality), obtained from following relation:

$$Q = \text{RQD}/J_n \cdot J_r/J_a \cdot J_w/\text{SRF}$$

Table 3.8 The input parameters to RMi

Uniaxial compressive strength of intact rock (σ_c)	**value** in Mpa (from lab. tests or assumed from handbook tables)
Block volume (Vb)	**value** in m³ (from observations at site or on drill cores, etc.)
Joint condition factor (jC)	**jC = jR x jL / jA** (ratings of jR, jA and jL from the tables below)

jR (joint roughness factor, which is composed of large scale and small scale undulations, similar to Jr in the Q-system)

(The ratings in***bold italic*** are similar to Jr)		Large scale waviness of joint plane				
		Planar	Slightly undulating	Undulating	Strongly undulating	Stepped or interlocking
Small scale smoothness of joint surface	Very rough	2	3	4	6	6
	Rough	1.5	2	3	4.5	6
	Smooth	1	1.5	2	3	4
	Polished or slickensided*)	0.5	1	1.5	2	3
	For filled joints jR = 1 For irregular joints a rating of jR = 6 is suggested					

**) For slickensided surfaces the ratings apply to possible movement along the lineations*

jA (joint alteration factor, which ratings are based on Ja in the Q-system)

Contact between joint walls	CLEAN JOINTS:	Healed or welded joints	filling of quartz, epidote, etc.		jA = 0.75
		Fresh joint walls	no coating or filling, except from staining (rust)		1
		Altered joint walls	- one grade higher alteration than the rock		2
			- two grades higher alteration than the rock		4
	COATING or THIN FILLING OF:	Frictional materials	sand, silt calcite, etc. without content of clay		3
		Cohesive materials	clay, chlorite, talc, etc.		4
Partly or no wall contact	THICK FILLING OF:	Frictional materials	sand, silt calcite, etc. (non-softening	jA = 4	8
		Hard, cohesive materials	clay, chlorite, talc, etc.	6	5-10
		Soft, cohesive materials	clay, chlorite, talc, etc.	8	12
		Swelling clay materials	material exhibits swelling properties	8 - 12	13 - 20
				Thin filling (< 5 mm)	Thick filling

jL (joint size factor, which is composed of the length and continuity of the joint)		Continuous joints	Discont. joints *)
Bedding or foliation partings	length < 0.5 m	jL = 3	jL = 6
Joints	with length 0.1- 1 m	2	4
	with length 1- 10 m	1	2
	with length 10- 30 m	0.75	1.5
(Filled) joint, seam or shear **)	length > 30 m	0.5	1

**) Discontinuous joints end in massive rock **) Often a singularity and should in these cases be treated separately*

where:

1. RQD is the modified core recovery percentage of a drilling test (Table 3.10). When no cores are available to determine the RQD, this parameter can be estimated using following empirical relation: RQD = 115−3.3J_v where J_v is the number of discontinuities be volume unit (1 m^3);
2. J_n is the number of discontinuity sets (Table 3.10);

Table 3.9 Geomechanical Rock Mass classification system SRC: (a) SRC basic, (b) adjustment to ratings to account for the surface data, and (c) adjustment to ratings to account for construction factors (Gonzales de Vallejo 2003)

(a)							
(1) Intact rock strength							
Point-load test (MPa)	>8	8-4	4-2	2-1	Not applicable		
Uniaxial compressive strength (MPa)	>250	250-100	100-50	50-25	25-5	5-1	<1
Rating	20	15	7	4	2	1	0
(2) Spacing or RQD							
Spacing (m)	>2	2-0.6	0.6-0.2	0.2-0.06	<0.06		
RQD (%)	100-90	90-75	75-50	50-25	<25		
Rating	25	20	15	8	5		
(3) Conditions of discontinuities	Very rough surfaces	Slightly rough surfaces	Slightly rough surfaces	Slickensided surfaces	Slickensided surfaces		
	Not continuous joints	Not continuous joints	Not continuous joints	continuous joints	Continuous joints		
	No separation	Separation > 1mm	Separation 1mm	Joints open 1-5 mm	Joints open < 5mm		
	Hard joint wall	Hard joint wall	Soft or weathered joint walls	Gouge materials	Gouge materials millimeterthick		
Rating	30	25	20	10	0		
(4) Groundwater inflow per 10-m tunnel length (l/min)	None	<10	10-25	25-125	>125		
General conditions	Dry	Slightly moist	Occasional seepage	Frequent seepage	Abundant seepage		
Rating	15	10	7	4	0		
(5) State of stresses							
Competence factor[a]	>10	10-5	5-3	<3	-		
Rating	10	5	-5	-10			
Tectonic structures	Zones near thrusts/faults of regional importance	Compression	Tension				
Rating	-5	-2	0				
Stress relief factor[b]	>200	200-80	80-10	<10	Slopes		
					200-80	79-10	<10
Rating	0	-5	-8	-10	-10	-13	-15
Neotectonic activity	None unknown	Low	High				
Rating	0	-5	-10				
(6) Rock mass classes							
Class number	I	II	III	IV	V		
Rock quality	Very good	Good	Fair	Poor	Very Poor		
Rating	100-81	80-61	60-41	40-21	≤20		

[a] Uniaxial intact rock strength/vertical stress.

[b] Ratio between the age of the last main orogenic deformation affecting the rock mass (in years x 10-3) and maximum overburden thickness (in meters).

Table 3.9 (continued)

(b) The total rating from Table (a) must be adjusted fro the surface data	
Spacing and RQD	
Compression fractures = 1.3	
Tension fractures = 0.8	
weathering degree $\geq$ IV=0.8	
weathering degree III=0.9	
weathering degree I or 11 = 1.0	
For depths <50 m = 1.0	
The maximum score is 25 points	
Conditions of discontinuities	
Compression fractures: + 5	
Tension fractures: 0	
Not applicable for depths <50 m	
The maximum score is 30 points	
Groundwater	
Compression fractures : + 5	
Tension fractures: 0	
Not applicable for depths <50 m	
(c) The total rating from Table a must be adjusted for the following factors:	
Excavation methods	
Tunneling boring machines, continuos miner,cutter machines, roadheaders, etc.	+5
Controlled blasting, presplitting, soft blasting, etc.	0
Poor-quality blasting[a]	-10
Stand up time[b]	
Class I	0
Class 11	
<10 d	0
>10 -< 20 d	-5
>20 d	-10
Class III	
<2 d	0
>2 -< 5 d	-5
>5 -< 10 d	-10
>10 d	-20
Classes IV and V	
<8 h	0
>8 -<24 h	-10
>24 h	-20
Distance to adjacent excavation[c]	
AEF<2.5	-10
Portals, accesses and areas withsmall overburden thickness[d]	
PF<3	-10
Rock durability to weathering[e]	
Rock of high durability (low clay content)	0
Rock of low durability (high clay content)	-5
Rock of very low durability (very high clay content)	-10

Table 3.9 (continued)

Discontinuity orientations[g]						
Strike perpendicular to hmnel axis				Strike parallel to tunnel axis		Dip 0-20° at any direction
Drive with dip		Drive against dip				
Dip 45-90°	Dip 20-45 "	Dip 45-90°	Dip 20-45°	Dip 45-90°	Dip 20-45°	Unfavourable
(very favourable)	(favourable)	(fair)	(unfavourable)	(very unfavourable)	(fair)	
0	-2	-5	-10	-12	-5	-10

[a] Conventional blasting: 0.

[b] Based on Bieniawski's (1989) graphic representation of the stand-up time and the unsupported span, the ratings are applied in relation to the maximum stand-up time. d: days, h: hours.

[c] AEF is the adjacent excavation factor, defined as the ratio between the distance to an adjacent excavation (in meters) from the excavation under design and the span of the adjacent excavation (in meters).

[d] PF is the portal factor, defined as the ratio between the thickness of overburden and the span of the excavation, both in meters.

[e] Durability can be assessed by the slake durability test, or indirectly by the clay content.

[f] After Bieniawski (1978).

3. J_r represents the roughness of discontinuity surfaces making a distinction between closed and open joints. If more discontinuities sets are present with different J_r values, the lowest value is considered, as the more unfavourable mass conditions for stability are taken as a reference (Table 3.10);
4. J_a refers to the weathering and filling of discontinuities, always making a distinction between closed and open joints. If more discontinuities sets are presents, the value representing the worst situation is considered, e.g. the highest (Table 3.11);
5. J_w is a reduction factor linked to the presence of water inside discontinuities (Table 3.12);
6. SRF is a reduction factor taking into consideration the stresses inside the rock mass. Four situations are possible: presence of weak zones interrupting the mass continuum; rigid massive rocks; squeezing rocks with plastic behaviour; swelling rocks (Table 3.13).

Referring to this classification, it can be observed that the first product of the relation is expressed in a unit volume, the second provides information on the shear strength along the joints and the last one refers to the active stress on the rock mass.

Once the value of Q is obtained (ranging from 0.001 to 1,000), the belonging to one of the rock classes distributed along a logarithmic scale is determined (Fig. 3.12). Thanks to that classification it is also possible to assess quickly the possible need to predispose supporting measures during the excavation. To that aim, the "Equivalent Dimension" has to be defined, resulting from the ratio between the diameter of the cavity and a safety value ESR that depends on the type of tunnel (Table. 3.14).

Using a graphic (Fig. 3.13) with Q on the x-axis and "equivalent dimension" on the y-axis, it will be cleared if supporting measures (bolts, mesh, steel arches, shot concrete, rock anchors, nails etc.) will be needed during the excavation.

Table 3.10 *RQD*, J_n and J_r parameters. (Barton et al. 1974)

Description	Value	Notes
1. Rock quality designation	*RQD*	
A. Very poor	0–25	1. Where RQD is reported or measured as < 10 (including 0), a nominal value of 10 is used to evaluate Q
B. Poor	25–50	
C. Fair	50–75	
D. Good	75–90	2. RQD intervals of 5, i.e. 100, 95, 90 etc. are sufficiently accurate
E. Excellent	90–100	
2. Joint set number	J_n	
A. Massive, no or few joints	0.5–1.0	
B. One joint set	2	
C. One joint set plus random	3	
D. Two joint sets	4	
E. Two joint sets plus random	6	
F. Three joint sets	9	1. For intersections use (3.0 x J_n)
G. Three joint sets plus random	12	
H. Four or more joint sets, random, heavily jointed, "sugar cube", etc.	15	2. For portals use (2.0 x J_n)
J. Crushed rock, earthlike	20	
3. Joint roughness number	J_r	
a. Rock wall contact		
b. Rock wall contact before 10 cm shear		
A. Discontinuous joints	4	
B. Rough and irregular, undulating	3	
C. Smooth undulating	2	
D. Slickensided undulating	1.5	1. Add 1.0 if the mean spacing of the relevant joint set is greater than 3 m
E. Rough or irregular, planar	1.5	
F. Smooth, planar	1.0	
G. Slickensided, planar	0.5	2. $J_r = 0.5$ can be used for planar, slickensided joints having lineations, provided that the lineations are oriented for minimum strength
c. No rock wall contact when sheared		
H. Zones containing clay minerals thick enough to prevent rock wall contact	1.0	(nominal)
J. Sandy, gravely or crushed zone thick enough to prevent rock wall contact	1.0	(nominal)

For various aspects, this classification may be considered as the most complete among the existing ones both for the range of the parameters considered and for the wide range of numerical values the Q index may have.

Table 3.11 J_a parameter (Barton et al. 1974)

Description	Value	Notes	
4. Joint alteration number	J_a	$Ø_r$ degrees (approx.)	
a. Rock wall contact			
A. Tightly healed, hard, non-softening, impermeable filling	0.75		1. Values of $Ø_r$, the residual friction angle, are intended as an approximate guide to the mineralogical properties of the alteration products, if present
B. Unaltered joint walls, surface staining only	1.0	25–35	
C. Slightly altered joint walls, non-softening mineral coatings, sandy particles, clay-free disintegrated rock, etc.	2.0	25–30	
D. Silty-, or sandy-clay coatings, small clay-fraction (non-softening)	3.0	20–25	
E. Softening or low-friction clay mineral coatings, i.e. kaolinite, mica. Also chlorite, talc, gypsum and graphite etc., and small quantities of swelling clays. (Discontinuous coatings, 1–2 mm or less)	4.0	8–16	
b. Rock wall contact lower than 10 cm shear			
F. Sandy particles, clay-free, disintegrating rock etc.	4.0	25–30	
G. Strongly over-consolidated, non-softening clay mineral fillings (continuous < 5 mm thick)	6.0	16–24	
H. Medium or low over-consolidation, softening clay mineral fillings (continuous < 5 mm thick)	8.0	12–16	
J. Swelling clay fillings, i.e. montmorillonite, (continuous < 5 mm thick). Values of J_a depend on percent of swelling clay-size particles, and access to water	8.0–12.0	6–12	
c. No rock wall contact when sheared			
K. Zones or bands of disintegrated or crushed rock and clay (see G, H and J for clay conditions)	6.0–12.0	6–24	
N. Zones or bands of silty- or sandy-clay, small clay fraction, non-softening	5.0		
O. Thick continuous zones or bands of clay (see G.H and J for clay conditions)	10.0–13.0	6.0–24.0	

Still, there is a serious problem regarding the evaluation of the SRF parameter, as a quite precise determination of the stress field is usually very complex and expensive. In a publication, Bieniawski (1979) analyzed a number of case histories of

Table 3.12 J_w parameter (Barton et al. 1974)

Description	Value	Notes	
5. Joint water reduction	J_w	approx. water pressure (MPa)	
A. Dry excavation or minor inflow i.e. < 5 l/m Locally	1.0	< 0.1	
B. Medium inflow or pressure, occasional outwash of joint fillings	0.66	0.1–0.25	
C. Large inflow or high pressure in competent rock with unfilled joints	0.5	0.25–1.0	1. Factors C to F are crude estimates; increase J_w if drainage installed
D. Large inflow or high pressure	0.33	0.25	
E. Exceptionally high inflow or pressure at blasting, decaying with time	0.2–0.1	> 1	2. Special problems caused by ice formation are not considered
F. Exceptionally high inflow or pressure	0.1–0.05	> 1	

underground excavations and deduced that following relation exists between quality values of rock masses obtained by means of RMR classifications and Q-System (to be used carefully):

$$\text{RMR} = 9 \ln Q + 44$$

After him, other authors found similar correlations:

$$\text{RMR} = 13.5 \log Q + 43 \qquad \text{(Rutledge, 1978)}$$

$$\text{RMR} = 38 + 8.7 \ln Q \qquad \text{(Kaiser and Gale, 1985)}$$

$$\text{RMR} = 15 \log Q + 50 \qquad \text{(Barton, 1995)}$$

In this way, both classifications can be used separately and confronted, thus limiting the errors.

Barton defines a normalised value of Q, given by:

$$Q_c = Q \cdot \frac{\sigma_{\text{ci}}}{100}$$

where:

Q is the previously defined Barton's parameter

σ_{ci} is the uniaxial compressive strength of the intact rock expressed in MPa.

Author in his recent works (Barton 2013) suggests that the Q_c-formulation can be used to define easily some of the most important parameters:

- rock mass modulus:

$$E_{rm} = 10 Q_c{}^{1/3} [\text{GPa}]$$

Table 3.13 SRF parameter (Barton et al. 1974)

Description			Value	Notes
6. Stress Reduction Factor			SRF	
a. Weakness zones intersecting excavation, which may cause loosening of rock mass when tunnel is excavated				
A. Multiple occurrences of weakness zones containing clay or chemically disintegrated rock, very loose surrounding rock any depth)			10.0	1. Reduce these values of *SRF* by 25–50 % but only if the relevant shear zones influence do not intersect the excavation
B. Single weakness zones containing clay, or chemically disintegrated rock (excavation depth < 50 m)			5.0	
C. Single weakness zones containing clay, or chemically disintegrated rock (excavation depth > 50 m)			2.5	
D. Multiple shear zones in competent rock (clay free), loose surrounding rock (any depth)			7.5	
E. Single shear zone in competent rock (clay free), (depth ofexcavation < 50 m)			5.0	
F. Single shear zone in competent rock (clay free), (depth ofexcavation > 50 m)			2.5	
G. Loose open joints, heavily jointed or 'sugar cube', (any depth)			5.0	
b. Competent rock, rock stress problems				
	σ_c/σ_1	$\sigma_t\ \sigma_1$		2. For strongly anisotropic virgin stress field (if measured): when $5 < \sigma_1/\ \sigma_3 < 10$, reduce σ_c To $0.8\sigma_c$and σ_t to 0.8 σ_t. When $\sigma_1/\ \sigma_3 > 10$, reduce σ_c and σ_t to 0.6 σ_c and 0.6 σ_t. where σ_c = unconfined compressive strength, and σ_t = tensile strength (point load) and σ_1 and σ_3 are the major and minor principal stresses
H. Low stress, near surface	> 200	> 13	2.5	
J. Medium stress	200–10	13–0.66	1.0	
K. High stress, very tight structure (usually favourable to stability, may be unfavourable to wall stability)	10–5	0.66–0.33	0.5–2	
L. Mild rockburst (massive rock)	5–2.5	0.33–0.16	5–10	
M. Heavy rockburst (massive rock)	< 2.5	< 0.16	10–20	3. Few case records available where depth of crown below surface is less than span width
c. Squeezing rock, plastic flow of incompetent rock under influence of high rock pressure				
N. Mild squeezing rock pressure			5–10	
O. Heavy squeezing rock pressure			10–20	
d. Swelling rock, chemical swelling activity depending on presence of water				
P. Mild swelling rock pressure			5–10	
R. Heavy swelling rock pressure			10–15	

- shear strength angle:

$$\phi = arctg\left(\frac{J_r}{J_a}\cdot\frac{J_w}{1}\right)\ [°]$$

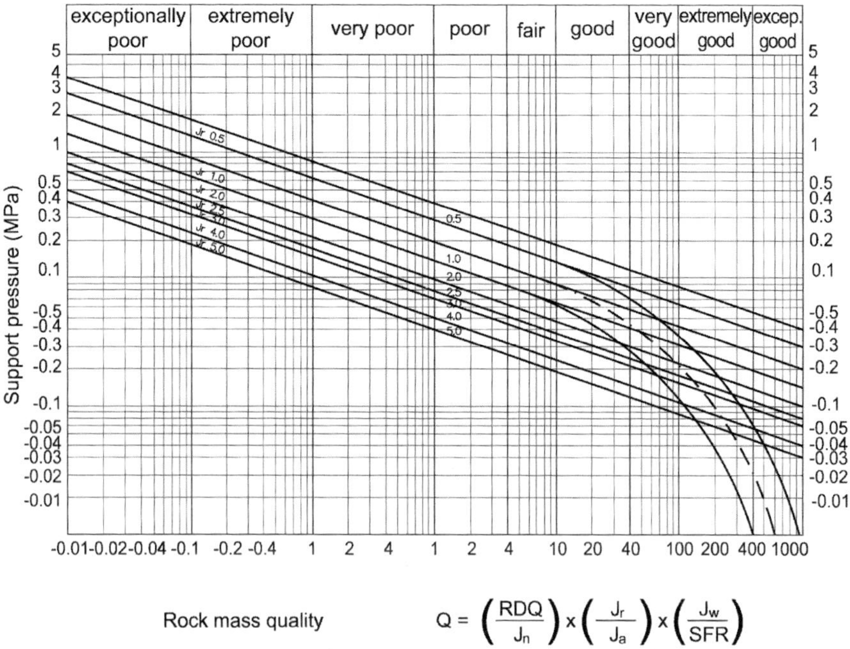

Fig. 3.12 Q parameter and rock quality. (Barton et al. 1974)

Table 3.14 Value of the coefficient ESR (Excavation Support Ratio) according to the type of underground cavity. (Barton et al. 1974)

Excavation Category		ESR
A	Temporary mine openings	3–5
B	Permanent mine openings, water tunnels for hydro-electric projects, pilot tunnels, drifts and headings for large excavations	1.6
C	Storage rooms, water treatment plants, minor road and railway tunnels, surge chambers and access tunnels in hydro-electric project	1.3
D	Underground power station caverns, major road and railway tunnels, civil defense chamber, tunnel portals and intersections	1.0
E	Underground nuclear power stations, railway stations, sports and public facilities, underground factories	0.8

- cohesion:

$$c = \left(\frac{\text{RQD}}{J_n} \cdot \frac{1}{\text{SRF}} \cdot \frac{\sigma_c}{100}\right) [\text{MPa}]$$

- global "rock mass strength":

$$\sigma_{cm} = 5 \cdot \gamma \cdot {Q_c}^{1/3} [\text{MPa}]$$

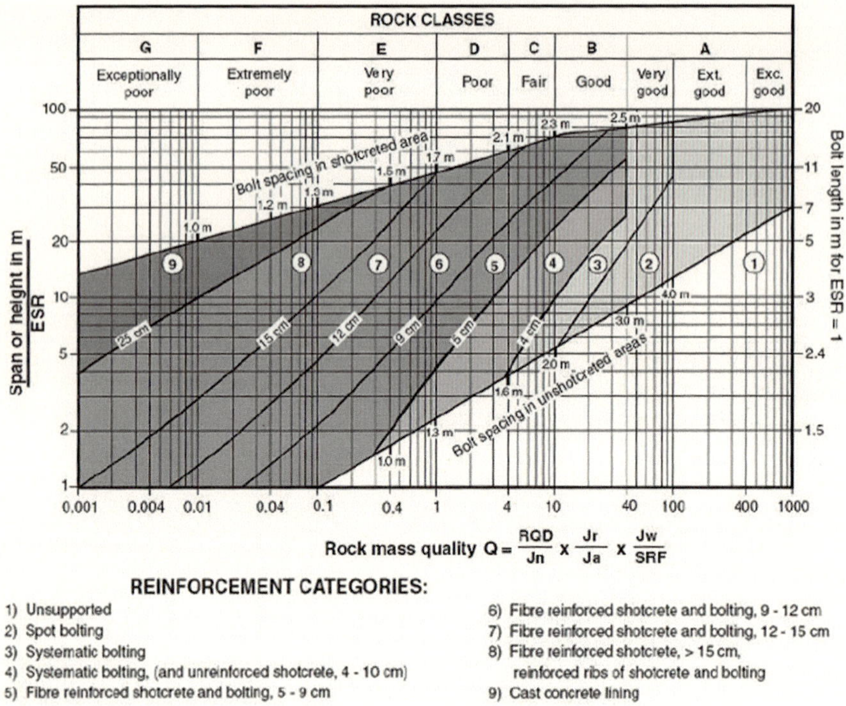

Fig. 3.13 Guide-lines of Q-based permanent, single-shell support. (Barton et al. 1974)

- Lugeon parameter:

$$L(Lugeon) = \frac{1}{Q_c}\left[\frac{l}{min}\right]$$

- cavity convergence

$$\Delta = \frac{\text{SPAN}(m)}{Q_c}[mm]$$

- Velocity (V_p) of primary waves obtained in seismic refraction tests (Q-system normally requires a high number of boreholes, and permeability tests. The advantage of linking Q to the V_p consists in reduction of the number of these tests and possibility of extrapolation of the point data to an larger area.):

$$V_p = 3.5 + \log Q_c \left[\frac{km}{s}\right]$$

In conclusion, according to Barton, the Q-system provides an important design tool which is simple to use and can be applied in potentially enormous variability of geology and structural geology (in fact the range of Q values is very high, approximately 10^6).

3.5.4 Q_{TBM} Classification System

Q-system was extended to a new Q_{TBM} system for predicting penetration rate (PR) and advance rate (AR) for tunneling using tunnel boring machine (TBM) in 1999 (Barton 1999). The method is based on the Q-system and average cutter force in relations to the appropriate rock mass strength. Orientation of joint structure is accounted for, together with the rock material strength. The abrasive or nonabrasive nature of the rock is incorporated via the cutter life index (CLI). Rock stress level is also considered. As a consequence, the $\boldsymbol{Q_{\mathrm{TBM}}}$ classification is strongly based on the above described Q-system, but it has additional rock-machine-rock mass interaction parameters:

$$Q_{\mathrm{TBM}} = Q \times \frac{\mathrm{SIGMA}}{F^{10}/20^{9}} \times \frac{20}{\mathrm{CLI}} \times \frac{q}{20} \times \frac{\sigma_\theta}{5}$$

where Q is the index of the Q-system by Barton calculated considering the RQD oriented along the tunnelling direction, SIGMA is the rock mass strength (MPa) found with a complicated equation, F is the average cutter load (t), CLI is the cutter life index, σ_θ is the quartz content in percentage terms, σ_θ is the induced biaxial stress on tunnel face (MPa).

The Q_{TBM} value can be used for estimating TBM penetration rates and advance rates for different rock conditions, both for prediction and for back analysis.

3.6 Hoek-Brown Constitutive Model for Rock Mass

As above described, quality index obtained by means of geomechanics classifications allow an approximate evaluation of the rock mass or soil behaviour, as well as its strength and deformation parameters.

On the contrary, among the different constitutive models allowing to model the rock mass behaviour using analytical or numerical calculation methods, the Hoek and Brown (1980) constitutive model is the most widely used at present. These authors defined a method to attribute a plasticity domain to a given rock mass. The starting point is represented by the properties of the intact rock constituting each single block of the rock mass and the influence of discontinuities on the behaviour of the rock mass.

The yielding criterion of Hoek and Brown is characterized by a non-linear plasticity domain, whose parameters are experimentally deduced by the merging of in situ experiences and lab tests.

With respect to the original formulation of 1980, this criterion was modified and improved during the years (e.g. Hoek and Brown 1988; while Marinos and Hoek (2006) describe the evolution of the criterion), also for its application to weak rocks and complex formations.

First of all, the Authors defined a specific yielding criterion for the intact rock that considers the influence of petrographic and textural characters of the rock and

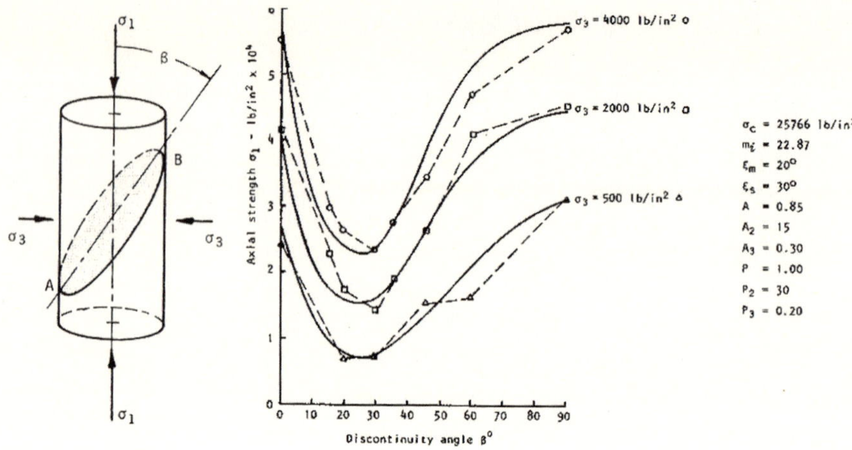

Fig. 3.14 Variation of the strength of a rock specimen containing an inclined discontinuity. (Hoek and Brown 1994)

its strength:

$$\sigma_1' = \sigma_3' + \sigma_{ci}\left(m_i \frac{\sigma'_3}{\sigma_{ci}} + 1\right)^{0.5}$$

where:

σ_1' and σ_3'= respectively major and minor principal stresses at failure,

σ_{ci} = uniaxial compressive strength of the intact rock; in case of materials characterised by discontinuities (e.g. schistosity) the strength varies in function of the angle between major loading direction and discontinuity direction (Fig. 3.14),

m_i = coefficient of intact rock, linked to the petrographic characters and texture (Table 3.15), it is obtained by means of a minimum square error method on the basis of a set of tri axial tests; to this aim the relation between the principal stresses at plasticity for intact rock is used.

In 2000, the Australian authors Mostyn and Douglas proposed a modification of the criterion introduced by Hoek and Brown. They published a statistical study on a big data base of lab test results and demonstrated that the approximation of experimental results can be improved by far using the following equation:

$$\sigma_1' = \sigma_3' + \sigma_{ci}\left(m_i \frac{\sigma'_3}{\sigma_{ci}} + 1\right)^{\alpha}$$

With

$$\alpha = 0.4032 + 1.08585/(1 + \exp(m_i/7.455))$$

Table 3.15 Values of coefficient m_i, obtained by testing, the values given in brackets are estimated (Hoek et al. 1995)

Rock type	Class	Group	Texture			
			Course	Medium	Fine	Very fine
SEDIMENTARY	Clastic		Conglomerate (22)	Sandstone 19	Siltstone 9	Claystone 4
				← Greywacke (18) →		
	Non-Clastic	Organic		← Chalk 7 →		
				← Coal (8 - 21) →		
		Carbonate	Breccia (20)	Sparitic Limestone (10)	Micritic Limestone 8	
		Chemical		Gypstone 16	Anhydrire 13	
METAMORPHIC	Non Foliated		Marble 9	Hornfels (19)	Quartzite 24	
	Slightly foliated		Migmatite (30)	Amphibolite 31	Mylonites (6)	
	Foliated*		Gneiss 33	Schists (10)	Phyllites (10)	Slate 9
IGNEOUS	Light		Granite 33 Granodiorite (30) Diorite (28)		Rhyolite (16) Dacite (17) Andesite 19	Obsidian (19)
	Dark		Gabbro 27 Norite 22	Dolerite (19)	Basalt (17)	
	Extrusive pyroclastic type		Agglomerate (20)	Breccia (18)	Tuff (15)	

Moreover Mostyn and Douglas (2000) suggest that the m_i parameter depends on the ratio between compression (σ_{ci}) and tension (σ_t) strength of material:

$$\frac{\sigma_{ci}}{|\sigma_t|} - 1 \leq m_i \leq \frac{\sigma_{ci}}{|\sigma_t|}$$

Furthermore using the modulus ratio MR (Table 3.16) proposed by Deere (1968) (modified by the authors based in part on this data set and also on additional correlations from Palmström and Singh (2001)) it is possible to estimate the intact rock modulus from:

$$E_i = MR \cdot \sigma_{ci}$$

Table 3.16 Guidelines for the selection of modulus ratio MR values based on Deere (1968) and Palmstrom and Singh (2001)

	Class	Group	Texture			
			Course	Medium	Fine	Very fine
Sedimentary	Clastic		Conglomerates 300–400 Breccias 230–350	Sandstones 200–350	Siltstones 350–400 Greywackes 350	Claystones 200–300 Shales 150–250[a] Marls 150–200
	Non-Clastic	Carbonates	Crystalline Limestone 400–600	Sparitic Lime-stones 600–800	Micritic Lime-stones 800–1000	Dolomites 350–50
		Evaporites		Gypsum (350)[b]	Anhydrire (350)[b]	
		Organic				Chalk 1000 +
Metamorphic	Non Foliated		Marble 700–1000	Hornfels 400–700 Metasand-stone 200–300	Quartzites 300–450	
	Slightly foliated		Migmatite 350–400	Amphibolites 400–500	Gneiss 300–750[a]	
	Foliated*			Schists 250–1100[a]	Phyllites/Mica Schist 300–800[a]	Slates 400–600[a]
Igneous	Plutonic	Light	Granite + 300–550 Granodiorite + 400–450	Diorite + 300–350		
		Dark	Gabbro 400–500 Norite 350–400	Dolerite 300–400		
	Hypabyssal		Porphyries (400)[b]		Diabase 300–350	
	Volcanic	Lava		Rhyolite 300–500 Andesite 300–500	Dacite 350–450 Basalt 250–400	
		Pyroclastic	Agglomerate 400–600	Volcanic breccia (500)[b]	Tuff 200–400	

[a] Highly anisotropic rocks: the value of MR will be significantly different if normal strain and/or loading occurs parallel (high MR) or perpendiculare (low MR) to a weakness plane. Uniaxial test loading direction should be equivalent to field application

+ Felsic Granitoids: Coarse Grained or Altered (high MR), fined grained (low MR)

[b] No data available, estimated on the basis of geological logic

Then, Hoek et al. (2002) define a specific failure criterion for rock masses:

$$\sigma_1' = \sigma_3' + \sigma_{ci}\left(m_b \frac{\sigma'_3}{\sigma_{ci}} + 1\right)^{\alpha}$$

where: σ_1, σ_3, σ_{ci} as given above;

m_b, s, α are coefficients that depend on the features of the rock mass. Those coefficients can be calculated using following formulas:

$$m_b = m_i e^{\frac{\mathrm{GSI}-100}{28-14\cdot D}}$$
$$s = e^{\frac{\mathrm{GSI}-100}{9-3\cdot D}}$$
$$\alpha = 0.5 + \frac{1}{6} \cdot (e^{-\mathrm{GSI}/15} - e^{-20/3})$$

where:

- m_i is a coefficient typical of intact rock as given above.
- GSI (Geological Strength Index) is a quality index of the rock mass ranging from 5 to 100, as a function of geomechanical conditions and of the weathering degree. This parameter can be calculated thanks to correlations linking it to Beniawski RMR index and Barton Q index. In particular, GSI = RMR'-5 being RMR' = BRMR for the dry rock, that is RMR with R6 = 0 and R5 = 15. These correlations are not reliable for very low-quality rock masses, with GSI < 25, for which GSI is obtained by means of the abaci shown in Figs. 3.15 and 3.16.
- D is the disturbance factor (Fig. 3.17) taking into account the impact of excavation technologies on the rock masses considered and/or of deformations undergone by the rock mass after of before the excavation. It ranges from 0 (undisturbed masses) to 1 (very disturbed masses).

One of the main problems linked to the Hoek and Brown approach is its implementation in numerical applications that is made difficult by the domain curvature, in particular the part concerning low confinement stresses. For this reason, the Authors provide indication on how to correctly use a Mohr-Coulomb linear failure criterion; this is done by determining equivalent parameters able to fit the Hoek and Brown curve.

Therefore, it is necessary to determine equivalent angles of friction and cohesion for each rock mass and stress range. This is done by fitting an average linear relationship to the curve generated by solving the Hoek and Brown's equation for a range of minor principal stress values defined by $\sigma_t < \sigma_3 < \sigma_{3\max}$, as illustrated in Fig. 3.18. The fitting process involves balancing the areas above and below the Mohr-Coulomb plot. This results in the following equations for the angle of friction ϕ' and cohesive strength c':

$$c' = \frac{\sigma_{ci}\left[(1+2\alpha)s + (1-\alpha)\, m_b \sigma'_{3n}\right](s + m_b\, \sigma'_{3n})^{\alpha-1}}{(1+\alpha)(2+\alpha)\sqrt{1 + (6\alpha m_b (s + m_b \sigma'_{3n})^{\alpha-1})/((1+\alpha)(2+\alpha))}}$$

$$\phi' = \sin^{-1}\left[\frac{6\alpha m_b (S + m_b\, \sigma'_{3n})^{\alpha-1}}{2(1+\alpha)(2+\alpha) + 6a m_b (s + m_b \sigma'_{3n})^{\alpha-1}}\right] \quad \text{with} \quad \sigma'_{3n} = \sigma'_{3\max}/\sigma_{ci}$$

where the symbols have the same meaning as the one described for Hoek et al. (2002) formulation.

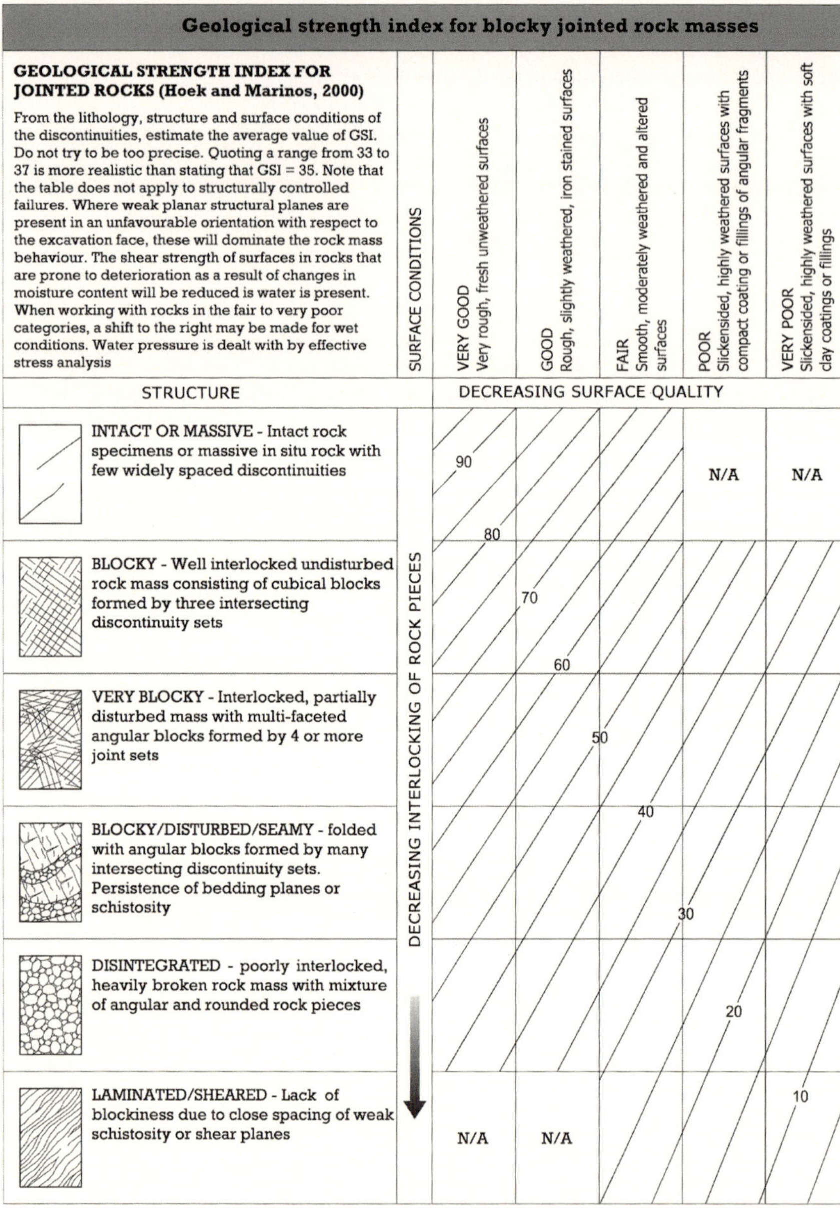

Fig. 3.15 Abacus to estimate the GSI on the base of the fracturing degree and of weathering conditions of the rock mass. It has to be noted that in their note on the top, the Authors advise not to expect to obtain precise GSI values from this graph, but to consider ranges of variability. Furthermore, the Authors reaffirm that the criterion cannot be applied to very anisotropous materials, where slidings are ruled by structural weaknesses of the rock mass (modified from Hoek's corner http://www.rocscience.com/education/hoeks_corner)

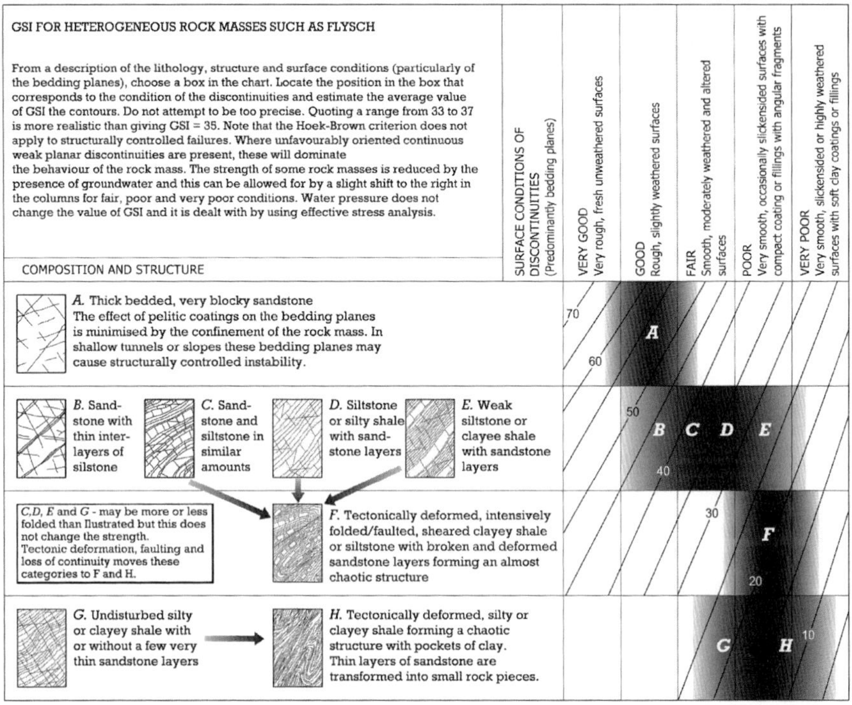

Fig. 3.16 Abacus for the estimation of GSI in case of complex formations. The parameters of an heterogeneous mass are obtained averaging those of the rocks composing the mass with the coefficients in Table 3.17 (modified from Hoek's corner http://www.rocscience.com/education/hoeks_corner)

The value of σ'_{3max} is the upper limit of confining stress over which the relationship between the Hoek-Brown and the Mohr-Coulomb criteria is considered and for both deep and shallow tunnels it can be determined as:

$$\sigma'_{3\max} / \sigma'_{cm} = 0.47 \cdot \left(\frac{\sigma'_{cm}}{\gamma \cdot H} \right)^{-0.94}$$

where γ is the unit weight of the rock mass, H the depth of the tunnel below surface and σ_{cm} is a global "rock mass strength" which considers the overall behaviour of the rock mass rather than the detailed failure propagation process. This strength can be estimated as:

$$\sigma'_{cm} = \sigma_{ci} \cdot \frac{(m_b + 4s - \alpha \cdot (m_b - 8s)) \cdot (m_b/4 + s)^{\alpha - 1}}{2 \cdot (1 + \alpha) \cdot (2 + \alpha)}$$

Hoek and Brown (1997) suggested an elastic-brittle-plastic, elastic-strain softening and elastic-perfectly plastic behaviours for very good, average and very poor quality rock masses, respectively, and recommended the use of dilatancy angle (Ψ) values related to the friction angle (ϕ); the values suggested by Hoek and Brown (1997) for Ψ are:

Appearance of rock mass	Description of rock mass	Suggested value of D
	Excellent quality controlled blasting or excavation by Tunnel Boring Machine results in minimal disturbance to the confined rock mass surrounding a tunnel.	D = 0
	Mechanical or hand excavation in poor quality rock masses (no blasting) results in minimal disturbance to the surrounding rock mass. Where squeezing problems result in significant floor heave, disturbance can be severe unless a temporary invert, as shown in the photograph, is placed.	D = 0 D = 0.5 No invert
	Very poor quality blasting in a hard rock tunnel results in severe local damage, extending 2 or 3 m, in the surrounding rock mass.	D = 0.8

Fig. 3.17 Guidelines for estimating disturbance factor D (modified from http://www.rocscience.com/education/hoeks_corner)

- for very weak rocks (GSI < 30): $\Psi = 0$
- for jointed rocks with 30 < GSI < 50: $\Psi = \phi'/8$
- for jointed rocks with 50<GSI<70: $\Psi = \phi'/8 \div \phi'/4$
- for jointed rocks with 70<GSI<90: $\Psi = \phi'/4$

Based on the Hoek-Brown criteria also the deformation modulus of the rock mass can be estimated (Hoek and Diedrichs 2006):

$$E_{rm}(\text{MPa}) = 100{,}000\left(\frac{1 - D/2}{1 + e^{((75+25D-\text{GSI})/11)}}\right)$$

Finally, when using numerical models to study the progressive failure of rock masses, estimates of the post-peak or post-failure characteristics of the rock mass are required. In some of these models, the Hoek-Brown failure criterion is treated as a yield criterion and the analysis is carried out using the plasticity theory. Based upon experience in numerical analysis, the Authors suggest the post-failure characteristics illustrated in Fig. 3.19.

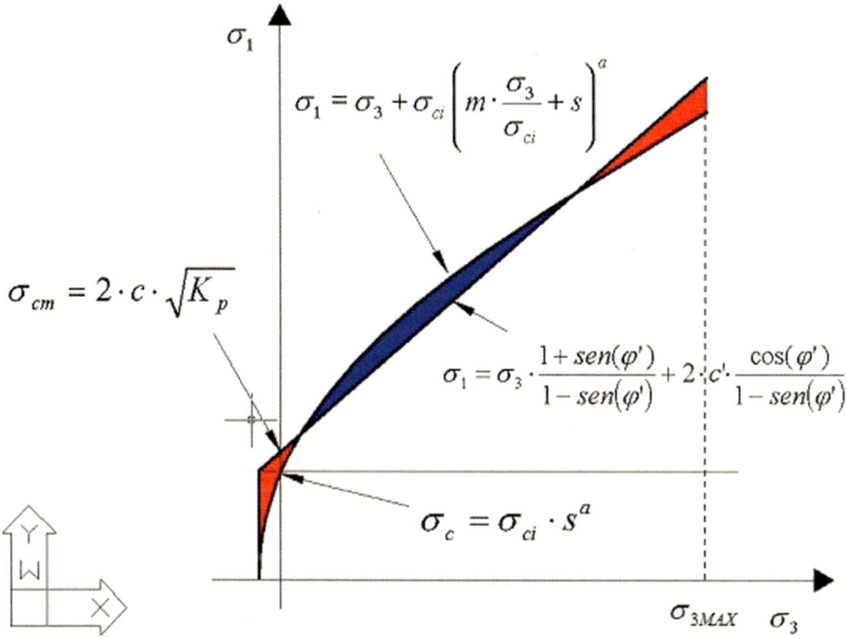

Fig. 3.18 Relationships between major and minor principal stresses for Hoek-Brown and equivalent Mohr-Coulomb criteria

Table 3.17 Suggested proportions of parameters σ_{ci} and m_i for estimating rock mass properties for flysch

Flysch type see Fig. 3.16	Proportions of σ_{ci} and m_i for each rock type to be included in rock mass property determination
A and B	Use values for sandstone beds
C	Reduce sandstone value by 20 % and use full values for siltstone
D	Reduce sandstone value by 40 % and use full values for siltstone
E	Reduce sandstone value by 40 % and use full values for siltstone
F	Reduce sandstone value by 60 % and use full values for siltstone
G	Use values for sandstone or shale
H	Use values for sandstone or shale

In a numerical model, in order to describe properly the post-peak rock mass behaviour, a reduction of Hoek-Brown parameters m_b and s is needed. The determination of residual strength parameters of the rock mass can be carried out, according to the generalized Hoek-Brown criterion, through the determination of the residual value of GSI ($\mathrm{GSI_r}$) by means of the relationship proposed by Cai et al. (2007):

$$\frac{\mathrm{GSI_r}}{\mathrm{GSI}} = e^{-0.0134 \cdot \mathrm{GSI}}$$

Starting from $\mathrm{GSI_r}$, values of $\mathrm{m_r}$ and $\mathrm{s_r}$ (and then φ'_r and c'_r) are then calculated.

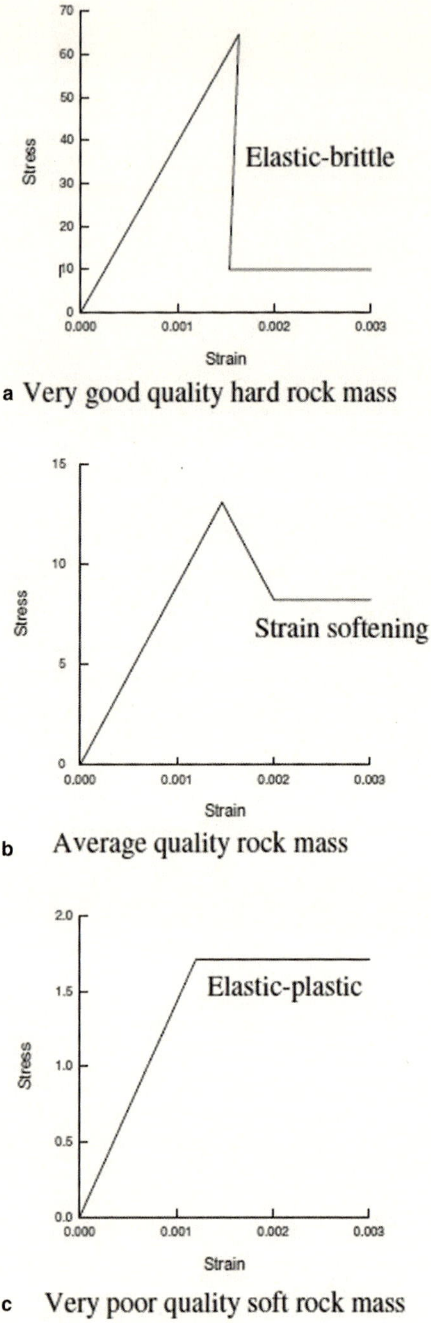

Fig. 3.19 Suggested post failure characteristics for different quality rock masses

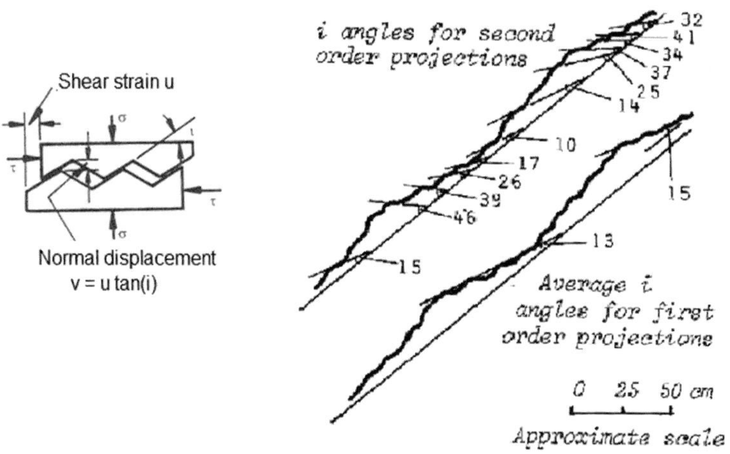

Fig. 3.20 **a** Idealized scheme of the dilatancy phenomenon along a rough joint subject to shear stress (Hoek and Bray 1981). **b** Patton's measurements of *i* angles for first and second order projections on rough rock surface

3.7 Strength of Discontinuities

If a rock mass is interested by discontinuities that can influence its behaviour substantially, it is essential to define the intrinsic resistance of the discontinuities.

The shear strength of discontinuities can either be determined by means of in situ or laboratory shear tests or estimated using empirical methods based on the geomechanical survey of discontinuities, e.g. on the description of their physical and geometrical features.

Starting from those data, empirical criteria have been introduced to evaluate the mechanical behaviour of discontinuities.

The better known and most widely used criteria to determine the shear strength of discontinuities are:

- Patton criterion
- Barton equation
- Ladanyi and Archanbault criterion

3.7.1 Patton Criterion

The roughness on a discontinuity surface is characterized by an angle *i* representing the angle formed by the asperity on the surface of the discontinuity (Fig. 3.20a). That angle, added to the base friction angle ϕ_b, (smooth discontinuity surface made of the same material) gives the total value of the shear strength angle along the surface:

$$\phi_p = \phi_b + i$$

The criterion does not take cohesion into account, therefore the result is:

$$\tau = \sigma_n \ \mathrm{tg}\ (\phi_b + i)$$

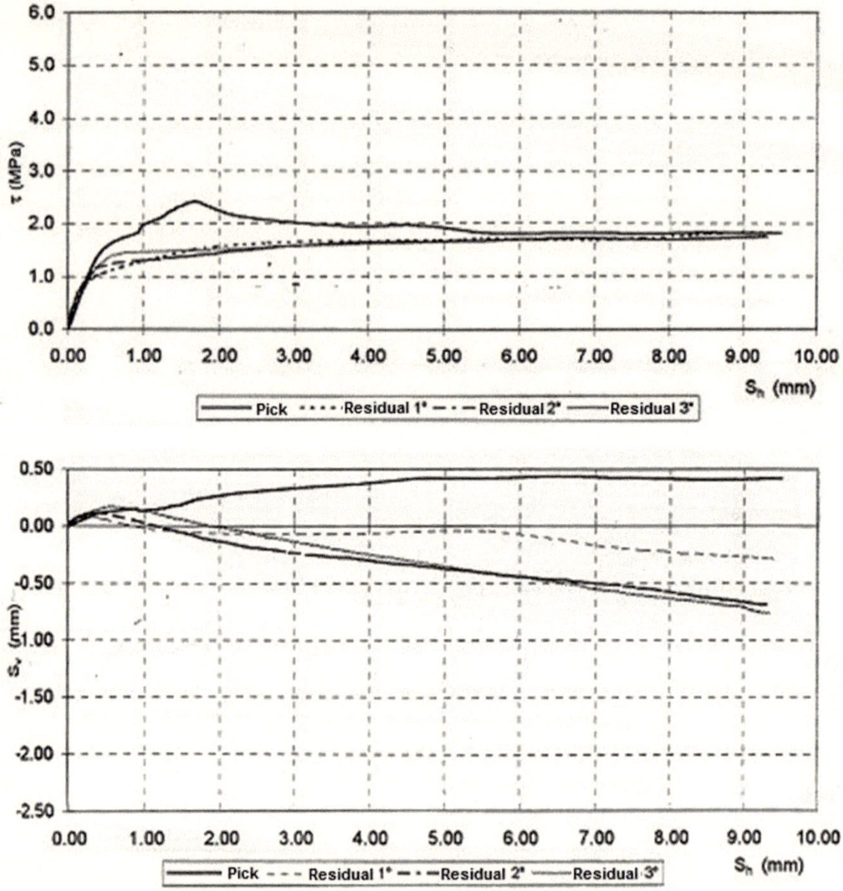

Fig. 3.21 Dilatancy phenomenon along a rough joint subject to direct shear test. S_h and S_v are respectively the horizontal and vertical shear displacements

It is clear that the presence of roughness determines an increase in the shear strength ("interlocking" effect) (Fig. 3.20).

If a shear stress is applied on a discontinuity surface, asperities have to be overcome in order to allow the relative shift of the two discontinuity surfaces. This can happen through the dilatancy (aperture or separation) phenomenon on the discontinuity walls, causing an increase in the volume of the sample (Figs. 3.20a, 3.21).

3.7.2 Barton Equation

Barton (1973) proposes an empiric equation to calculate the shear strength on rough not cemented joints (Fig. 3.22):

$$\tau = \sigma_n \text{ tg } [\phi_b + \text{JRC Log}(\text{JCS}/\sigma_n)]$$

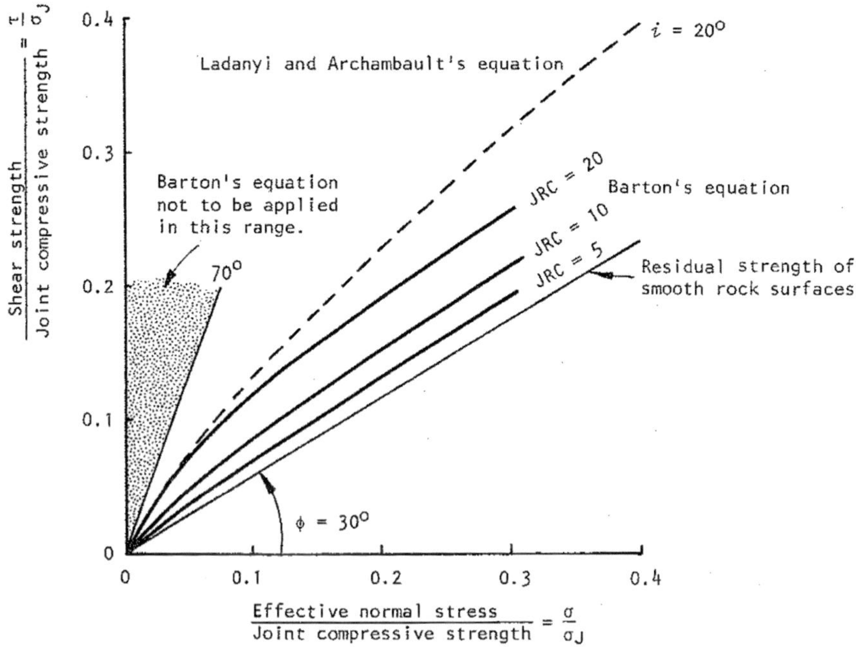

Fig. 3.22 Envelope lines based on Barton failure criterion for different JRC values compared to the envelope line of Ladanyi and Archambault's criterion. (Hoek and Bray 1981)

where:

- JRC is the Joint Roughness Coefficient, ranging from 0 (joint with planar and smooth surface) to 20 (joint with undulating and rough surface) that can be measured with the shape tracer.
- the JCS (Joint Wall Coefficient Strength) is a coefficient expressing the Joint Wall Compression Strength along the joint surface σ_j, that can be determined with the Point Load Strenght Test or with the Schmidt test hammer directly on the fracture surface.
- ϕ_b (base friction angle) is only function of the rock type and can be obtained with the Tilt Test (sliding test on a smooth sloping surface).
- σ_n is the nominal stress on the joint surface, it depends on the depth of the joint (for superficial joints, Barton suggests to set $\sigma_n = 0.1$ MPa).

Barton equation is valid and experimentally verified if $\sigma_n/\sigma_j = 0.01 \div 0.3$. If $\sigma_n/\sigma_j < 0.01$, the ratio σ_j/σ_n tends to infinite and the equation loses validity.

The shear strength along rough discontinuities is also influenced by the composition and the width of the material present inside the discontinuities.

To this purpose, Goodman (1970) demonstrated that if the width of the filling is equal or higher than the maximum undulation width, the shear strength along the discontinuity is a function of the cohesion and of the shear strength angle of the filling (Fig. 3.23).

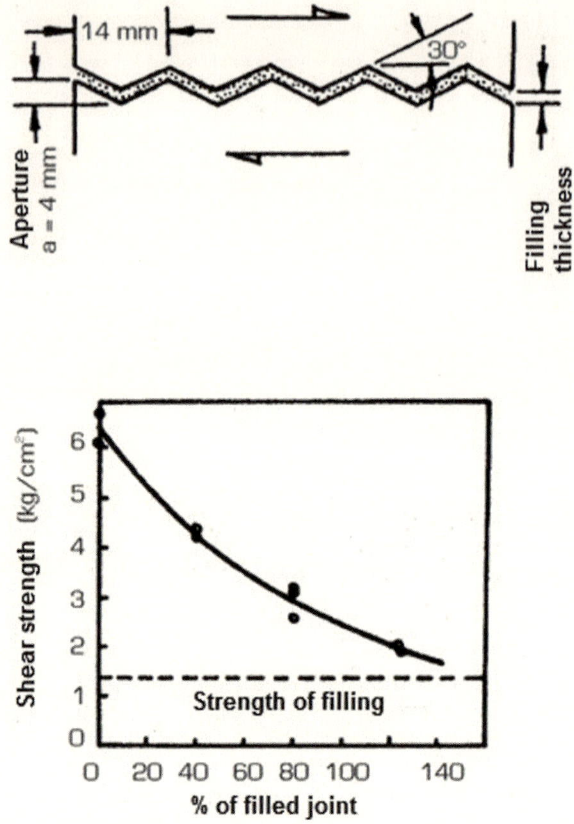

Fig. 3.23 Filling influence on the shear strength of a discontinuity

3.7.3 *Ladanyi and Archanbault Criterion*

The relationship linking the normal stress to the tangential stress according to the Ladanyi and Archanbault criterion is (Fig. 3.24):

$$\tau = \frac{\sigma(1 - a_s)\left(\frac{v}{u} + \text{tg}\,\phi\right) + a_s \tau_r}{1 - (1 - a_s)\left(\frac{v}{u}\,\text{tg}\,\phi\right)}$$

where:

$a_s =$ parameter determining that part of the discontinuity surface where intact rock is affected by the failure;

$v/u =$ dilatancy (relationship between the shift along the direction of the normal stress component and the shift along the direction of the shear strength component).

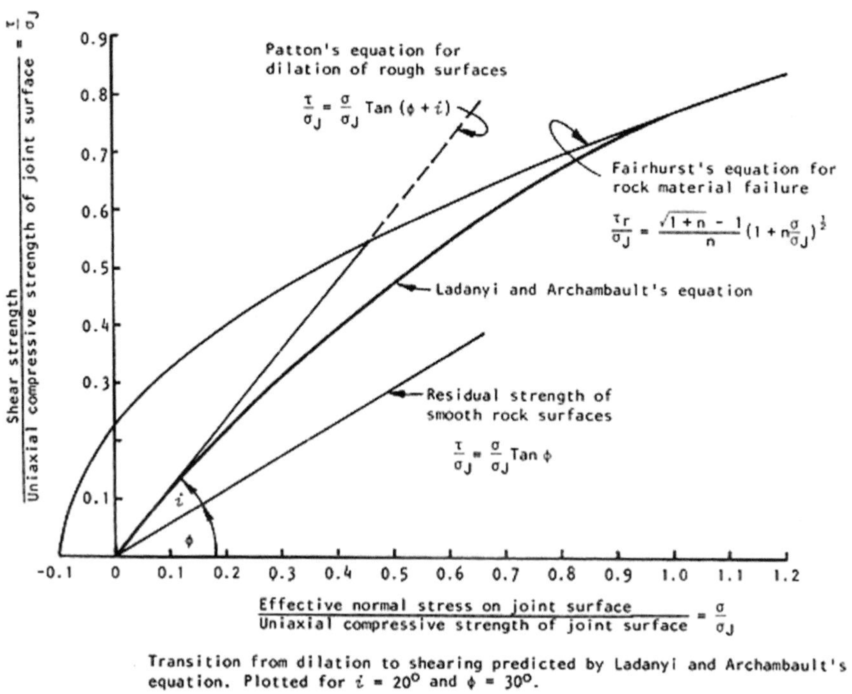

Fig. 3.24 Comparison among different failure criteria (where in Fairhurst equation n is the ratio between compressive and tensile strength of intact rock mass $n = \sigma_{ci}/\sigma_t$). (Hoek and Bray 1981)

There are two limit conditions:

1. there are no asperity cuts for low values of nominal stress (e.g. parts of intact rock) and $a_s = 0$, $v/u = \text{tg}\,(i)$; in that case, the formulation is reduced to Patton equation. Therefore, in these conditions, the Patton criterion is recommended for its simplicity.
2. the shear failure only occurs in intact rock for high values of nominal stress, $a_s = 1$ e $\tau = \tau_r$(e.g. a value comparable to the one obtainable from triaxial compression tests on intact rock samples characterised by not null values of tensile stress).

As shown in Fig. 3.24 the models of Ladanyi-Archambault and Barton give similar results for low load values and for JRC = 20 and $i = 20°$; otherwise, if the normal stress increases the discrepancy between the two criterions grows. In fact, as σ_n grows the relation of Ladanyi-Archambault tends to residual values ($\tau \rightarrow \tau_r$), while the Barton's one conduce to pure shear strength ($\tau \rightarrow \sigma_n(\text{tg}\,\phi_b)$). Moreover it can be seen (Fig. 3.24) that for low load (low normal stress) values of Ladanyi-Archambault criterion corresponds also to the Patton's equation (in fact in this conditions the asperities are not damaged so that $a_s \rightarrow 0$ and $v/u \rightarrow \text{tg}i$ and the Ladanyi-Archambault becomes equal to the Patton's one).

Ladanyi-Archambault criterion can be seen also as a transition between Patton's criterion (which corresponds to the maximum dilatancy) and a nonlinear envelope of intact rock yielding criterion (Fairthurst's equation).

References

Barton N (1999) TBM performance estimation in rock using Q_{TBM}. Tunnels & Tunneling Intern 31(9)

Barton N, Lien R, Lunde J (1974) Engineering classification of rock masses for the design of tunnel support. Rock Mech 6(4):189–236

Barton N (2013) Shear strength criteria for rock, rock joints, rock fill and rock masses: Problems and some solutions. J Rock Mech Geotech Eng 5:249–261

Barton NR (1973) Review of a new shear strength criterion for rock joints. Eng Geol 7:287–332

Bieniawski ZT (1989) Engineering rock mass classifications. Wiley, New York

Bieniawski ZT, Caleda B, Galera JM, Alvares MH (2006) Rock Mass Excavability (RME) index. Seoul, In ITA World Tunnel Congress

Cai M, Kaiser PK, Tasaka Y, Minami M (2007) Determination of residual strength parameters of jointed rock masses using the GSI system. Int J Rock Mech Mining Sci 44(2):247–265

Deere DU (1968) Geological considerations. Rock Mechanics in Engineering Practice, eds. K.G. Stagg and O.C. Zienkiewicz. Wiley, London, pp 1–20

Gonzalez de Vallejo LI (2003) SRC rock mass classification of tunnels under high tectonic stress excavated in weak rocks. Eng Geol 69:273–285

Gonzalez de Vallejo LI (2006) Geoingegneria (p. 716). Pearson Ed. Prentice Hall, Milan

Hoek E, Bray J (1981) Rock slope engineering (p. 402). Institution of Mining and Metallurgy, London

Hoek E, Brown ET (1988) The Hoek-Brown failure criterion—a 1988 update. Proceedings of the 15th Canadian rock mechanics symposium, pp 31–38

Hoek E, Brown ET (1994) Underground Excavations in Rock (p. 527). Inst. of Mining and Metallurgy, Chapman & Hill, London.

Hoek E, Brown ET (1997) Practical estimates or rock mass strength. Int J Rock Mech Mining Sci Geomech Abstr 34(8):1165–1186

Hoek E, Carranza-Torres CT, Corkum B (2002) "Hoek-Brown failure criterion-2002 edition". Proceedings of the fifth North American rock mechanics symposium 1, pp 267–273

Hoek E, Kaiser PK, Bawden WF (1995) Support of underground excavations in hard rock. A.A. Balkema, Rotterdam, p. 215

Marinos P, Hoek E (2006) The construction of the Egnatia Highway through unstable slopes in northern Greece. Proc. XI Conference on Rock Mechanics, Turin, November 2006.

Mostyn G, Douglas K (2000) Strenght of intact rock and rock masses (Melbourne). Proc Geo Engg:1389–1421

Palmstrom A (1996) Characterizing rock masses by the RMi for use in practical rock engineering: Part 1: The development of the Rock Mass index (RMi). Tunn Under Space Technol 11(2):175–188

Palmström A, Singh R (2001) The deformation modulus of rock masses—comparisons between in situ tests and indirect estimates. Tunn Under Space Technol 16(3):115–131

Chapter 4
Underground Excavation Analysis

4.1 Introduction

The knowledge of the underground geological complexity is essential to design and construct underground works. As already seen in Chap. 3, the results that emerged from the studies and geognostic surveys allow the reconstruction of a reference conceptual model of the excavated area and the definition of design parameters, as well as the constitutive model of the ground. Then, on the basis of those inputs, it is possible to forecast the stress-strain behaviour of the rock mass at the cavity aperture. The reference conceptual model is also a decision-making tool that is required to design, carry out risk analysis and optimise the realisation and operating costs of the work.

Starting from the geotechnical behaviour, suitable calculation procedures (i.e. empirical procedure, characteristic lines, numeric modelling etc.) can be used to forecast the deformation phenomena and the kinematisms developing after the excavation. This means that it will be possible to forecast a substantial stability or instability situation for the intercepted materials, after the excavation. In the latter case, expected deformations will have to be estimated as precisely as possible, as well as the typologies and the entities of those instabilities, in order to make more suitable design choices.

Generally, the results of geological, geotechnical and geomechanical characterisation will allow to choose:

- A more suitable excavation technique to contain deformation and settlement risks, the latter mainly in urban areas; stabilisation works that can improve the technical characteristics of the material and contain deformation processes
- Confinement structures and works, lining in the first and in the final phase

It has to be said that geological structures are generally complex and, in most cases, cannot be observed directly. As a consequence, a fully reliable forecast of geological, geotechnical and hydrogeological conditions of the work is not usually possible. Furthermore, a good approach for its quantification in the risk analysis process is required (Chap. 5). In this view, the conceptual model is meant to evolve after its formulation according to the new data available during the construction phase

P. Gattinoni et al., *Engineering Geology for Underground Works*,
DOI 10.1007/978-94-007-7850-4_4,

(through monitoring); new data can be coherently integrated or they can justify a revision that could lead to a new formulation of the reference model.

This chapter is meant to provide a short overview of the leading analysis methodologies used to foresee the behaviour of the ground during and after the excavation. The presentation of the treated topics provides the first comprehension elements in terms of physical stability of the excavation process, without discussing more strictly the analytic aspects. The treatise of different analysis methods is preceded by some introduction paragraphs providing the base concepts concerning the following aspects: the choice of the medium (equivalent continuum or discontinuous medium) to use as a reference, the convergence-confining relationship, and the difference between low and high overburden.

4.2 Discontinuous Medium and Equivalent Continuum

As already underlined in Sect. 1.4 (Chap. 1), it is important to establish if the ground mass can be regarded as a discontinuous medium (rather fractured rocks) or an 'equivalent continuum' (soils, intact rocks, heavily fractured rock masses etc.) before analysing its behaviour. In the first case, the analysis methods will consider the interactions between rock blocks separated by discontinuities and then model the behaviours of both single rock volumes and joints. In the second case, it will be possible to use a modelling approach that considers representative parameters and constitutive models for the rock mass (as a whole) as those valid for the soils (e.g. Mohr–Coulomb, Druker–Prager, Cam Clay etc.) or for rock masses (see, for example Chap. 3: Hoek–Brown model). Figure 4.1 provides some information on how to choose a more suitable modelling approach for rock mass according to the fracturing characteristics of the medium and the analysis scale.

In case of intact rock, a general isotropic behaviour can be considered and the Hoek–Brown failure criterion (Sect. 3.6) can be used. In case of heavily jointed rock mass, the Hoek–Brown criterion for jointed rock mass (Sect. 3.6) allows to describe the behaviour of interlocking angular pieces. On the contrary, for massive rock mass with a few sets of discontinuities defining the mechanical behaviour of rock mass, the behaviour of the whole is anisotropic. It depends on the number, orientation, and shear strength of discontinuities. Therefore, the Hoek–Brown failure criterion has to be used with extreme care, or different criteria have to be considered to describe the intact rock and joint planes (Sect. 3.7).

4.3 Convergence and Confinement

Convergence (movements of the cavity perimeter due to rock mass deformations after excavations) and *confinement* (stresses on the cavity perimeter) are further fundamental concepts.

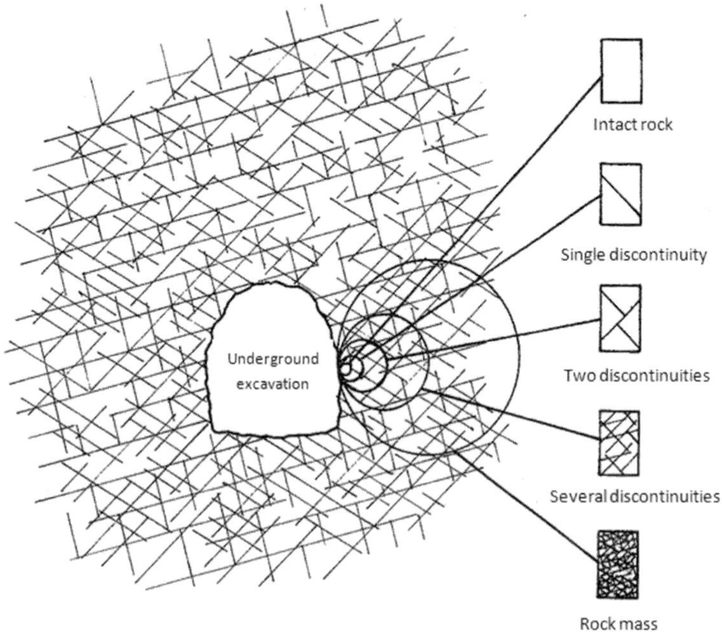

Fig. 4.1 Idealised diagram showing the transition from intact rock to a heavily jointed rock mass with increasing sample size. (Modified from Hoek and Brown 1980)

The realization of underground excavations is a process implying the progressive removal of material (ground) from a mass undergoing an initial compression stress state (σ_{r0}, $\sigma_{\theta 0}$, σ_{z0} at time t_0, with reference to the polar coordinate in space, see Fig. 4.9 for reference) that is more or less isotropic, followed by the application of improvement/confinement works and/or lining (not always strictly necessary) in one or more phases. In case of isotropic stress state ($\sigma_{r0} = \sigma_{\theta 0} = \sigma_{z0} = \sigma_0$ at the beginning) the stress σ_0 can be generally considered coincident with the lithostatic load ($\sigma_0 = \gamma H$, where H is the depth of the excavation axis with respect to the land surface and γ is the specific weight of the material).

The soil removal causes a change in the stress field of the mass and, therefore, its deformation. As the removal process is progressive, the change in the stress field induced by the excavation (from σ_0 for t_0 to $\sigma_{r\infty} \neq \sigma_{\theta\infty} \neq \sigma_{z\infty}$ for t_∞) and the following deformation depends on time (they are not instantaneous).

In case of peculiar geotechnical and boundary conditions (no rheology), the dependence of the stress-strain status in time, in a generic cross section orthogonal to the advancing direction, can be represented by its dependence on a geometric variable: the distance from the excavation face. In particular, the excavation implies a radial displacement u_r of the perimeter toward the cavity centre (convergence) both behind and beyond the excavation face (Fig. 4.2). It can be considered as the consequence of a progressive reduction of a fictitious radial pressure p_r on the perimeter of the excavation (confinement). That reduction takes into account the three-dimensional effects, both behind and beyond the advancing face, of the progressive core removal.

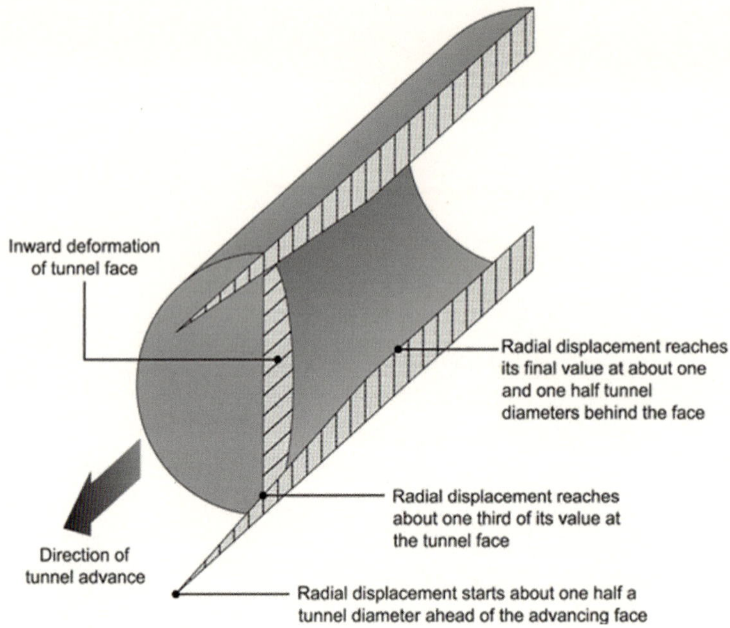

Fig. 4.2 Pattern of radial deformation in the roof and floor of an advancing tunnel. (Hoek 2013)

In particular, the zero value of the fictitious radial pressure represents the condition far behind the tunnel face, with no support. Reduction of fictitious pressure at the boundary is assumed to depend on the parameter λ (Panet 1995) that grows from 0 (representing the stress condition of the cross section far beyond the face) to 1 (representing the stress conditions of excavated sections far behind the face) (Fig. 4.3). Depending on the ground behaviour, λ at tunnel face (λ_f) assumes values approximately equal to 0.3 (elastic case) or greater than 0.3 (elasto-plastic case) (Figs. 4.4 and 4.5).

It follows that, for example (Fig. 4.6), at the time of application of the lining (in time $t_i > t_f$) the fictitious radial pressure will be $p_{ri} < p_{rf} < 0{,}7\sigma_0$ (with p_{rf} = fictitious radial pressure at the face).

The supports are initially unloaded, so, at the application moment t_i, the fictitious radial pressure is equal to p_{ri} on the ground (corresponding to a displacement u_{ri}) and the real radial pressure is $\sigma_{ri} = 0$ on the lining.

The equilibrium will be reached after a further reduction Δp_r of p_r equivalent to an increase $\Delta\sigma_r$ of the pressure on the lining and of the real pressure on the excavation perimeter, and to a further displacement of the excavation perimeter and of the lining.

It can be noticed that, before excavation (at time t_0) the stress σ_0 is equal to fictitious pressure p_{r0}. Moreover at the equilibrium $\sigma_{r-\text{equilibrium}}$ on the lining is equal to fictitious pressure $p_{r-\text{equilibrium}}$. In these two conditions the values σ and p_r are equivalent.

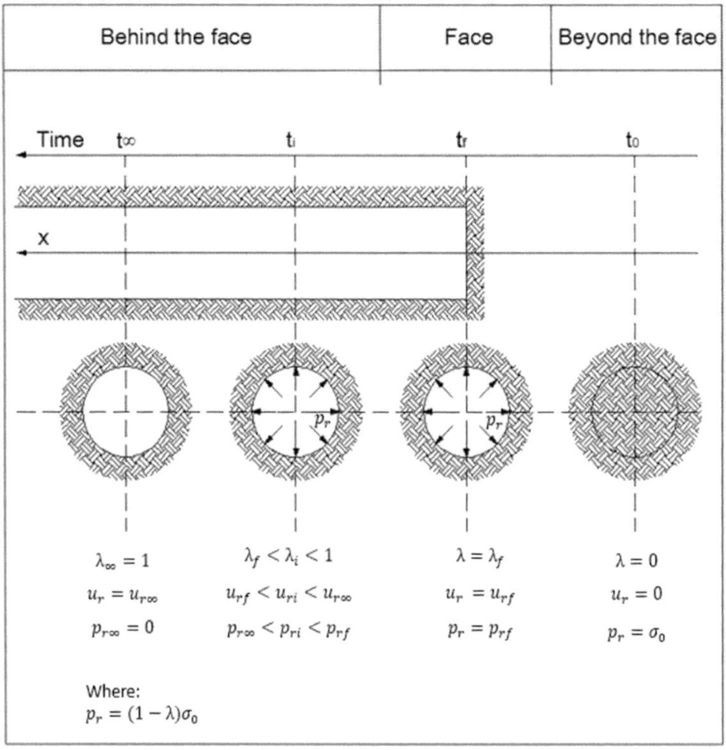

Fig. 4.3 Change of the deconfinement rate according to the distance from the face. (Panet 1995, modified)

Soil is a material with strongly non-linear and often irreversible (plastic) behaviour. Moreover, in particular conditions, almost all materials (both rocks and soils, with the exception of granular soils) show a quite important rheological behaviour (dependent on time).

Let us consider a ground element subject to in situ stress (σ_{r0}, $\sigma_{\theta 0}$, σ_{z0}) on the perimeter of the excavation. σ_r decreases (unloading, decompression) as the excavation advances up to becoming null right behind the face (if the cavity is not lined), whereas σ_θ increases up to a few times σ_0 (loading, compression) according to the initial stress conditions, and it can be assumed that σ_z remains practically the same.

The progressive increase of the difference between the two main stresses (deviator) from the initial stress state to that at time t_i determines the deformation of the element. According to the initial conditions, the final deviator size, the change modalities of the deviator, the stress-strain behaviour can be of different type (hardening, softening, dilatant, viscous) and may or may not generate the failure of the material (brittle or ductile).

Hardening behaviour means that plastic deformations have to be supported by increasing the deviator, whereas in the ideal case of perfect plasticity they occur with

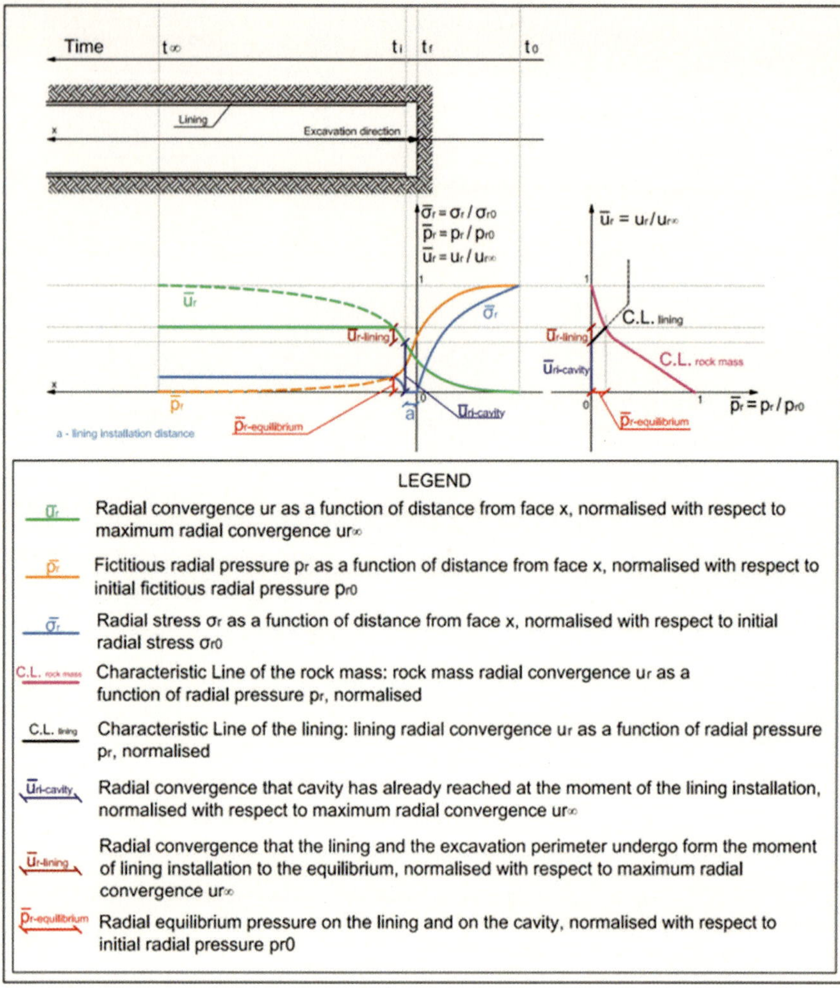

Fig. 4.4 Change of the confinement rate ($pr/\sigma 0 = 1 - \lambda$) according to the normalized distance from the face (x/r). Elastic case; r excavation radius; x distance from the face. (Gesta 1993)

constant deviator. Softening behaviour means a progressive reduction of strength as the deformation increases. Dilatancy means a progressive and partial de-structuring (it may begin with micro-fractures or relative movements among grains and get more important up to the inner dislocations of the matrix) with an increase in volume during plastic deformation. Viscous behaviour means the progression of deformations in time, also if the deviator remains constant. Failure is a complete de-structuration of the material. Failure can be brittle, if sudden (small deformations), or ductile, if it follows important deformations. Failure is always preceded by dilatancy (except in the case of loose sands and soft clays). Moreover, except in the case of soils with no cohesion (granular soils), failure may also be delayed with regards to the moment

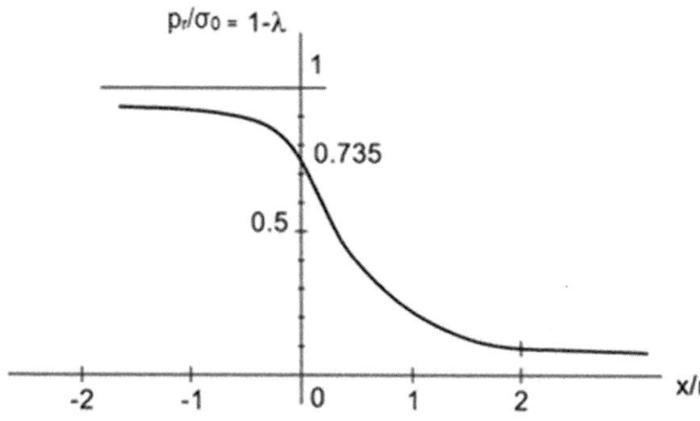

Fig. 4.5 Change of the confinement rate ($p_r/\sigma_0 = 1-\lambda$) according to the normalized distance from the face (x/r). Elastoplatic case ($N_s > 1$) with cohesive soil; r excavation radius, x distance from the face, N_s stability number defined as $N_s = \sigma_0/C_u$, where C_u is the undrained cohesion. (Gesta 1993)

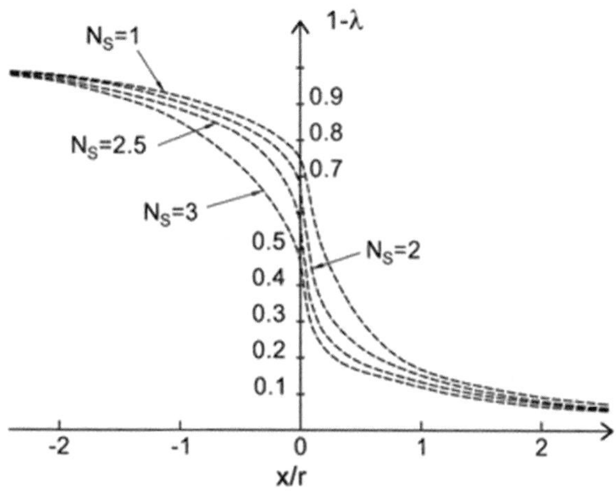

Fig. 4.6 Behaviour of rock mass and lining according to the excavation face position; Characteristic Lines of rock mass and lining

the critical stress is reached, in other words, a particular stress may not cause the failure in the short term, but provokes it in the long term, as a consequence of the deformation increase and of the dilatancy due to creep.

In case of an underground work, the soil failure is identified with the collapse (closure of the cavity).

The lining application can be theoretically avoided if the geotechnical and boundary conditions allow to reach stability, e.g. the increasing process of the deviator is completed and displacements no longer increase in time and they are still compatible with the underground work functionalities and with the limit deformations of the ground. The lack of lining is actually very rare because, beyond the respect of the work functionalities and the containment of deformation below the critical threshold, local instability phenomena, that usually require some kind of confinement measures, have to be considered.

When a lining is installed (obviously when needed), no fall can occur if the lining itself does not reach the collapse.

Lining is any kind of measure that can provide a confining pressure aimed at guaranteeing the stabilisation of the cavity.

Usually, the structure behaviours can be schematised as perfectly as elastic-plastic with more or less brittle failure.

As a consequence, they can offer an increase in confining pressures only within their elastic limit. If the lining is needed to guarantee the cavity stability and if the structures reach yielding limit during the development of deformations, following two cases may be possible :

- The cavity reaches stability with practically stable confinement pressure
- Sooner or later the tunnel will collapse, unless the structures are reinforced

Radial displacements causing plastic deformation phenomena and structure failure are smaller than those causing the failure of the surrounding ground.

It is clear that the deformation of the tunnel has to be always compatible with its functionality; therefore, the lining has to deform in a controlled way, that means as it was forecast in the design phase.

4.4 Underground Works at Shallow and Great Depth

As already discussed in Chap. 1, the types of geological problems that can be met during the excavation of an underground work are different in case they refer to works at shallow or great depth.

Indicatively it is possible to define an underground excavation "at shallow depth" when the increase of the mass stress around the cavity does not imply sensible and extended plastic deformation phenomena in the ground. In other words, depth is shallow when the lithostatic load γH is not bigger than the compression strength of the rock mass σ_{cm},

$$\sigma_{cm} = 2 \cdot c \cdot \sqrt{k_p},$$

k_p being the passive earth pressure coefficient. In these conditions, it can be assumed that cavity convergence is quite modest and that it fades shortly behind the excavation face.

It is obvious that the upper limit of 'shallow depths' is very low (even null) for soils where the mass strength is very small, whereas it will get higher and higher moving from weak rocks to hard rock masses. Generally, in soils and weak or very weathered rocks, underground excavation can be very insidious with low overburden. If the rock masses have better qualities, the excavation behaviour under low overburden is essentially ruled by the structural features of the rock mass. Therefore, the analysis has to be performed with calculation methods taking into account the geometry and the discontinuity features, as the Block Theory and the numerical Distinct Elements Methods.

Underground works are considered, at great depth, when the deviatoric stress on the cavity boundary is much higher than the mass strength ($\gamma H >> \sigma_{cm}$). Typically, this condition is true for soils with limited overburden, whereas it generally occurs in rock masses with high overburden. In these cases, not only the already discussed

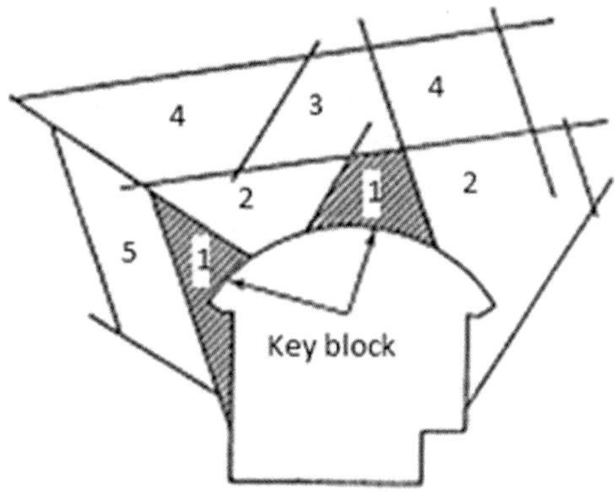

Fig. 4.7 Example of key block in the application of Block Theory

aspects related to the stability linked to the presence of discontinuities have to be considered, but it is also necessary to estimate the entity of the cavity deformation and the interaction among the rock masses and the confinement structures.

4.5 Analysis Methods of the Excavation Behaviour

4.5.1 Block Theory

The Block Theory (Shi and Goodman 1985) studies the stability of excavations within the rock masses with quite regular discontinuity networks.

It is a three-dimensional calculation method, that uses the approach of analytic geometry to study thoroughly the discontinuity systems of a rock mass and their mutual interaction, to identify the 'key blocks' whose stability determines the stability of the whole excavation (Fig. 4.7).

'Key blocks' are those rock wedges whose movement creates a space toward which other wedges—previously confined—can move, thus triggering a series of progressive falls. Therefore, the stability is guaranteed by an adequate stabilisation of the key blocks.

The block theory can be applied only if following assumptions are valid:

1. All joint surfaces are assumed to be perfectly planar.
2. Joint surfaces will be assumed to extend entirely through the volume of interest. that is, no discontinuities will terminate within the area of a key block.
3. Blocks defined by the system of joint faces are assumed to be rigid. This means that block deformation and distortion will not be considered.
4. The discontinuities and the excavation surface are assumed to be input parameters.

The development and application of the method can be carried out both with a vectorial calculation and a graphic procedure (with equal angle stereographic projection).

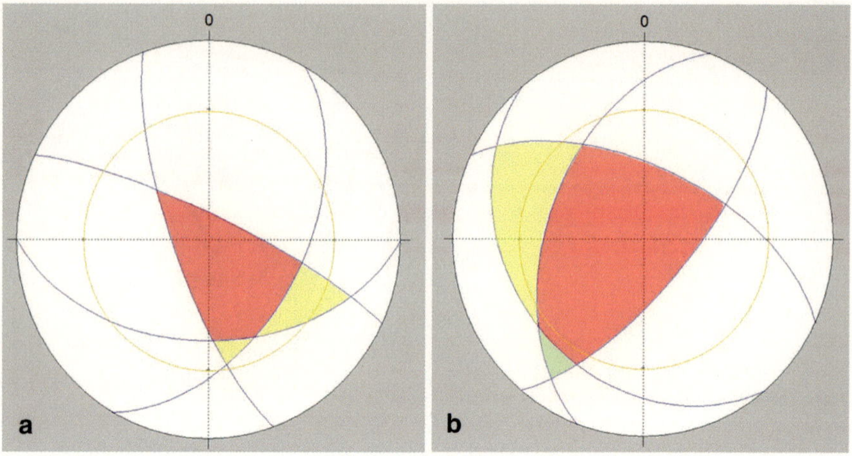

Fig. 4.8 Examples of stereographic analysis of key blocks. Discontinuity sets are represented in *blue*, the friction angle circle in *orange*. Falling blocks are represented in *red*, sliding blocks in *yellow* and stable blocks in *green*

Equal angle stereographic projections allow the identification of removable blocks and it is also possible to understand if the wedges formed by discontinuity intersections may trigger fall or sliding phenomena.

In particular, the wedge can only fall and the sliding is not possible if the great circles representing the joint planes form a closed figure that surrounds the centre of the net (Fig. 4.8a).

If the figure formed by the intersection of the great circles falls to one side of the centre of the net, failure can only occur by sliding on one of the joint surface or along one of the lines of intersection.

In this case, an additional condition is necessary: either the plane or the line of intersection along which sliding occurs should be steeper than the friction angle. This condition is satisfied if the figure, formed by the intersection of discontinuities, falls within the circle representing the friction angle and the circumference of the net.

When the entire figure falls outside the friction angle, the wedge is stable, because the gravitational weight of the wedge is not high enough to overcome the frictional resistance of the plane or planes on which the sliding would occur (Fig. 4.8b).

4.5.2 Characteristic Lines

To define the behaviour of the rock mass when excavated and the overall response of the tunnel when advancing, it is necessary to carry out the analysis of some properties representative of the cavity deformation (u_r = radial convergence), of plastic deformation phenomena around it (R_{pl} = plastic radius or F_{pl} = Plastic zone

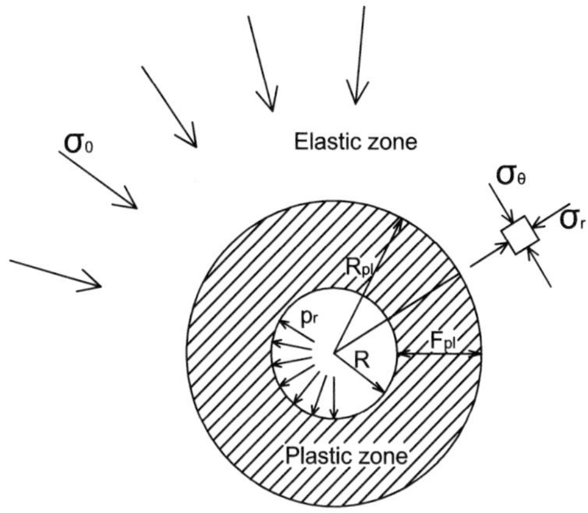

Fig. 4.9 Graphical representation of the variables related to the ground reaction curves

thickness) and of confinement pressures (p_r) required to reach stability (Fig. 4.9). The characteristic lines (or ground reaction curves or convergence-confinement curves) can be used to that aim. The method is shortly described below.

In a ground reaction curve, a mutual bond between the radial pressure on the boundary of the cavity (p_r) and radial displacement (u_r) (convergence) is defined (Fig. 4.10). The characteristic line of the rock mass starts from a point, A, corresponding to the initial stress state; the A–B stretch corresponds to the elastic behaviour of the mass, whereas in the B–C stretch the elastic limit is overcome and the convergence increases more than linearly with the reduction of inner confinement pressure.

Generally, the characteristic line for low confinement pressures can show two different behaviours (Fig. 4.10): The first one is characterised by a continuous increase of deformations, like in the Fig. 4.10 (case 1), with an asymptote defined by a value of radial pressure p_{cr} (in case of no cohesion); in the second behaviour (case 2), the curve reaches the y-axis in correspondence of a finite deformation, that is the convergence does not increase any further also with null confinement pressure: in this second case, theoretically the cavity could reach stability without lining. Figure 4.10 shows the characteristic lines of some types of lining that will be treated more in detail in Sect. 7.8 (0–p_1: elastic lining; 0–p_2: elastic lining reinforced in a second phase, 0–p_3: elastic-plastic lining).

Analytic formulations of the characteristic lines and their solutions in closed form (for example, those by Lombardi (1973), Ribacchi and Riccioni (1977), Panet and Guenot (1982), Hoek and Brown (1980)) provide a particularly useful support for the design.

The ground reaction curves can be drawn assuming, for the rock, an elasto-plastic constitutive model with softening and not associated flow rule.

Other curves can also be analysed to determine the behaviour of the rock mass during the excavation (Fig. 4.11):

- Convergence–Distance from face
- Distance from face–Fictitious radial pressure

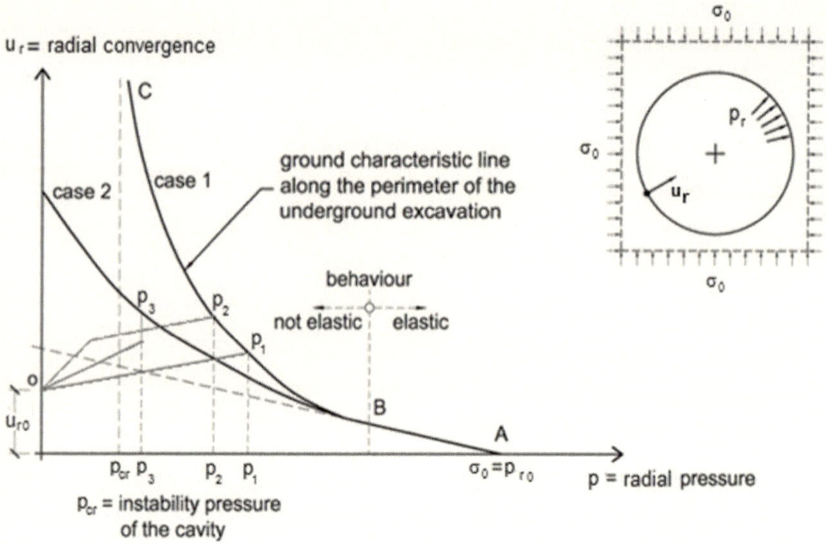

Fig. 4.10 Characteristic line of the ground at the cavity border. (Macori and Benussi 1982)

- Radial pressure–Plastic zone thickness
- Distance from face–Plastic zone thickness

The curve 'Convergence–Distance from face' can be obtained using a simplified analytic procedure that uses the following relations (Nguyen-Minh and Guo 1993; Panet and Guenot 1982) :

$$u_{rf} = 0.3 \cdot u_{r\infty}$$

$$\frac{c(x)}{c_\infty} = 1 - \left[\frac{0.84 \cdot R_{pl\infty}}{x + 0.84 \cdot R_{pl\infty}}\right] = 1 - \left[\frac{1}{1 + x/(0.84 \cdot R_{pl\infty})}\right]^2$$

where

x Distance from face
u_{rf} Convergence at face
$u_{r\infty} = u_r(x = \infty)$ Absolute convergence to infinity
$c\,(x) = c_r\,(x) = \left(u_r\,(x) - u_{rf}\right)$ Relative convergence
$c_\infty = (u_{r\infty} - u_{rf})$ Relative convergence to infinity
$R_{pl\infty} = R_{pl}(x = \infty)$ Total plastic radius (to infinity)

The curves 'Distance from face–Fictitious radial pressure' and 'Distance from face–Plastic zone thickness' are extrapolated from the other three curves.

It is also possible to plot the characteristic line close to the face, for example shifting the characteristic line of the cavity to the point given by the coordinates (strength of 1/2 core–convergence at the face). The latter characteristic line is particularly important for the design of the face improvement measures (see Sect. 4.6).

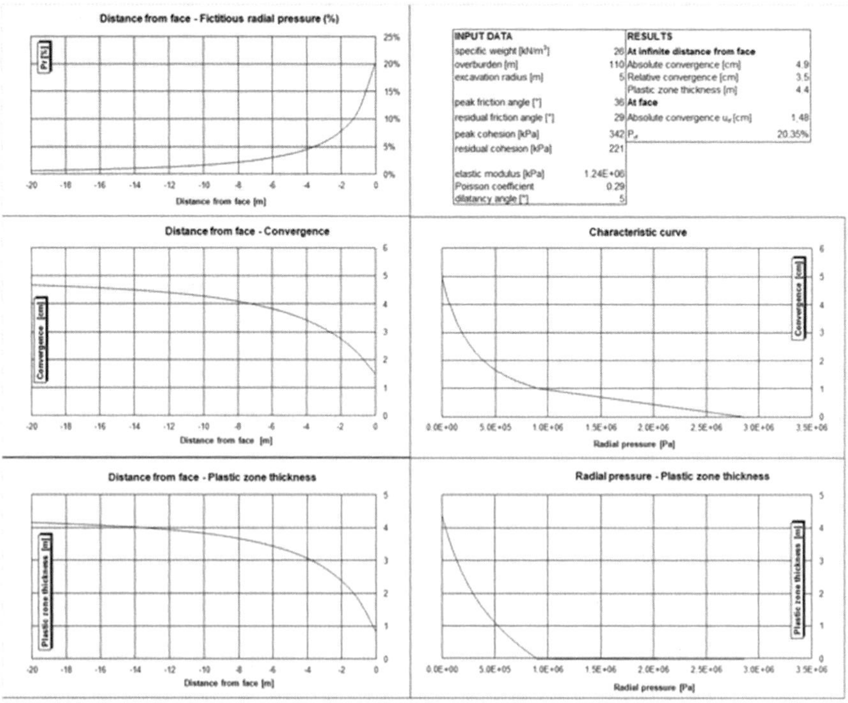

Fig. 4.11 Example of characteristic lines

The results obtained with the characteristic lines method allow the assessment of the behaviour of the ground to excavation, according to, for example, the convergence at the face and the Plastic zone thickness at the face (as it is explained in detail in Sect. 4.6).

4.5.3 *Numerical Methods*

Numerical methods allow the detailed modelling of the ground behaviour after excavation and the analysis of the influence of different factors and parameters that rule the realisation phase. This allows the setting of design criteria adequate for the excavation or to make decisions if an instability problem or important deformations occur.

In general, numerical methods start from the discretisation of the rock mass by means of different approaches: finite elements or differences and distinct elements.

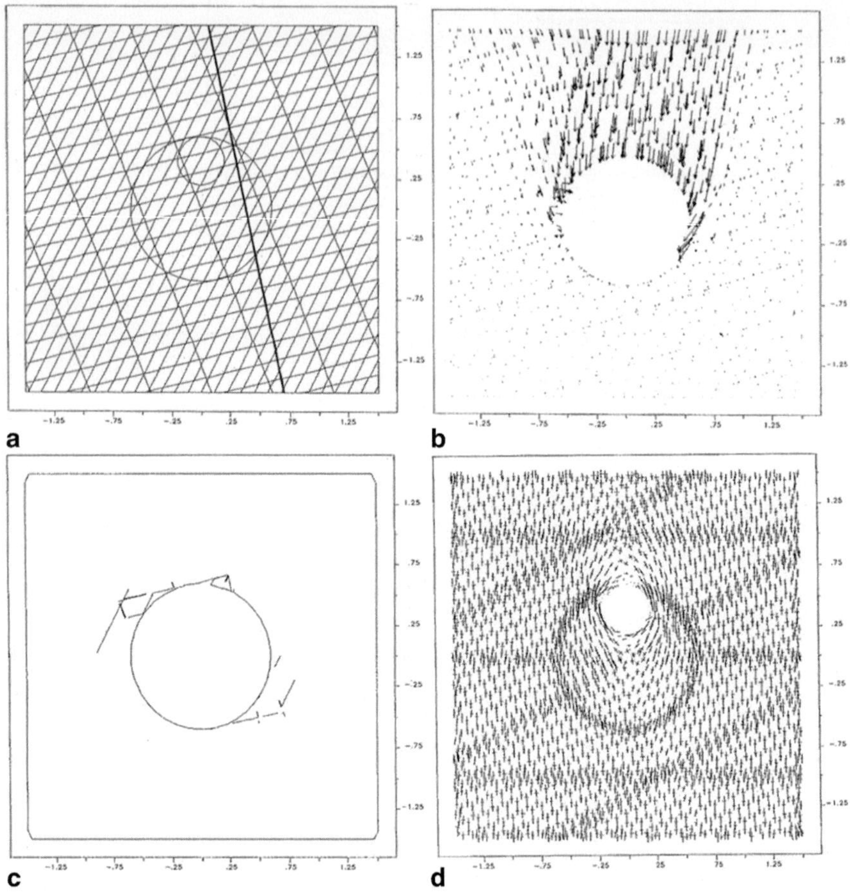

Fig. 4.12 Example of distinct elements modelling using the UDEC code: **a** modelling domain, **b** blocks displacement, **c** shear stress increment along joints and **d** principal stress orientation

4.5.3.1 Distinct Elements Method

The distinct elements method is a numerical approach apt to simulate the discontinuous behaviours of fractured media (Fig. 4.12). Adopting this approach, the inner part of the rock mass is divided into geometrically simple elements, each one having specific properties. The collective behaviour and interaction of these simplified elements models the more complex overall behaviour of the rock mass. Therefore, this kind of approach is particularly useful in ground conditions conventionally described as blocky, in which intersecting joints form wedges of rock that may be regarded as rigid bodies. For such conditions it is usually necessary to model many joints explicitly. Each block is considered as a unique free body that may interact at contact locations with surrounding blocks.

The main input parameters are represented by the mesh defining the analysis domain, the geomechanical properties of each element of the mesh and the discontinuity surfaces, the boundary conditions and the loads. The elaboration takes place through a series of calculation steps, where each step is characteristic of an operation phase. Therefore, the calculation makes it possible to follow the stress-strain evolution of the medium. The results of each calculation step consist in the numerical values of the displacements and the stresses of barycentre for each element of the mesh (Figs. 4.12b and 4.12c).

4.5.3.2 Finite Elements or Finite Differences Methods

They differ from the previously described distinct element methods because they consider the rock mass as a continuum. As far as the output results are concerned, the finite elements method is usually indistinguishable from the finite differences method; thus, they will be treated here as one and the same.

Those methods are aimed at the designing of underground works in soil or rock masses that can be assimilated to an equivalent continuum, considering that they allow taking into account both the stress and lithological anisotropies and the effect of possible improvement and/or confinement works in the cavity.

The physical problem is modelled numerically by dividing the entire problem region into elements. Obviously, the outer boundaries of the model domain must be placed at an adequate distance from the excavations.

The main input parameters are represented by the mesh defining the analysis domain, the geomechanical properties of each element, the boundary conditions and the loads.

Similarly to what happens for the distinct elements method, the elaboration occurs through a series of calculation steps, where each step is characteristic of an operation phase. Therefore, the calculation makes it possible to follow the corresponding stress-strain evolution in each constructive phase. But, in these methods, the algorithm requires the respect of the congruence of the displacement nodes of adjoining elements. The results of each calculation step consist in the numerical values of the displacements and stresses in correspondence of each node of the mesh.

4.6 Squeezing and Time-Dependent Behaviour

Squeezing is characterised by time-dependent deformations associated with plastic behaviour, caused by a particular combination of material properties and overstress conditions in the rock mass around the tunnel.

Some effects of squeezing are evident immediately after excavation (for example, the convergence that occurs even when the excavation does not advance), but normally long-term effect are prevalent, including continued ground movements and/or gradual build up of load on the tunnel support system.

Prediction of squeezing conditions is very important to define a stable support system of the tunnel. Generally, it is possible to say that (Barla 2002) :

- Squeezing behaviour is associated with poor rock mass deformability and strength properties in relation to initial stress.
- Squeezing behaviour implies the occurrence of yielding around the tunnel. The onset of yielding zone in the tunnel surrounding causes a significant increase in tunnel convergence and face displacements (extrusion). These can have generally large increase in time and represent the more significant aspects of the squeezing behaviour.
- Orientations of discontinuities such as bedding planes and schistosities play a very important role in the onset and development of large deformations around tunnels, and also on the squeezing behaviour. In general, if the main discontinuities strike parallel to the tunnel axis, the deformation will be enhanced significantly, as observed in terms of convergence during face advance.
- The pore pressure distribution and the piezometric head can also influence the rock mass stress-strain behaviour. Drainage measures causing a reduction in piezometric head both in the tunnel surround and beyond of the tunnel face often help to reduce ground deformations.
- Construction techniques for excavation and support (i.e. the excavation sequences and the number of excavation stages adopted, including the stabilization methods used) may influence the overall stability conditions of the excavation.
- Large deformations associated with squeezing may also occur in rocks susceptible to swelling. Although the factors causing either behaviour are different, it is often difficult to distinguish between squeezing and swelling, as the two phenomena may occur at the same time and induce similar effects. For example, in overconsolidated clays, the rapid stress-relief due to the tunnel excavation causes an increase in deviatoric stresses with simultaneous onset of negative pore pressure. In undrained conditions, the ground stresses may be such as not to cause squeezing. However, due to the negative pore pressure, swelling may occur with a more sudden onset of deformations under constant loading. Therefore, if swelling is restrained by means of early invert installation, a stress increase may take place with probable onset of squeezing.

Some methods that can be effectively used for a quick qualitative estimation of the risk of squeezing are presented briefly in the following paragraphs.

4.6.1 *Singh et al. (1992) Empirical Approach*

The line separating non-squeezing and squeezing conditions obtained by the authors was defined by the study of a number of case histories in which the overburden H and rock mass quality Q (Barton) were analysed. The line has the following equation:

$$H = 350 \cdot Q^{\frac{1}{3}} [m].$$

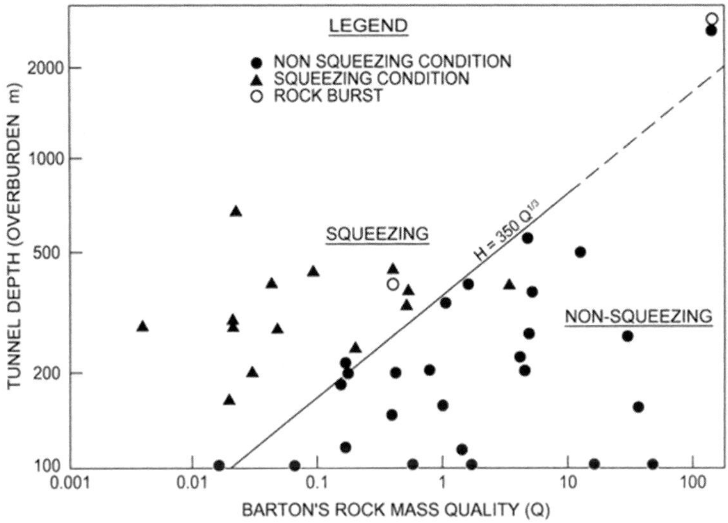

Fig. 4.13 Approach to predict squeezing. (Singh et al. 1992)

The analysed cases showed that squeezing occurs above this line (Fig. 4.13).

4.6.2 Goel et al. (1995) Empirical Approach

The method proposed by Goel et al. (1995) is similar to the one discussed above, but it is based on rock mass number N, defined as stress-free Q :

$$N = (Q)_{\mathrm{SRF}=1}$$

This formulation allows to avoid the problems and uncertainties in obtaining the correct rating of the parameter SRF in Barton et al. (1974) Q.

The analysed cases have been plotted on a log-log diagram having N and $H.\ B^{0.1}$ axes (where H is the tunnel depth, and B the tunnel span or diameter); the results obtained are shown in Fig. 4.14.

It can be seen that squeezing and non-squeezing cases are delimited by line:

$$H = (275 \cdot N^{0.33}) \cdot B^{-1}[m]$$

so that for squeezing conditions

$$H >> (275 \cdot N^{0.33}) \cdot B^{-1}[m].$$

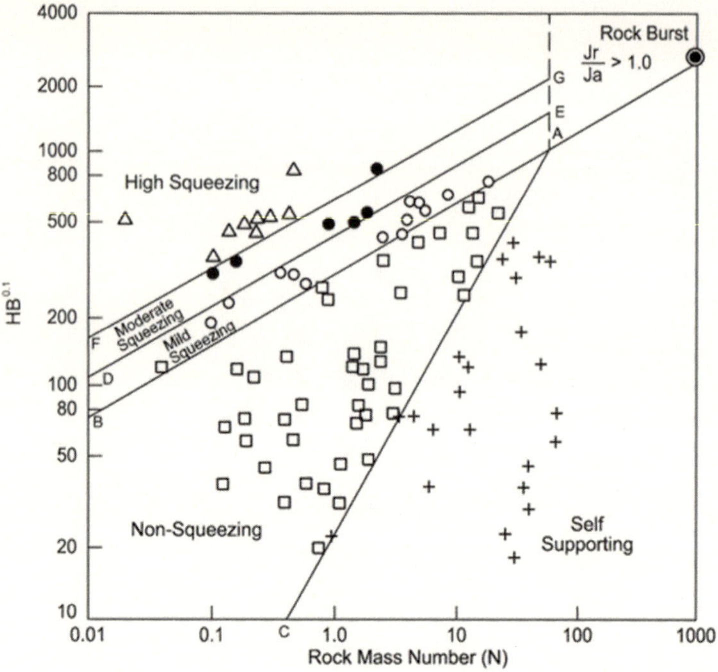

Fig. 4.14 Approach to predict squeezing conditions. (Modified from Goel 2000)

4.6.3 Hoek and Marinos (2000) Semi-Empirical Method

Hoek and Marinos (2000) demonstrated that, connecting the tunnel strain with the ratio of rock mass strength to in situ stress, a basis for estimating the potential risk of tunnel instability can be provided. In this context, strain is defined as the percentage ratio of tunnel wall deformation to tunnel radius.

Figure 4.15 shows that strain increases asymptotically when the ratio of rock mass strength to in situ stress falls below 0.2. This reveals the onset of severe instability, and without adequate support both the tunnel and the face would collapse.

If the internal support pressure $p_i$upon the strain of the tunnel ($\sigma_r = p_i$) and the face ($\sigma_3 = p_i$) is considered, it is possible to find the following approximate relationships for the strain of the tunnel $\varepsilon_t$and the face ε_f and the ratio of support pressure to in situ stress :

$$\varepsilon_t\% = 0.15\left[1-\left(\frac{p_i}{\sigma_0}\right)\right] \times \frac{\sigma_{cm}}{\sigma_0}^{-\left[3\left(\frac{p_i}{\sigma_0}\right)+1\right]/\left[3.8\left(\frac{p_i}{\sigma_0}\right)+0.54\right]}$$

$$\varepsilon_f\% = 0.1\left[1-\left(\frac{p_i}{\sigma_0}\right)\right] \times \frac{\sigma_{cm}}{\sigma_0}^{-\left[3\left(\frac{p_i}{\sigma_0}\right)+1\right]/\left[3.8\left(\frac{p_i}{\sigma_0}\right)+0.54\right]}$$

where σ_0 is the in situ stress and σ_{cm} is the rock mass strength.

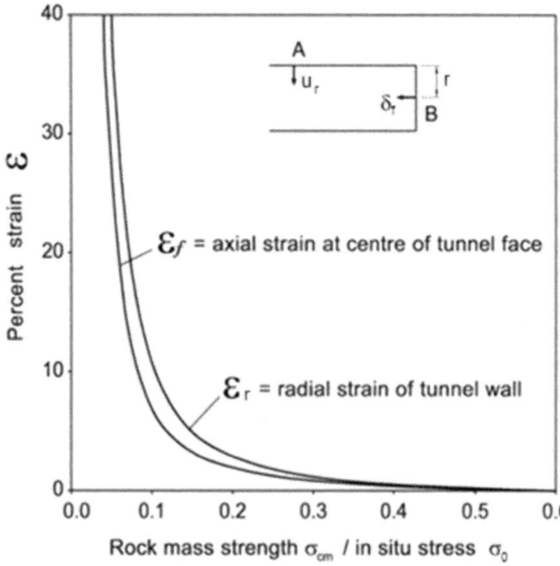

Fig. 4.15 Relationship between Rock Mass Strength σ_{cm} to in situ stress σ_0 and percentage strain for unsupported tunnels (strain ε_t is defined as a percentage ratio of radial tunnel wall displacement to tunnel radius, while strain ε_f is a percentage ratio of axial face displacement to tunnel radius; note that this analysis is applicable to a circular tunnel subjected to equal horizontal and vertical in situ stresses)

The curve defined by the first equation can be used to provide a set of approximate guidelines on the degree of difficulty that can be encountered for different levels of strain. Since these strain levels are associated with specific ranges of the ratio of rock mass strength to in situ stress, the curve given in Fig. 4.16 can be used to provide a first estimate of tunnel squeezing problems (Hoek and Marinos 2000) .

4.6.4 Jehtwa et al. Method (1984)

The criterion proposed by Jehtwa et al. (1984) for the estimation of the squeezing degree is based on the ratio between the mass strength and lithostatic pressure :

$$N_c = \frac{\sigma_{\text{cm}}}{\sigma_0}$$

where

σ_0 Pressure due to overburden (equal to $H\gamma$)

σ_{cm} Unconfined compressive strength of the rock mass that can be obtained by applying yielding criterion by Hoek et al. (2002) or, more approximatively, by applying the coefficient proposed by John (1971) to the compression strength σ_{ci} of the intact rock (Table 4.2).

The expected squeezing degree is obtained on the basis of this ratio (Table 4.3).

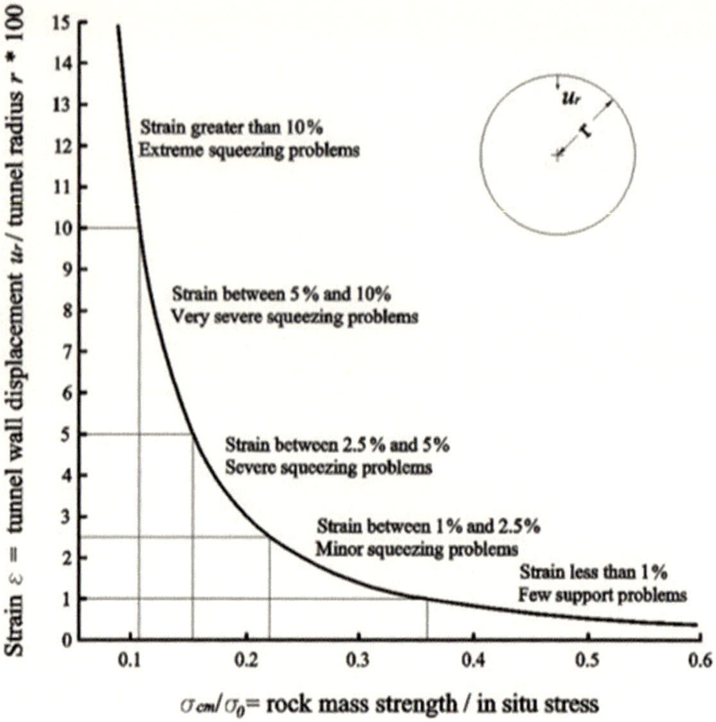

Fig. 4.16 Tunnelling problems associated with different levels of strain. (Hoek and Marinos 2000)

4.6.5 *Bhasin Method (1994)*

In the method proposed by Bhasin (1994), the behaviour of weak rocks at the excavation is defined starting from the stability factor N_t:

$$N_t = \frac{2\sigma_0}{\sigma_{cm}}$$

where

σ_0 Pressure due to the overburden ($H\gamma$)

σ_{cm} Unconfined compressive strength of the rock mass that can be obtained by applying the yielding criterion by Hoek et al (2002) or, more approximatively, by applying the coefficient proposed by John (1971) to the compression strength σ_{ci} of the intact rock (already presented in Table 4.2).

The expected squeezing degree is obtained on the basis of stability factor (Table 4.4).

In the graph of Figure 4.17, the stability factor N_t is shown on the y-axis and the mass strength on the x-axis; in the same graph, a limit curve for the excavation stability is also shown. This curve is determined on the basis of real cases and represents the progressive decrease of N_t when the rock mass strength decreases.

Table 4.1 Indications about the type of support according to the deformation ε_t of the cavity. (Hoek and Marinos 2000)

	Strain ε %	Geotechnical issues	Support types
A	Less than 1 %	Few stability problems and very simple tunnel support design methods can be used. Tunnel support recommendations based upon rock mass classifications provide an adequate basis for design	Very simple tunneling conditions, with rockbolts and shotcrete typically used for support
B	1–2.5 %	Convergence confinement methods are used to predict the formation of a "plastic" zone in rock mass surrounding a tunnel and of the interaction between the progressive development of this zone and different types of support	Minor squeezing problems which are generally dealt with by rockbolts and shotcrete; sometime light steel sets or lattice griders are added for additional security
C	2.5–5 %	Two-dimensional finite elements analysis, incorporating support elements and excavation sequence, is normally used for this type of problem.Face stability is generally not a major problem	Severe squeezing problems requiring rapid installation of support and careful control of construction. Heavy steel sets embedded in shotcrete are generally required
D	5–10 %	The design of the tunnel is dominated by face stability issues and, while two-dimensional finite elements analysis are generally carried out, some estimates of the effect of forepoling and face reinforcement are required	Very severe squeezing and face stability problems. Forepoling and face reinforcement with steel sets embedded in shotcrete are generally required
E	More than 10 %	Severe face instability as well as squeezing of the tunnel make this an extremely difficult three-dimensional problem for which no effective design methods are currently available. Most solution are based on experience	Extreme squeezing problems. Forepoling and face reinforcement are usually applied and yielding support may be required in extreme cases

Table 4.2 Reduction coefficient of σ_{ci} for the calculation of mass strength σ_{cm} proposed by John (1971)

Rock mass	Reduction coefficient of σ_{ci}
Massive rock, slightly stratified	0.8
Rock mass with slight stratification and low alteration	0.6
Rock mass with high stratification, fracturing and alteration degree	0.4
Very fractured rock mass	0.2

4.6.6 *Panet Method (1995)*

The convergence–confinement method simulates the excavation of a tunnel using a plane strain model. This criterion can be applied for soft rocks where the excavation of a tunnel with high overburden triggers the cavity convergence and a reduction

Table 4.3 Squeezing degree forecast using Jehtwa et al. method (1984)

N_c	Squeezing degree
< 0.4	Highly squeezing
0.4–0.8	Moderately squeezing
0.8–2.0	Mildly squeezing
> 2.0	Non-squeezing

Table 4.4 Squeezing degree forecast using Bhasin's method (1994)

Stability factor N_t	Squeezing degree
< 1	Non-squeezing
1–5	Mild to moderate squeezing
> 5	Highly squeezing

of the stresses acting on the tunnel boundary. The reduction from the initial stress value (σ_0) to a fictitious internal pressure p_r (simulating the supporting effect on the cavity) can be expressed by following equation:

$$p_r = (1 - \lambda)\sigma_0.$$

As a consequence, the amount of stress relief is expressed by stress relief factor (deconfinement rate) λ:

$$0 \leq \lambda \leq 1.$$

The face stability can be assessed as a function of the critical value of the stress relief factor λ_e, for which plastic deformations begin, that can be evaluated by Mohr–Coulomb or by Hoek–Brown criterion or else, as a function of the value of N. In

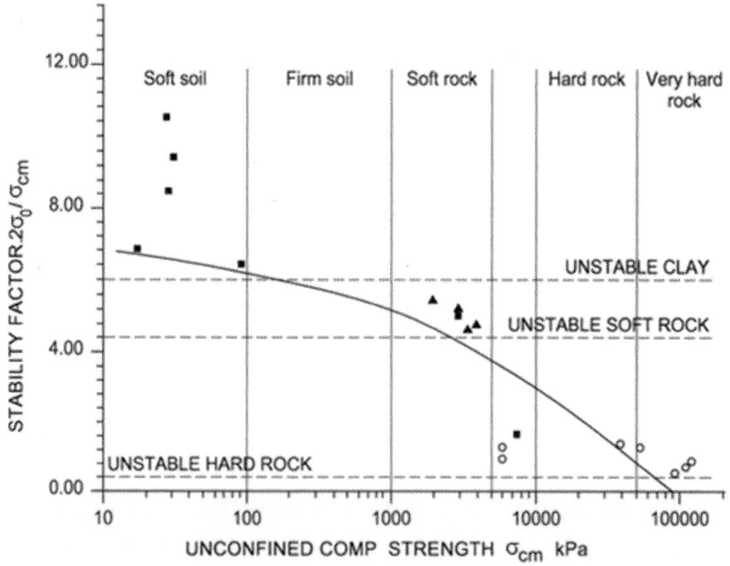

Fig. 4.17 Stability curve N_t–σ_{cm}. (Bhasin 1994)

particular, with reference to the yielding criterion by Mohr–Coulomb, that index is a function of the passive earth pressure coefficient k_p and of the stability factor N according to the equation:

$$\lambda_e = \frac{1}{k_p + 1}\left[k_p - 1 + \frac{2}{N}\right]$$

whereas, with reference to the failure criterion by Hoek and Brown, it is a function of the parameter m and of the stability criterion N according to the equation:

$$\lambda_e = \frac{1}{4N}(\sqrt{m^2 + 8mN + 16s} - m)$$

$$N = \frac{2\sigma_0}{\sigma_{cm}}$$

where

σ_0 Pressure due to the overburden
σ_{cm} Unconfined compressive strength of the rock mass that can be obtained by applying the yielding criterion Hoek et al. 2002
m and s Constants of the yielding criterion by Hoek et al. (2002).

If $N < 1$, the mass strength is never reached by the pressures acting around the cavity; therefore, this is a case of elastic conditions. If, on the contrary, yielding deformations occur ($N > 1$), three conditions are possible according to Panet:

- $1 < N < 2$: the excavation face is stable and the deformations in the face area remain in the elastic field; plastic deformation appears behind the face.
- $2 < N < 5$: part of the excavation face presents plastic deformation, whereas the area behind the face is completely plastified.
- $N > 5$: the excavation face is unstable, the plasticisation interests the area beyond the excavation face.

When the plastic limit is reached, the author suggests following limit ranges of the deconfinement rate, corresponding to specific stability conditions of the face:

- $0.6 < \lambda_e < 1.0$: the excavation face is stable; pressures reach the limit value of the mass strength behind the face.
- $0.3 < \lambda_e < 0.6$: the excavation face is stable in the short term; pressures at the excavation face reach the maximum strength value near the cavity border first, and then toward the core.
- $\lambda_e < 0.3$: the excavation face is unstable, therefore, prior improvement measures are required.

4.7 Rock Burst

Rock burst and squeezing (previous described) are the two main modes of underground instability caused by the overstressing of the ground, the first occurring in brittle hard rocks, the second in ductile soft rocks.

Depending on the magnitude and typology of instability, rock burst is also known as *spalling* or *popping*, also a variety of other names are in use, among them are *splitting* and *slabbing*. Generally, these phenomena take place at great depths in hard low fractured rocks (brittle behaviour), but they can also be induced at shallower depth where high horizontal stresses are acting. Selmer-Olsen (1964) and Muir Wood (1979) underline the significant impact arising from the great differences between horizontal and vertical stresses. Selmer-Olsen (1964, 1988) has experienced that, in the hard rocks in Scandinavia, such anisotropic stresses might cause spalling or rock burst in tunnels located inside valley sides steeper than 20° and with the top of the valley reaching height higher than 400 m above the level of the tunnel.

Rock burst can consist of sudden failures associated with high energy release that can cause the projection of rock volumes from the tunnel wall whose dimensions range from small rock fragments to slabs of several cubic metres. However, they cause significant problems and reduced safety for the tunnel crew during excavation.

The relationship between vertical stress $p_z = \sigma_0$ and the matrix strength σ_{ci} is an index of the probabilities that rock burst or phenomena of increasing intensity (spalling and slabbing) might occur in rock masses with good geomechanical qualities; Fig. 4.18 (Hoek and Brown 1980) shows a graph vertical stress p_z/ matrix strength σ_{ci} of excavations carried out in quartzite with very good geomechanical quality. Following this approach, Hoek and Brown (1980) have classified the rock burst activity as:

$p_z/\sigma_{\mathrm{ci}} = 0.1$ Stability
$p_z/\sigma_{\mathrm{ci}} = 0.2$ Spalling
$p_z/\sigma_{\mathrm{ci}} = 0.3$ Severe spalling—slabbing
$p_z/\sigma_{\mathrm{ci}} = 0.4$ Need of important stabilisation measures
$p_z/\sigma_{\mathrm{ci}} > 0.5$ Cavity collapse (rock burst)

Similarly, Russenes (1974) has shown the relations between rock burst activity, tangential stresses in tunnel surface and the point load strength of the rock (Fig. 4.19).

4.8 Face Stability Assessment

This chapter summarizes some methods that can be used to define the behaviour of the face and the overall response of the tunnel during the excavation. These methods are based on "benchmark" quantities that allow an immediate evaluation of the behaviour.

In general, the following conditions may occur:

- Stable face: no measures are required at the face
- Short-term stable face: stabilisation works at the face are suggested (fiberglass elements, self-drilling steel bars, injections, etc.)
- Unstable face: requires the implementation of stabilisation works at the face

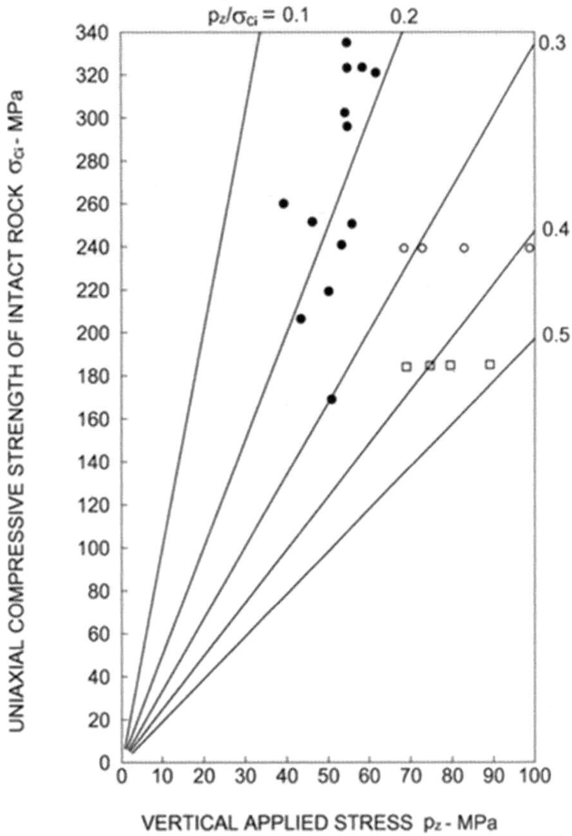

Fig. 4.18 Rock burst assessment according to the ratio p_z/σ_{ci} (Hoek and Brown 1980). *Points* and *squares* represent different site tests

From the analytical point of view, the stabilisation measures of the face can be reproduced either by considering a confining pressure on the face σ_3 or an improvement of the mechanical properties of the core material (cohesion increase).

4.8.1 *Shallow Overburden*

4.8.1.1 Undrained Behaviour of Cohesive Soils

In cohesive soils, it can be assumed that the excavation speed is so fast that undrained behaviours can be taken as a reference to check the face stability. The limit value of the stabilisation pressure of the face σ_3 is evaluated according to the expression by Assadi Sloan (1991):

$$\sigma_3 = Q_\gamma \cdot \gamma \cdot \frac{D}{2} + Q_c \cdot C_u$$

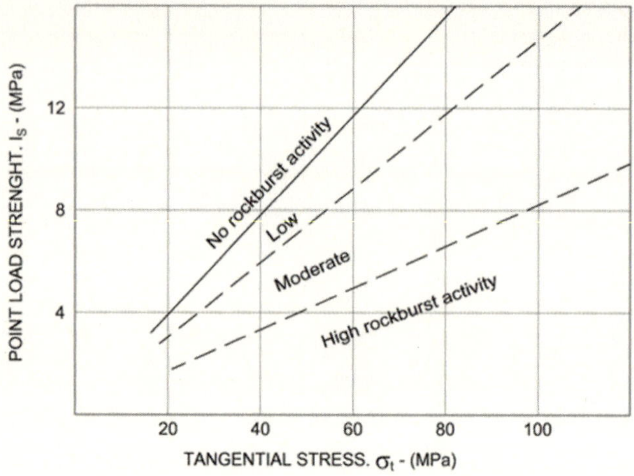

Fig. 4.19 The level of rock burst related to the point load strength of the rock and the tangential stress ($\sigma_t = \sigma_\theta$) in the tunnel surface calculated from Kirsch's equations (in case of an isotropic state of stress: $\sigma_t = 2\gamma H$). (From Nilsen and Thidemann 1993, based on the data from Russenes (1974))

where

$$Q_\gamma = \frac{2 \cdot z + D}{D}$$

$$Q_c = 2ln\frac{2 \cdot z + D}{D}$$

γ Specific weight of the soil
z Overburden referring to the crown of the tunnel
D Equivalent diameter of the excavation
σ_3 Stabilisation pressure at the face
C_u Undrained cohesion

The values of $\sigma_3 \leq 0$ show that the cavity is stable also when no confinement pressure is present.

Adopting the definition of the stability factor given by Broms and Bennemark (1967) and obtained by the following expression:

$$N = \frac{(\sigma_s + \gamma \cdot (z + D/2) - \sigma_3)}{C_u}.$$

where σ_s is the pressure due to the overloads on the surface, whereas the other symbols still have the above-written meanings, the authors state that the face is stable if:

$$N \leq 6 - 7.$$

That value can also be compared with the limit values given by Broms and Bennemark (1967), Attewell and Boden (1971), Davis et al (1980), Léca and Dormieux (1990).

If both conditions lead to the conclusion that the face is unstable ($\sigma_3 > 0, N > 7$), then it is possible to determine the value of σ_3 that stabilizes the face. In particular, applying the first condition, $\sigma_3 \neq 0$ has to be found that leads to a stability factor $N \leq 5$; or, starting from the second condition, the $\sigma_3 > 0$ will have to be assessed according to the same value of Q_γ but with Q_c equal to the maximum value between the two values that follow:

$$Q_c = max \begin{cases} 4ln\frac{2z+D}{D} \\ 2 + 2ln\frac{2z+D}{D} \end{cases}$$

4.8.1.2 Grain Material with Drained Behaviour

According to Leca and Dormieux (1990), for granular soil where no water table is present and for which the failure behaviour can be expressed with the Mohr–Coulomb criterion with c' and φ', the stability condition of the excavation face can be assessed using the following expression:

$$\sigma_3 = -c' + \cdot cot\phi' + Q_\gamma \cdot \gamma \cdot \frac{D}{2} + Q_s \cdot (\sigma_s + c' \cdot cot\phi')$$

where

- Q_γ and Q_s Adimensional coefficients of the Leca and Dormieux method (1990) defined according to the friction angle and the ratio overburden/radius (H/r) Fig. 4.21.
- γ Specific weight of the soil
- D Equivalent diameter of the excavation
- r = D/2 Equivalent radius of the excavation
- σ_3 Stabilisation pressure at the face
- σ_s Pressure due to overload on the surface
- c' Drained cohesion
- φ' Drained friction angle

That expression is obtained considering the three-dimensional situation presented in Fig. 4.20.

The values of Q_γ and Q_s can be assessed according to what is suggested by Leca and Panet (1988) and Leca and Dormieux (1990) in the diagram in Fig. 4.21.

The values of $\sigma_3 \leq 0$ show that the face is stable and also with no confinement pressure.

4.8.1.3 Stability of the Excavation Face by Tamez (1985)

The method proposed by Tamez (1985) can be applied to both granular and cohesive soils and can be adapted to different draining conditions. That method considers the

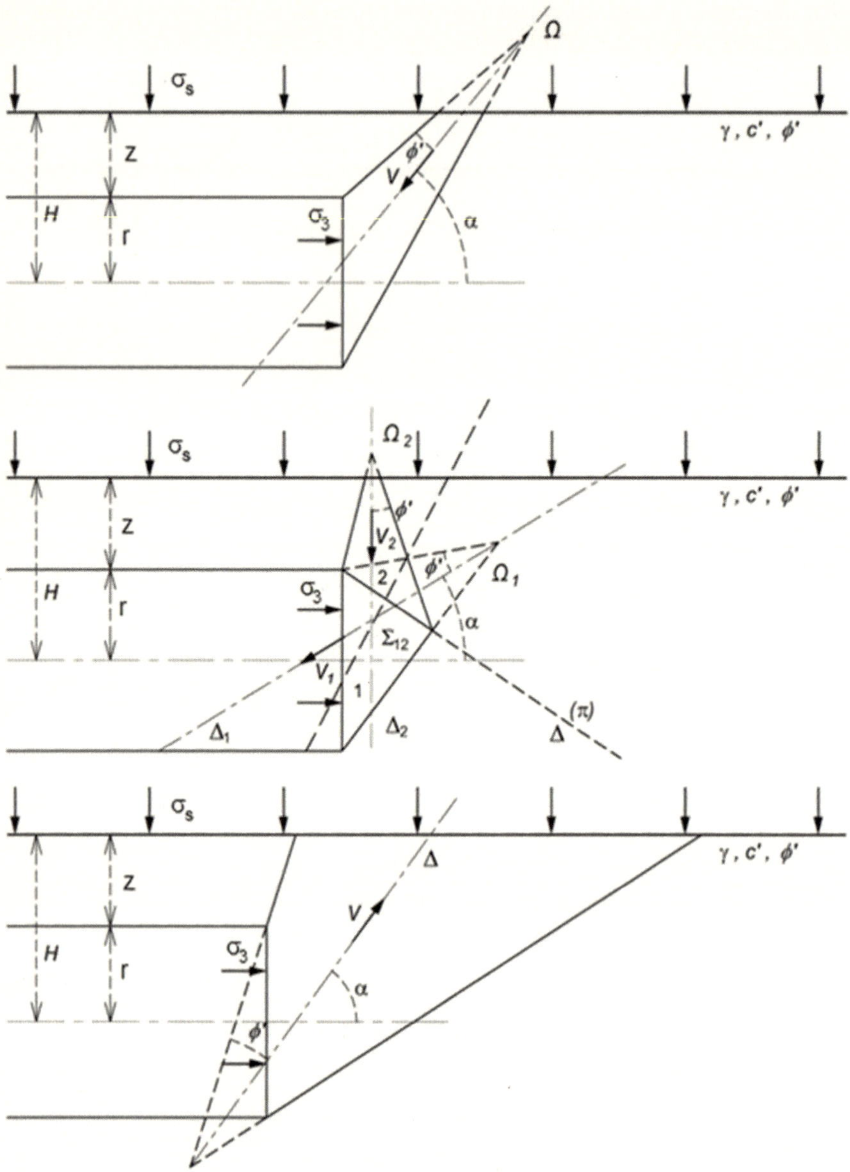

Fig. 4.20 Possible failure mechanisms. (Leca and Dormieux 1990)

equilibrium of the prism loading on the face (Fig. 4.22) and defines a safety factor for the stability as a ratio between stabilising and destabilising forces, according to following equation :

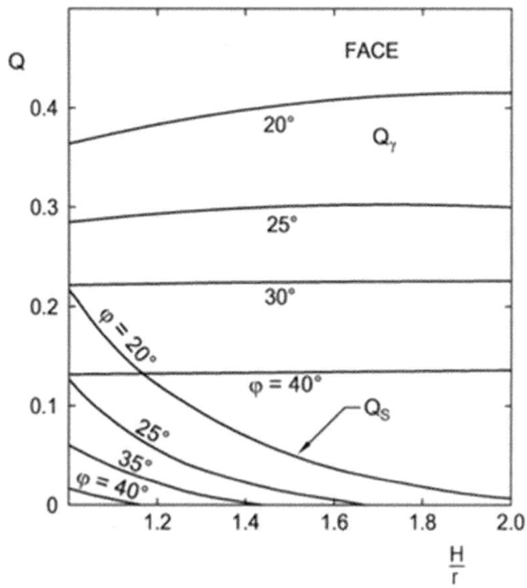

Fig. 4.21 Definition of the numerical values of coefficients Q_γ and Q_s of the Leca and Dormieux method (1990)

$$FSF = \frac{\left[\frac{2\cdot(\tau_{m_2}-\tau_{m_3})}{(1+\frac{a}{l})^2} + 2\cdot\tau_{m_3}\right]\cdot\frac{h_1}{b} + \frac{2\cdot\tau_{m3}}{(1+\frac{a}{l})\cdot\sqrt{k_a}}\cdot\frac{h_1}{D} + \frac{3.4\cdot c'}{(1+(\frac{a}{l})^2)\cdot\sqrt{k_a}}}{[1+\frac{2D}{3z(1+\frac{a}{l})^2}]\cdot[\gamma\cdot z - \sigma_3]}.$$

where

a Length of the stretch where lining has not been installed yet (Fig. 4.21)
l = $D\cdot \mathrm{tg}(45-\varphi/2)$
b Width of the tunnel;
D Height of the tunnel
h_1 Height of Protodyakonov's curve (defined in the following pages)
c' Cohesion
k_a Active earth pressure coefficient,
γ Specific weight of the soil,
z Overburden at the crown of the tunnel,
σ_3 Stabilisation pressure at the face.

The parameter h_1, defined as the height of Protodyakonov's curve:

$$h_1 = \begin{cases} B/2f & \text{if } h_1 < z \\ z & \text{otherwise} \end{cases}$$

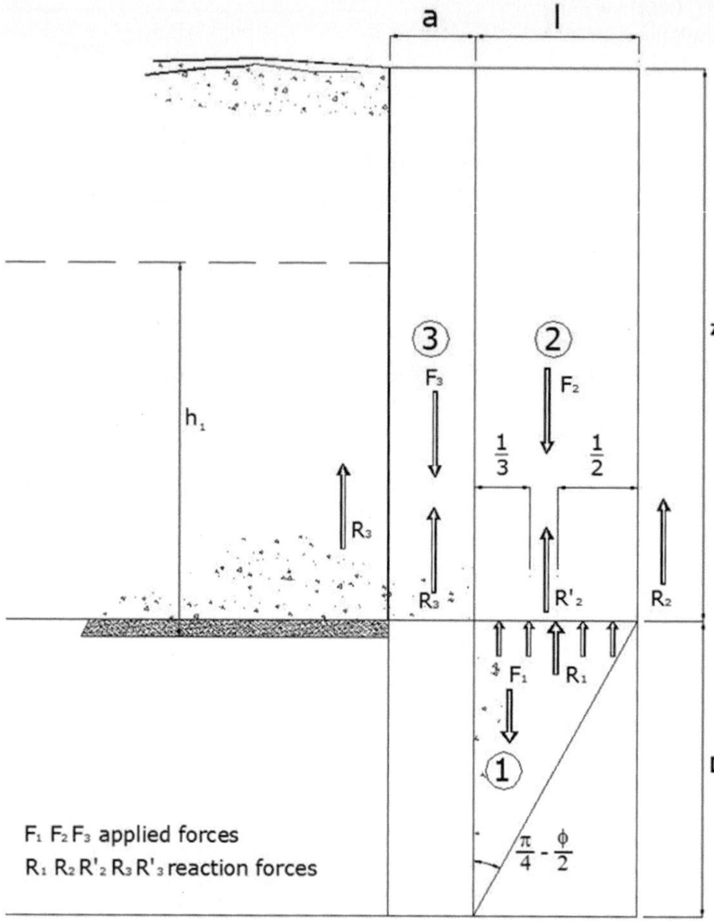

Fig. 4.22 Geometrical scheme of Tamez (1985) method

B= $b + 2 \cdot D \cdot \text{tg}(45 - \varphi/2)$
f Protodiakonov factor, defined as follows:

$$f = \sigma_c/100 \quad \text{for rock}$$
$$f = c'/\sigma c + \text{tg}\phi \quad \text{for soil}$$

σ_c being the ground compression strength.τ_{m2} and τ_{m3} are the shear strength values of the ground along the faces of slices (2) and (3); they depend on the tunnel depth. More precisely, if the tunnel is deep (e.g. for a ratio between overburden z and excavation diameter D higher than 3), following equations are valid:

$$\tau_{m3} = c' + \{0.25 \cdot [w\gamma + (z - h_1 - w) \cdot (\gamma - \gamma_w) - u]\} \cdot tg\phi$$

$$\tau_{m2} = c'_{\sigma_c} + \frac{k_0}{2} \cdot \left\{ w\gamma + (z - h_1 - w) \cdot (\gamma - \gamma_w) + 3.4 \cdot \frac{c'}{\sqrt{k_a}} - \frac{(\gamma - \gamma_w) \cdot D}{2} \right\}$$

w being the depth of the water table with respect to the land surface, γ_w the specific weight of water, u the neutral pressure of water, c' and φ the cohesion and the friction angle of the ground, respectively.

In case of shallow tunnel (e.g. for per $z/D \leq 3$), τ_{m2} *and* τ_{m3} have to be calculated using the following expressions :

$$\tau_{m3} = c'$$

$$\tau_{m2} = c' + \frac{k_0}{2} \cdot \left\{ 3.4 \cdot \frac{c'}{\sqrt{k_a}} - \frac{(\gamma - \gamma_w) \cdot D}{2} \right\}$$

where k_0 is the rest thrust coefficient.

The safety factor of the prism corresponding to the stretch with no lining (prism 3) is given by:

$$FS_3 = \frac{2.\tau_{m3}}{\gamma \cdot z - \sigma_3} \cdot \frac{h_1}{b} \cdot \left[1 + \frac{b}{a} \right],$$

σ_3 being the stabilisation pressure applied on the excavation face.

Ultimately, it is assumed that:

$$FS = \min(FSF; FS_3).$$

In general, *FS* has to be higher than 1.3 to guarantee stability. Where this does not happen with no stabilisation pressure (σ_3) applied to the excavation face, the value of the pressure that assures stability has to be found. The pressure value is then transformed in a number of equivalent supports at the face.

4.8.2 High Overburden

4.8.2.1 Face Stability as a Function of Characteristic Strength of Rock Mass

Since the instability of the face is induced by the same conditions leading the squeezing, the criteria of Bhasin (1994) and Hoek-Marinos (2000) already discussed in Sect. 4.6 can be used for the evaluation of the face stability:

- In terms of previously defined stability factor N:
  - $N < 1$: no plasticity occurs
  - $N < 5$: low plasticity occurs
  - $N > 5$: high plasticity occurs
- Using Fig. 4.17 (Bhasin 1994)
- Using Fig. 4.16 (Hoek-Marinos 2000)

4.8.2.2 Face Stability with Convergence–Confinement Method

Face stability can be seen as a function of convergence. In this case, the method proposed by Panet (1995) (described in Sect. 4.6) can be adopted for its evaluation. Following limits have to be considered:

- As a function of N:
  - $N < 2$: face is to be considered stable,
  - $2 < N < 5$: short term stability,
  - $N > 5$: face is to be considered unstable
- As a function of λ_e:
  - $0.6 < \lambda_e < 1$: face is to be considered stable,
  - $0.3 < \lambda_e < 0.6$: short-term stability
  - $\lambda_e < 0.3$: face is to be considered unstable

4.8.2.3 Face Stability as a Function of Shear Strength

The method proposed by Ellstein (1986) to assess the stability of the tunnel face estimates a safety factor evaluated in function of material characteristics in terms of c (peak cohesion) and φ (peak value of internal friction angle) and tunnel equivalent excavation diameter D.

For circular tunnels excavated in homogenous cohesive materials under phreatic surface (i.e. with groundwater table depth w higher than $z - D/2$) and excavated with a stabilisation pressure σ_3, the Ellstein formulation can be written as follows :

$$FSF = \frac{2 + \frac{2+\sqrt{2}}{1+a/D}}{k_0\left(\frac{\gamma-\gamma_w}{\gamma} + \frac{\gamma_w \cdot w}{\gamma \cdot z}\right) + \frac{D}{2z} + \frac{\gamma-\gamma_w}{6\gamma \cdot z} \cdot \frac{D}{z} - \frac{a}{6z} + \frac{\gamma_w}{\gamma}\left(1 - \frac{w}{z}\right) - \frac{\sigma_3}{\gamma \cdot z}} \cdot \frac{c}{\gamma . z}$$

where

a Length of the stretch where lining has not been applied yet,
c Cohesion,
γ e γ_w Specific weights of soil and of water respectively,
z Overburden at the crown of the tunnel,
σ_3 Stabilisation pressure at the face.
w Depth of the water table with regards to the land surface
k_0 The at rest thrust coefficient.
D Equivalent excavation diameter

The method is applicable to not very shallow tunnel, e.g. where following inequality is valid:

$$D \leq 3 \cdot z - \sqrt{9 \cdot z^2 - 24 \cdot z \cdot {}^{C}/\gamma} \, .$$

All conditions for which $Fs > 1.3$ are considered stable.

Table 4.5 Interpretation criteria for the ground reaction curve method

Face is stable	Short term stability of the face	Possible face instability	Face is unstable
$u_{rf} < 1\,\%R_{exc}$ $F_{plf} << R_{exc}$	$1\,\%R_{exc} < u_{rf} < 2\,\%R_{exc}$ $F_{plf} < R_{exc}$	$2\,\%R_{exc} < u_{rf} < 3\,\%R_{exc}$ $F_{plf} \geq R_{exc}$	$u_{rf} > 3\,\%\,R_{exc}$ $F_{plf} >> R_{exc}$

4.8.2.4 Face Stability in Relationship to the Tensional Field and Mechanical Characteristics of Rock Masses

The criterion is based on the comparison between stress conditions on the excavation perimeter and strength characteristics of rock masses. To that aim, Pelizza et al. (1993) defined a Competence Index R_0 as:

$$R_0 = \frac{S_{1(0)}}{S_t}$$

with

$$S_{1(0)} = 0.72 \cdot P_z + \sqrt{0.72 \cdot m \cdot \sigma_{ci} \cdot P_z + s \cdot \sigma_{ci}^2}$$
$$S_t = 1.67 \cdot P_z$$

where $P_z = \sigma_0$ is the vertical geostatic pressure, σ_{ci} is the uniaxial compressive strength of the intact rock whereas m and s are the parameters of the rock mass defined in Sect. 3.6 (coefficient of the model by Hoek and Brown).

On the basis of the directions given by the authors, it can be said that stability problems at the face are more likely to occur when $R_0 < 1$.

4.8.2.5 Face Stability with the Ground Reaction Curve Method

Analyses of the face stability at great depths can be carried out by the ground reaction curves method.

Criteria, described in Table 4.5, can be considered for the interpretation of the results obtained by the ground reaction curves. The limits used in the first place instance to assess the stability of the face are reported here (for the limits of admissibility and critical deformation limits, see for example Sakurai (1997)).

where

u_{rf} Convergence at the excavation face.
F_{plf} Plastic zone extension at the excavation face.
$R_{exc} = R_{eq}$ Equivalent radius, it is radius of a cavity which has same area of the studied excavation section.

Let us consider the case in which convergence u_{rf} is relevant, e.g. the face in unstable conditions; the pre-dimensioning of improvement works of the core, required to reduce u_{rf}, can be carried out drawing the characteristic line of the face. This curve can be obtained, for example (Fig. 4.22), shifting the characteristic line of

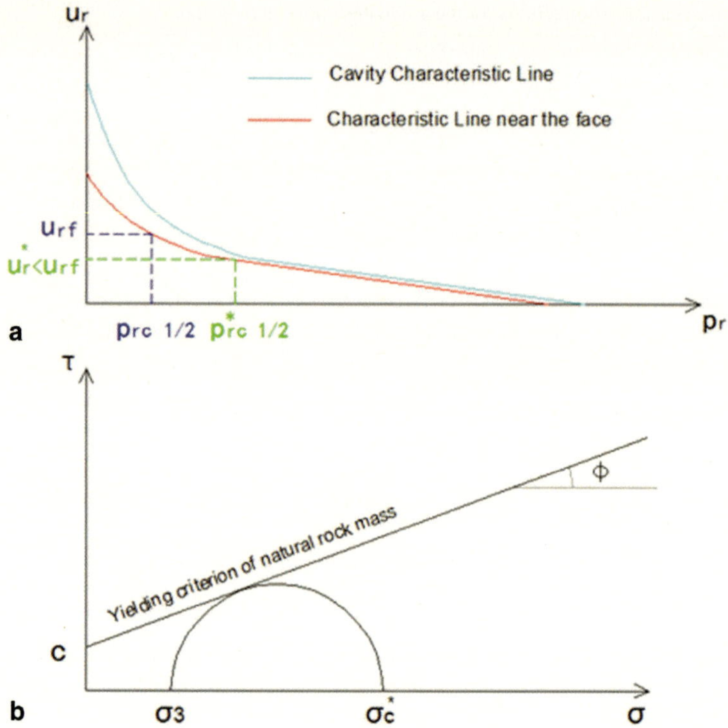

Fig. 4.23 **a** Characteristic lines of the cavity and the face, **b** Mohr–Coulomb yielding criterion of the rock mass (where σ_3 = stabilisation pressure on the face, τ = tangential stress, c cohesion, φ = shear angle, σ_c* = minimum rock mass strength which ensures face convergence u_{rf}*)

the cavity so that it goes through the point given by the coordinates (strength of half core–convergence at the face u_{rf}). The 'strength of half core' $p_{rc1/2}$ is a conventional fictitious radial pressure that simulates the confinement condition at the face (Lombardi 1973) given by the relation:

$$p_{rc1/2} = \frac{\sigma_{cm}}{2}$$

where σ_{cm} is the unconfined compressive strength of the rock mass.

As already said, in this conditions the u_{rf} is too high and it would lead to face instability, so it must be reduced to $u_{rf}* < u_{rf}$ (e.g. u_{rf}* can be defined as a reduced percentage of the excavation radius). As it can be seen in Fig. 4.23a, the pressure ($p_{rc1/2}$*) corresponding to u_{rf}* is higher than $p_{rc1/2}$; consequently, this pressure can be stood by the rock mass only if the strength on the core is increased.

The strength increase of the core can be obtained either by improving the mechanical properties of the material (e.g. by injections or freezing etc.) or applying a confining pressure on the face σ_3. In any case, from the definition of the half core strength $p_{rc1/2}$, the new rock mass strength that must be obtained is:

$$\sigma_c{}^* = 2 \cdot p_{rc1/2}{}^*.$$

The pressure at the face σ_3 needed to reach σ_c^* can be obtained geometrically from the Mohr–Coulomb yielding criterion (Fig. 4.23b), and it is expressed by the following relationship:

$$\sigma_3 = \frac{\sigma_c{}^* - 2c\sqrt{k_p}}{k_p}.$$

4.8.2.6 Face Stability Caquot Method

The traditional Caquot's solution (Carranza-Torres 2004) valid to assess the equilibrium and the radial pressure around a round cavity in an isotropic medium can be generalised in a single form that allows, according to the safety factor FS, the analysis of the round cavity (assessment of the safety of the lining) or the spherical analysis (assessment of the safety of the excavation face) through the variation of the index *k*.

That generalised form is :

$$\frac{\sigma_3}{\gamma \cdot r} = \left(\frac{\sigma_s}{\gamma \cdot r} + \frac{c}{\gamma \cdot r} \cdot \frac{1}{tg\phi}\right)\left(\frac{h}{r}\right)^{-k(N_\phi^{FS}-1)} - \frac{1}{k \cdot (N_\phi^{FS} - 1)} - 1 \cdot \left[\left(\frac{h}{r}\right)^{1-k(N_\phi^{FS}-1)} - 1\right] - \frac{c}{\gamma \cdot r} \cdot \frac{1}{tg\phi}$$

where

- σ_3 is the minimal inner pressure required for a stability situation
- γ is the dry volume weight of the material
- r is the equivalent radius of the excavation
- σ_s is the possible surface load
- c is the material effective cohesion
- φ is the angle of effective shear strength of the material
- h overburden at the cavity centre with respect to the land surface
- k is the index ruling the type of analysis (1 = round; 2 = spherical)

where $N\varphi^{FS}$ introduces the dependence from FS

$$N_\phi^{FS} = \frac{1 + sen\left(arctg\frac{tg\phi}{FS}\right)}{1 - sen\left(arctg\frac{tg\phi}{FS}\right)}.$$

4.9 Ground Water Influence

In the previous chapters it has already been highlighted that the presence of ground water may represent a hazard element for the realisation of an underground work, both for the equilibrium of the environment and for the work itself. All geotechnical works are strongly influenced by the presence of water, but for a work that is completely underground, this influence is even stronger.

The methodologies and the applicable techniques to limit and reduce the risks and the consequences of the presence of water in underground excavations are described in the following chapters. This chapter is focused on the description of the modalities used to analyse those problems. The modelling of the interaction between the system constituted by a ground saturated with water and the tunnel represents a very complex problem due to the many variables present, the lack of reliable data available during the design phase. Some simplified analysis methods will be briefly introduced as they can be used in the preliminary design phase.

Two main aspects linked to the presence of water have to be considered during the design of a tunnel under the water table:

(a) The determination of the flow rate that will be drained by excavation (both in transient state and at regime) and assessment of the effect that the realisation of the work will have on the natural hydrogeological setting
(b) The influence of groundwater (static or in seepage regime) on the cavity stability and on the static design of lining

4.9.1 Assessment of Tunnel Inflows

Tunnel water inflow is always an undesirable circumstance for hydrogeological-environmental reasons, with regards to possible surface settlements and also with respect to increasing realisation difficulties. The influence of an underground opening on the hydrogeological setting is a wide and complex aspect of the design that normally cannot be approached in a rigorous way. This chapter only provides some basic instruments for a preliminary tunnel inflow assessment. A rigorous assessment of the flow rates is only possible with numerical methods that require detailed input data, with a particular reference to the structure, the properties of the ground interested by the cavity and the boundary conditions.

An estimate of tunnel inflows can be obtained using geomechanical classifications (Gates 1997), through analytical formulations (Jacob and Lohman 1952; Kawecki 2000; Goodman et al. 1965) or the implementation of mathematic models (Dunning et al. 2004; Molinero et al. 2002), notwithstanding all the mistakes deriving from the uncertainty of the variables involved. Actually, it must be kept in mind that the results obtained are highly conditioned by the permeability value that depends on the fracturing degree and from the stress state, as well as on the hypothesis of isotropy

Table 4.6 Values of the J_w index when rock mass flow conditions change

Conditions	Drops/minute	Litre/minute	J_w
Dry	< 1		1
Wet	1–10		0.94
Drops	10–100		0.86
Dripping	> 100		0.76
Diffuse seepage		0.0075–0.075	0.66
Low flow		2.3–6	0.5
Average flow		6–60	0.33
High flow		> 60	0.2

and homogeneity that constitute the base of traditional formulations, which are often inadequate to describe correctly the draining process in rock masses.

A semi-quantitative technique to assess if a rock mass can be affected by a high groundwater flow was studied by Gates (1997), who developed a classification of the rock mass called HP (hydro-potential or hydrological potential). This classification mainly derives from Barton classification (Q, Barton et al. 1974) and expresses the HP value as:

$$\mathrm{HP} = \left[\frac{\mathrm{RQD}}{J_n}\right] \cdot \left[\frac{J_r}{J_k \cdot J_{af}}\right] \cdot J_w$$

where

RDQ, J_n and J_r Parameters used in Barton's classification
J_k Joint hydraulic connectivity
J_{af} Joint aperture factor (equal to 1 for closed joints, and 2, 5 for apertures up to 20 mm)
J_w Factor that identifies the hydraulic condition in the rock mass (Table 4.6)

If the HP value thus obtained is higher than 3, no relevant hydraulic circulation phenomena are forecast; on the contrary, if HP < 3, the rock mass is potentially interested by hydraulic circulation, whose flow (measured in litres/minute) is estimated as:

$$Q = 3.785(91.971e^{-2.3144\mathrm{HP}})$$

Concerning the complete saturated condition, the literature provides some analytical formulations that allow an approximate estimation of the tunnel inflow (Table 4.7).

As it can be seen in Table 4.7, the various analytical formulations are different according to:

- The flow conditions (steady or transient state)
- The relative position of the groundwater table with regards to land surface and tunnel depth
- The load conditions along the tunnel lining, the extension and the characteristics of the aquifer

Table 4.7 Analytic formulas for the tunnel inflow assessment

Formula	Author	Conditions
Steady state		
$Q = \frac{2\pi\ KL\ (H-h)}{\ln\left(\frac{2H-2h}{r}\right)}$	Goodman (1965)	Water table below land surface. Hydrostatic load constant along the tunnel border
$Q = \frac{2\pi\ KL\ (H-h)}{\ln\left(\frac{H-D-h}{r}+\sqrt{\left(\frac{H-D-h}{r}\right)^2-1}\right)}$	Lei (1999)	Water table above land surface. Hydrostatic load constant along the tunnel border
$Q = \frac{2\pi KL(H-h)}{\ln\left(\frac{2(H-D)}{r}\right)}$	Kolymbas and Wagner (2007)	Water table above land surface. Hydrostatic load constant along the tunnel border. Deep tunnel
$Q = \frac{2\pi KL(H-h)}{\ln(R/r_e)}\left(1+\frac{\ln(r_e/r_i)}{\ln(R/r_e)}\cdot\frac{K}{K_l})^{-1}\right)$	Ribacchi et al. (2002)	Water table below land surface. Hydrostatic load constant along the tunnel border. Tunnel lining
$Q = \frac{2\pi KL\ (A+D)}{\ln\left(\frac{(H-D)}{r}+\sqrt{\frac{(H-D)^2}{r^2}-1}\right)}$ *where* $A = (H-D)\frac{(1-\alpha^2)}{(1+\alpha^2)}$, $\alpha = \frac{1}{r}(H-D-\sqrt{(H-D)^2-r^2})$	Park et al. (2008)	Water table above land surface. Hydrostatic load along the tunnel border depending on the stage
$Q = 2\pi KL\frac{\lambda^2-1}{\lambda^2+1}\cdot\frac{(H-h)}{\ln\lambda}$ *where* $\lambda = \frac{(H-h)}{r}-\sqrt{\frac{(H-h)^2}{r^2}}-1$	El Tani (2003)	Water table below land surface. Hydrostatic load along the tunnel border depending on the stage. Extension for non horizontal water table
Transient state		
$Q(t) = \frac{4\ \pi KL\ [H\ (t)-h]}{\ln(2.25\ KLt/Sr^2)}$	Jacob and Lohman (1952)	Hydrostatic load constant along the tunnel border
$Q(t) = 2\pi\int_0^{vt}\frac{K[H(t)-h)]\cdot\theta(L-x)}{\ln\left[1+\sqrt{\frac{\pi K}{Sr^2}}\left(t-\frac{x}{v}\right)\right]}dx$	Perrochet al. (2005)	Hydrostatic load constant along the tunnel border. Extension for heterogeneous aquifer by Perrochet et al. (2007)

All the above cited formulas are based on the hypothesis of homogeneous and isotropic aquifer, horizontal water table and $r << H$ K hydraulic conductivity
L length of the tunnel, H depth of the tunnel centre from the water table, h hydraulic head into the tunnel, S specific storage coefficient, r tunnel radius (with lining: r_e is the external radius and r_i is the internal radius), R radius of influence, K_l tunnel lining hydraulic conductivity, D hydraulic load above land surface, t time, x spatial coordinate along the tunnel axis with the origin at the entry of the permeable zone, v drilling speed, $\theta(L-x)$ Heaviside step function (also named unit step function, when $(L-x) < 0$, $\theta(L-x) = 0$ and when $(L-x) > 0$, $\theta(L-x) = 1$)

Other equations were also defined for medium-depth tunnel in sedimentary rock masses (Gattinoni and Scesi 2010), which take into account the specific characteristics of the fracture network.

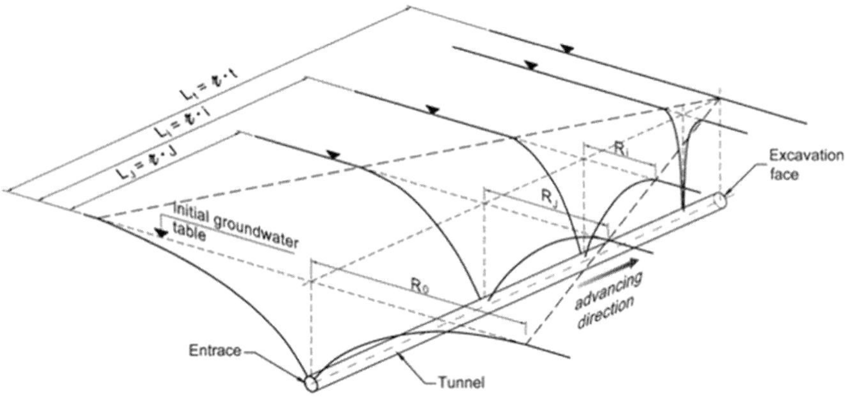

Fig. 4.24 Representation of the values of the radius of influence of the tunnel R_i along its axis. (Modified from Federic 1984)

4.9.1.1 The Draining Process from an Advancing Tunnel

The draining process from an advancing tunnel is a very complex problem, because it is three-dimensional and time depending; nevertheless, there are models allowing a simplified approach (Federic 1984). In particular, the problem is solved assuming that the flow direction is mainly horizontal (the area of the motion is laterally unlimited and it can be delimited at the bottom) or mainly vertical (the area of the motion is limited laterally and at the bottom).

The hypothesis considered in the simplified modelling are listed below:

- Straight tunnel with horizontal axis
- Water table originally static whose free surface is situated at height H above the tunnel centre
- The ground is considered a homogeneous, isotropic, undeformable medium, with porosity n and permeability k
- The flowing fluid is uncompressible and the Darcy law is valid
- The tunnel has no lining and the hydrostatic load on the profile is null

Following parameters can be determined (Fig. 4.24):

- The inflow rate into the tunnel for length unit, at the different distances and for different time steps ($q_i(t)$)
- The total tunnel inflow rate in different time steps ($Q(t)$)

The extension of the area interested by the water table drawdown at different distances and for different time steps ($R_i(t)$).

If the water table is known, the application of some reference scheme assuming mainly the horizontal (Fig. 4.25) or vertical (Fig. 4.26) flow direction can solve the problem. Generally, models with horizontal direction of the flow are applicable in a more correct way in the intermediate and final phases of the flow; models with

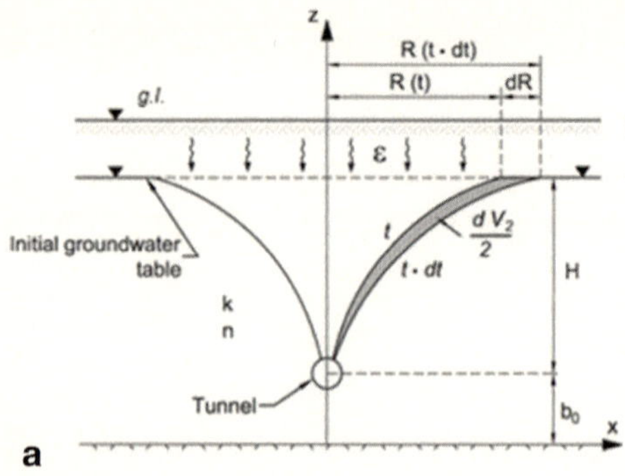

Fig. 4.25 Schematisation of the problem of the main horizontal motion: **a** finite aquifer, **b** for infinite aquifer. (Modified from Federic 1984)

Fig. 4.26 Schematisation of the problem for a mainly vertical flow, for an aquifer of limited extension. (Modified from Federic 1984)

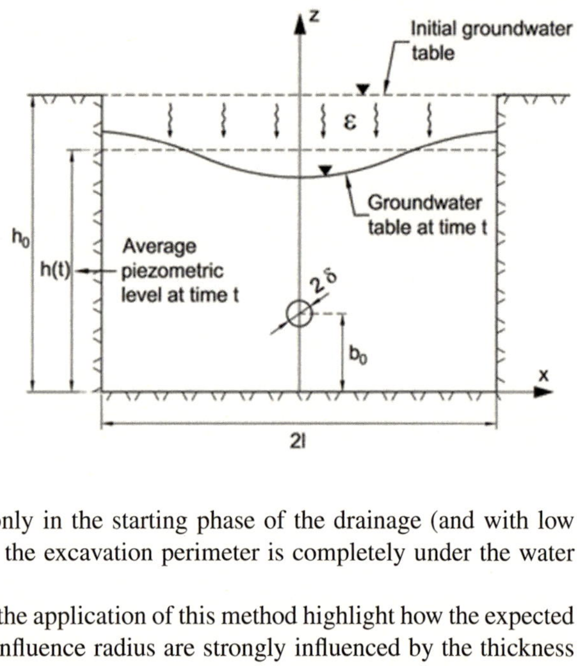

vertical flow are reliable only in the starting phase of the drainage (and with low values of k, assuming that the excavation perimeter is completely under the water table at any time).

The results obtained by the application of this method highlight how the expected tunnel inflow rate and its influence radius are strongly influenced by the thickness of the saturated zone that might be present below the tunnel. Moreover, for low permeability values, flow rates are overestimated as it assumed a water-table drawdown that reaches the centre of the tunnel.

It is clear that the described approximate methods have to be used only for a general estimate of the flow rate, considering the fact that many simplifying hypothesis have been adopted. In particular, all solutions are valid if the tunnel inflow can be drained completely.

Moreover, the results are strongly influenced by the value of permeability coefficient k and by the correspondence of the homogeneity and isotropy hypotheses to the

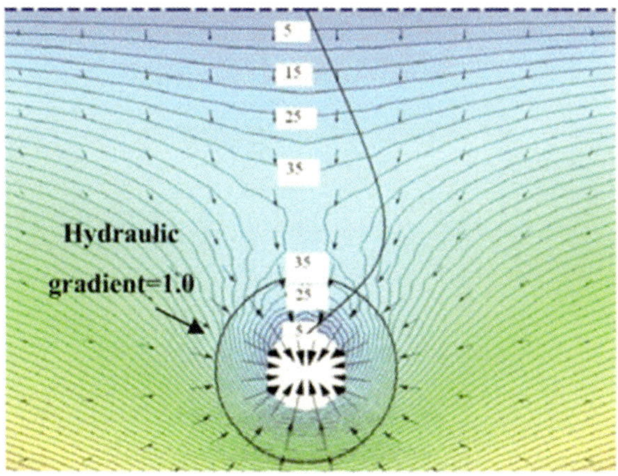

Fig. 4.27 Example of modelling results of drainage process induced by tunnelling. (Lee et al. 2007)

real condition. Moreover, permeability itself is strongly dependant on the stress and deformation field of the medium. The excavation of tunnel may imply two phenomena causing two opposite effects: the increase of shear stresses in radial direction induces a permeability reduction; on the contrary, in the yielded area, the reduction of effective stresses produces a permeability increase as an effect of dilatancy phenomena.

4.9.2 The Influence of Water on the Mass Behaviour

Some authors demonstrated that the presence of water during the realisation of a tunnel has important consequences on radial convergences as well (Lee et al. 2007). Actually, while the effective radial pressure decreases during the excavation, the seepage forces keep their value. Those forces increase with the hydraulic load and contribute to increase the convergences. Moreover, the yielded area extends with the increase of the seepage effect, having a negative influence on the cavity stability.

The entity of the seepage forces depends on the dimension of the area around the tunnel where those forces are calculated. According to the authors, from a practical point of view, the calculation area of seepage forces can be identified as the area with a hydraulic gradient higher than 1 (Figs. 4.27 and 4.28).

According to Lee et al. (2007), the value of seepage forces is not negligible with respect to the hydrostatic pressure. On the contrary, it changes according to the water-table level (ratio H/D) and the tunnel depth (ratio C/D). The values range between 40 % of initial hydrostatic pressure, in case of low water level (H/D $<$ 2), and 100 % of the same pressure for shallow tunnels (C/D = 2) characterised by high water level (H/D = 4). The seepage pressure ratio, plotted in the following figure as function of geometrical ratios C/D and H/D, represents the seepage pressure divided by the hydraulic pressure at the initial conditions before excavation (Fig. 4.29).

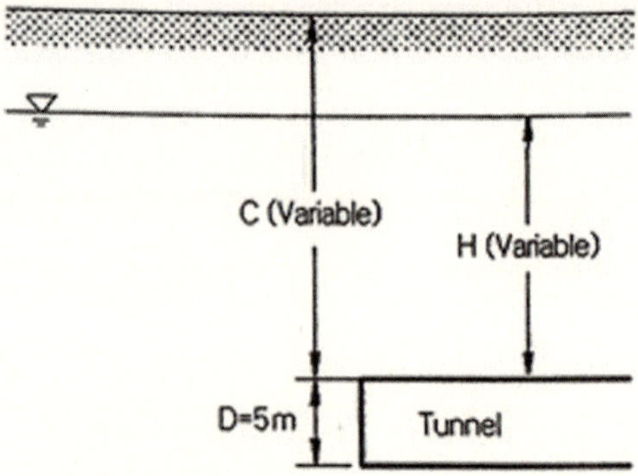

Fig. 4.28 Geometry to be considered for the analysis of drainage process in Fig. 4.27. (Lee et al. 2007)

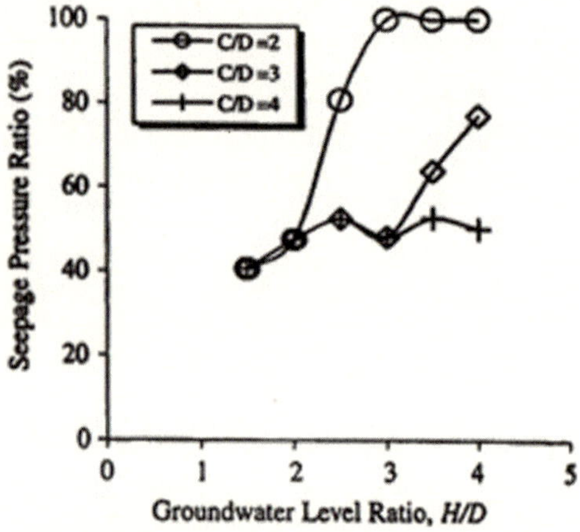

Fig. 4.29 Seepage pressure ratio at the tunnel crown. (Lee et al. 2007)

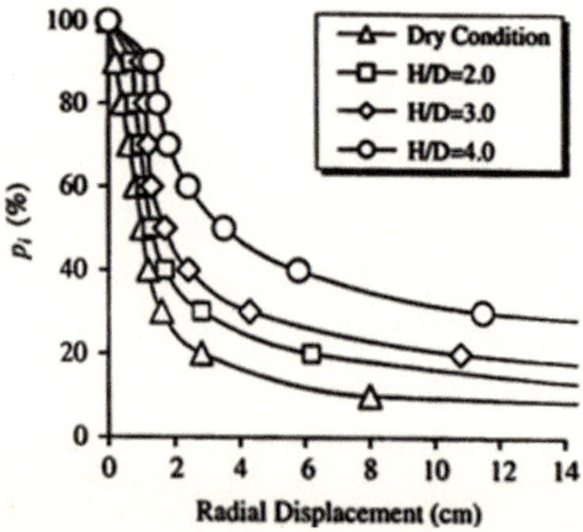

Fig. 4.30 Ground reaction curves both considering seepage forces and in dry condition. (Lee et al. 2007)

Therefore, influence of water can lead to a modification of the characteristic curves. The results provided by the authors suggest that, with equal pressure, the higher the water level above the tunnel (higher ratio H/D), the more important will be the convergence (Fig. 4.30).

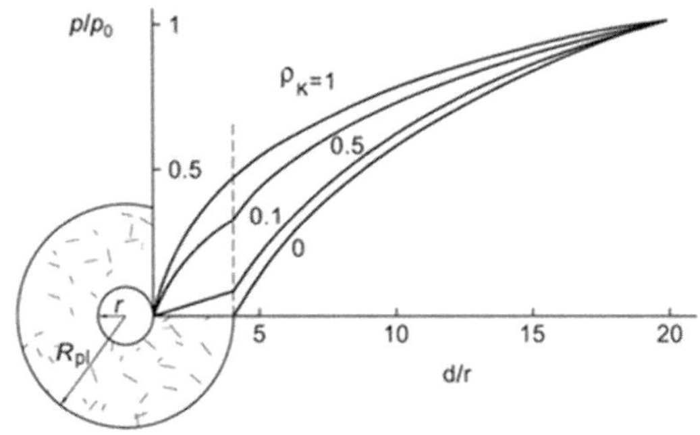

Fig. 4.31 Trend of the interstitial pressure p with respect to the initial pressure p_0 around the tunnel when the permeability ratio ρ_k changes (Vielmo 1973). R_{pl} is the extension of the area subject to plastic deformation phenomena, r is the tunnel radius and d the distance from the tunnel perimeter

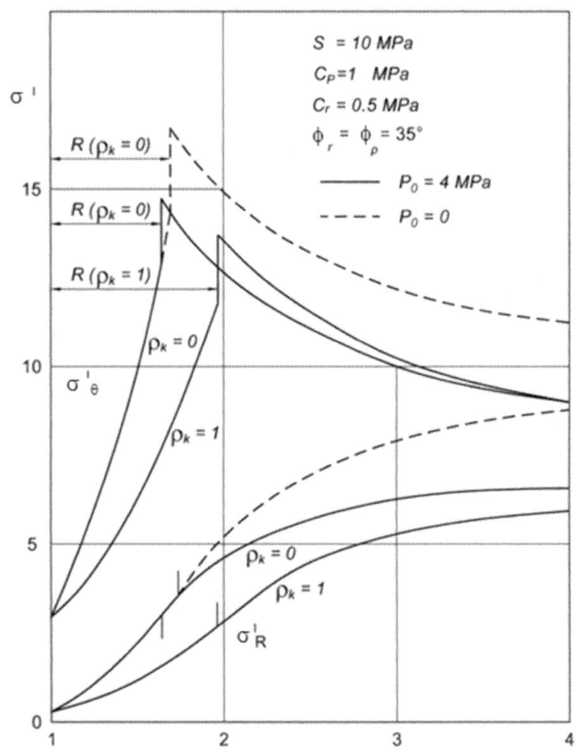

Fig. 4.32 Stress state around a circular tunnel in a soil with elasto-plastic behaviour (Lembo-Fazio and Ribacchi 1986). Effective stresses (continuous lines) σ_r' and σ_θ' are presented for two flow limit conditions, corresponding to permeability unchanged by plastic deformation phenomenon ($\rho_k = 1$) and total drainage in the yielded area ($\rho_k = 0$), respectively. The stress state is shown in the hypothesis of absence of groundwater (*dotted lines*), to enable the comparison. (C_p, C_r: peak and residual cohesion, $S = \sigma_{cm}$ rock mass compressive strength; $P_0 = \sigma_0$ initial vertical stress; φ_p, φ_r peak and residual friction angle)

It is clear that the exact solution of the problem of a tunnel under the water table can only be provided by a simultaneous numerical analysis of the flow process and stress-strain behaviours, in particular in those cases where permeability is strongly influenced by the stress state and, therefore, the static problem cannot be solved independently from the hydraulic one, and vice versa. Anyway, there are analytical approaches that allow solving the problem in a decoupled way but, considering the simplifying hypotheses, the use of such methods has to be evaluated for each single case.

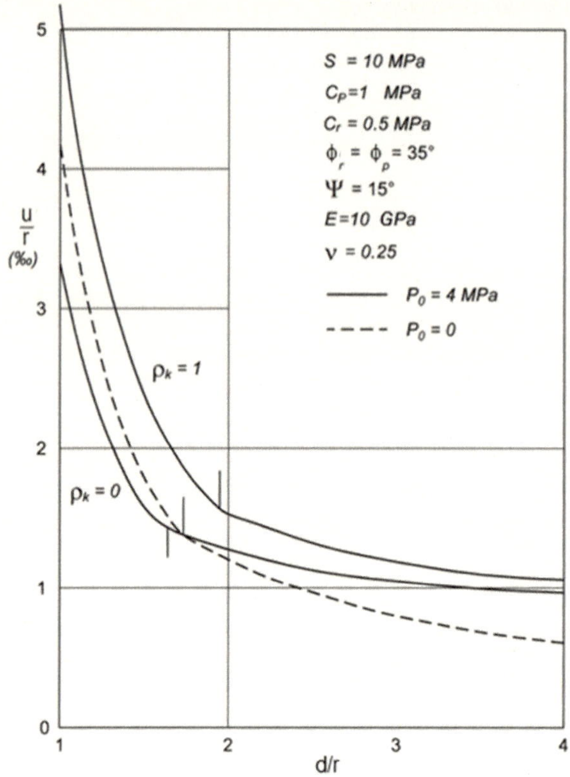

Fig. 4.33 Radial displacement around a circular tunnel in a soil with elasto-plastic behaviour (Lembo-Fazio and Ribacchi 1986). The displacements are shown in two flow limit conditions (*continuous lines*) and, to enable the comparison, also the displacement if no groundwater is present (*dotted line*). (C_p, C_r peak and residual cohesion, $S = \sigma_{cm}$ rock mass compressive strength; $P_0 = \sigma_0$ initial vertical stress; φ_p, φ_r peak and residual friction angle; ψ dilatancy angle; E elastic modulus; ν Poisson coefficient)

An analytical solution was provided, for example, by Lembo-Fazio and Ribacchi (1986). The development of the yielding area around the cavity implies the increase in permeability (k_{pl}) of the rock mass in that area, whereas permeability (k_{el}) in the elastic part of the rock mass remains almost constant and equal to the initial one. If the permeability ratio is defined as:

$$\rho_k = \frac{k_{el}}{k_{\rho l}},$$

the trend of the interstitial pressure around a cavity is defined by logarithmic equations whose trend is schematised in Fig. 4.31 for some ideal flow situations:

- $\rho_k = 1$ (unchanged permeability in the yielding area, e.g. $k_{pl} = k_{el}$)
- $\rho_k = 0$ (permeability in the yielding area is much higher than in the elastic area, e.g. $k_{pl} >> k_{el}$)
- $0 < \rho_k < 1$ (intermediate cases)

If the interstitial pressures p are known, using the principle of the effective stresses, it is possible to obtain the stress state around the cavity. The authors showed that the formula obtained for the calculation of the yielding radius and stresses in the yielding area when no water is present are still valid, although adequate substitutions have to be made (Lembo-Fazio and Ribacchi 1986). The results of the calculation can be represented in terms of minimal stress state around the tunnel (Fig. 4.32) and in terms of convergences (Fig. 4.33). These highlight that the convergence (and the extension of the plastic radius R_{pl}) may be sometimes smaller, sometimes higher with respect to those values estimated ignoring the presence of groundwater, according to the flow regime around the cavity .

References

Barla G. (2002) Tunnelling under squeezing rock conditions. In: Tunnelling Mechanics-Eurosummerschool, Innsbruck, 2001. Verlag Berlin, BERLIN, pp. 169–268. ISBN 3897228734

Barton N, Lien R, Lunde J (1974) Engineering classification of rock masses for the design of tunnel support. Rock Mech 6(4):189–236

Bhasin R (1994) Forecasting stability problems in tunnels constructed through clay, soft rocks and hard rocks using an inexpensive quick approach—Gallerie e Grandi Opere Sotterranee p 42, 14–17. In Cai M, Kaiser PK, Tasaka Y, Minami M (eds) Determination of residual parameters of jointed rock masses using the GSI system, International Journal of Rock Mechanics & Mining Sciences

Federic F (1984) Il processo di drenaggio da una galleria in avanzamento, R.I.G. 4/84, pp 191–208

Gattinoni P, Scesi L (2010) An empirical equation for tunnel inflow assessment: application to sedimentary rock masses. Hydrogeol J 18(2010):1797–1810

Gesta P (1993) Recommendations for use of convergence-confinement method. AFTES, Groupe de travail no.7: tunnel support and lining, 206–222

Goel RK (2000) Tunneling in squeezing ground conditions. Riv Italiana Geotecn 1:35–40

Goel RK, Jethwa JL, Paithakan AG (1995) An empirical approach for predicting ground conditions for tunneling and its practical benefits. Proc. 35th US Sym on Rock Mechanics, pp 431–435

Hoek E, Brown ET (1980) Underground excavations in rock. Institution of Mining and Metallurgy, London

Hoek E, Marinos P (2000) Predicting tunnel squeezing problems in weak heterogeneous rock masses. Tunn Tunn Int

Hoek E, Carranza-Torres CT, Corkum B (2002) "Hoek-Brown failure criterion-2002 edition"Proceedings of the fifth North American rock mechanics symposium 1, pp 267–273

Hoek E (2013) Rock mass properties. In Hoek's corner, Rocscienceinc http://www.rocscience.com/education/hoeks_corner

Jethwa JL, Singh B, Singh B (1984) Estimation of ultimate rock pressure for tunnel linings under squeezing rock conditions—a new approach. In: Brown ET, Hudson JA (eds) Design and performance of underground excavations. ISRM Symposium, Cambridge, pp 231–238

John M (1971) Properties and classification of rock with reference to tunnelling. CSIR pubblication MEG 1020, South Africa

Kolymbas D, Wagner P (2007) Groundwater ingress to tunnels—The exact analytical solution. Tunn Undergr Space Technol 22(1):23–27

Leca E, Panet M(1988)Application du calcul à la rupture à la stabilité du front de taille d'un tunnel. Revue Française de Géotechnique 43

Léca E, Dormieux L (1990) Upper and lower bound solutions for the face stability of shallow circular tunnels in frictional material. Géotechnique 40(4):581–606

Lee S-W, Jung J-W, Nam S-W, Lee I-M (2007) The influence of seepage forces on ground reaction curve of circular opening. Tunn Undergr Space Technol 22–1:26–36

Lembo-Fazio A, Ribacchi R (1986) Stato di sforzo e deformazione intorno ad una galleria, Primo Ciclo di Conferenze di Meccanica e Ingegneria delle Rocce: Recenti sviluppi e nuovi orientamenti nel campo delle gallerie, Politecnico di Torino 25–28 Novembre 1986

Lombardi G (1973) Dimensioning of tunnel linings with regard to constructional procedure. Tunn Tunn 5(4 July–August):340–351

Macori M, Benussi G (1982) Gallerie stradali—raccolta di memorie. Pubblicato da A.N.A.S. Estratto dalla rivista "Le Strade" casa editrice La Fiaccola—Milano

Muir Wood AM (1979) Ground behaviour and support for mining and tunnelling. Tunn Tunn, Part 1 in May 1979, pp 43–48, and Part 2 in June 1979, pp 47–51

Nguyen-Minh D, Guo C (1993) A ground support interaction principle for constant rate advancing tunnels. Proceedings of the Eurock '93, pp 171–177, Lisbon, Portugal, Balkema, Rotterdam

Nilsen B, Thidemann A (1993) Rock engineering. Hydropower Development, Publ. no. 9, Norwegian Institute of Technology, Division of Hydraulic Engineering, p 156

Palmström A (1995) Characterizing rock burst and squeezing by the rock mass index. Design and Construction of underground, New Delhi, 23–25 (February 1995)

Panet M (1995) Calcul des tunnels par la method convergence-confinement. Ponts et chausses. Paris 1995

Panet M, Guenot A (1982) Analysis of Convergence behind the face of a tunnel. International Symposium "Tunneling 82", Brighton

Ribacchi R, Riccioni R (1977) Stato di sforzo e di deformazione intorno ad una galleria circolare. Gallerie e grandi opere sotterranee 4:7–18

Ribacchi R., Graziani A, Boldini D (2002). Previsione degli afflussi d'acqua in galleria e influenze sull'ambiente. Meccanica e ingegneria delle rocce MIR, Torino, pp. 143–199

Russenes BF (1974) Analysis of rock spalling for tunnels in steep valley sides (in Norvegian). M.Sc. thesis, Norwegian Institute of Technology, Dept. of Geology, p 247

Sakurai S (1997) Lessons learned from field measurements in tunnelling—tunnelling and underground space technology, Pergamon, Oct–Dec. 1997

Selmer-Olsen R (1964) Geology and engineering geology (in Norwegian). Tapir, Trondheim, Norway, p 409

Selmer-Olsen R (1988) General engineering design procedures. Norwegian Tunnelling Today, Tapir, pp 53–58

Singh B, Jethwa JL, Dube AK, Singh B (1992) Correlation between observed support pressure and rock mass quality. Tun Undergr Space Technol 7:59–74

Shi G, Goodman RE (1985) Block theory and its application to rock engineering. Prentice-Hall, Inc., Englewood Cliffs, New Jersey

Chapter 5
Geological Risk Management

5.1 Introduction

Underground openings involve the resolution of technical problems whose complexity depends on the geological and environmental context. International experience has proved that the occurrence of an 'unforeseen geologic event' can produce a relevant increase in time and costs (Table 5.1). For this reason, several methods were developed for the management of geological and environmental risks in tunnelling scientific literature, especially with reference to financial aspect and hydrogeological risk (Gattinoni and Scesi 2006). In practice, geological risk management is a very complex procedure, both for the high number of variables and frequent lack of available data during the tunnel design. Moreover, the definition of a critical threshold is still an open problem.

The term 'geological risk' includes all issues relating to geology and hydrogeology as well as geotechnics and may be extended to 'antropic risks' and risks created by underground works on neighbouring infrastructures.

At a general level, the geological risk management can be defined as a set of activities aimed at maximizing safety and minimizing the cost and time increase for the realization of an underground work. It maximizes safety because adopting adequate mitigation criteria, a careful analysis and the knowledge and management of the risks allow the reduction of risks to an acceptable level and limit the increase of construction costs and time. This is possible because the chance of containing the realization costs of an underground work is higher in the initial planning and designing phases. On the contrary, this chance is considerably limited in the following designing phases (Fig. 5.1).

The risk management can lead to an assessment of the capability of designing and construction system to respond to both functional and environmental requirements. Generally speaking, risk management is divided in consecutive steps to be faced at different designing and realization phases (Fig. 5.2). The whole development of the risk management process can be formalized in a risk register, to be completed starting from the very first conception phases and up to the completion of the underground work.

P. Gattinoni et al., *Engineering Geology for Underground Works*,
DOI 10.1007/978-94-007-7850-4_5,

Table 5.1 Costs of damages occurred during tunnelling. (Source: Munich Re-Knowledge Management Topic Network Construction)

Date of damage	Project	Cause of damage	Damage (million US$)
1994	Great Belt tunnel Funen–Zeeland, Denmark	Fire	33
	'Heathrow Express Link', London, UK	Cave-in	141
	Underground Munich–Trudering, Germany	Cave-in	4
	Metro Taipei, Taiwan	Cave-in	12
1995	Metro Los Angeles, USA	Cave-in	9
	Metro Taipei, Taiwan	Cave-in	29
1999	Sewage tunnel, Hull, UK	Cave-in	55
	High-speed rail route Bologna–Florence, Italy	Cave-in	9
	Bolu tunnel Gurraisova–Gerede, Turkey	Earthquake	115
2000	Metro Taegu, South Korea	Cave-in	24
	High-speed rail route Bologna–Florence, Italy	Cave-in	12
2001	Metro Hong Kong 'Tseung-Kwan-O Line'	Typhoon	–
2002	High-speed rail route, Taiwan	Cave-in	30
	Socatop tunnel, Paris, France	Fire	8
2003	Metro Shanghai 'Pearl Line', PR of China	Cave-in	80
2004	Metro Singapore 'Circle Line'	Cave-in	–
2005	Metro Kaohsiung 'Orange Line', Taiwan	Cave-in	–
	Metro Barcelona, Spain	Cave-in	–
	Metro Lausanne, Switzerland	Cave-in	–
	Motorway Tunnel 'Lane Cove', Sydney, Australia	Cave-in	–
2007	Metro São Paulo, Brazil	Cave-in	–
		Total	> 600

– indicates that the project is still open

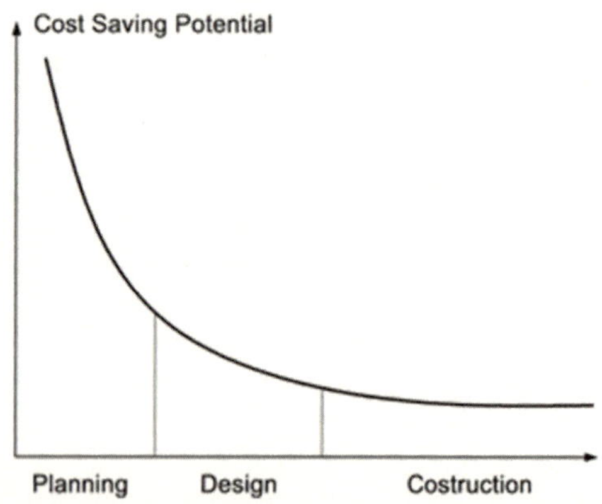

Fig. 5.1 Potential cost containment linked to different realization phases

In particular, the risk management associated with the construction of underground works follows a three-phase cyclic path (AFTES 2012):

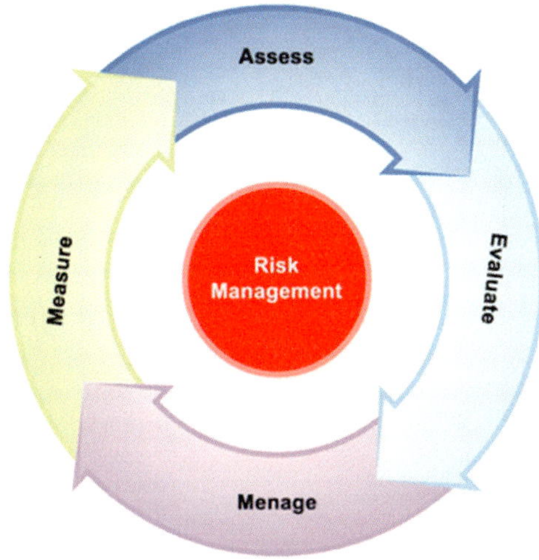

Fig. 5.2 Scheme of iterative process of risk management

(A) Review of Knowledge and Uncertainties

- Compilation of factual data
- Analysis of the data reliability
- Geological and geotechnical report and design assessment
- Register of geotechnical uncertainties (lists of the uncertainties)

(B) Risk Assessment. It includes the following sub-phases:

- *Identification of potential risks*: it consists of the individuation process of elements of hazard (e.g. a situation or a physical condition that may potentially originate a damage or cause undesired consequences, Table 5.2) and the evaluation of the respective causes and consequences.
- *Risk analysis*: this involves quantifying (or at least qualifying) the likelihood of hazardous events and the seriousness of their consequences in term of costs, lead times, worksite safety, environmental impact etc. First of all, the possible causes are identified and classified, and then their analysis follows, which usually includes the filling in of a risk matrix. The potentially damaging events are listed according to their occurrence probability and impact of the consequences. The list allows to move to the following designing phases of the measures apt to reduce and manage the consequences. The analysis can be quantitative, semi-quantitative or qualitative.
- *Risk evaluation*: it involves the comparison of the results of the previous analysis using some acceptability criteria. This makes it possible to determine which risk requires treatment to bring the damage down to an acceptable level. In this phase, it is necessary to define risk acceptability criteria.

(C) Risk Treatment. It involves the reduction of the level of risk, or even its elimination, by reducing the occurrence likelihood (e.g. with additional investigation),

Table 5.2 Most common geological hazard events in tunnelling

Potential hazard events	Controlling factors
Rock fall	Unfavourable structural conditions: primary discontinuity (cooling cracks, stratification and schistosity), secondary discontinuities (fractures and faults of tectonic origin, folds, karst dissolution); seismicity
Sidewall instability (rock burst, spalling, spitting and slabbing)	Unfavourable structural conditions; lithology (rock with brittle behaviour); tensional state (i.e. high depth, folds, overthrust etc) seismicity
Squeezing or swelling	Lithology (rock with ductile behaviour); tensional state; hydrogeological conditions
Face collapse	Lithology: cohesionless soils or loose rocks; structural conditions allowing sliding; hydrogeological conditions; seismicity; squeezing due to high-stress conditions
Groundwater (tunnel inflow, watertable drawdown, spring extinction)	Permeability; structural conditions; hydrogeological conditions
Gases in tunnel	Lithology; structural conditions
High temperature	High depth; proximity of magmatic bodies
Surface settlements and sinkholes	Lithology (loose and weak rocks); low depth; hydrogeological conditions

and limiting the consequences (e.g. modifying position, orientation, layout, profile, and construction methods). Once these measures have been applied, the level of risk is re-evaluated and compared to the acceptability criteria. This iterative analysis process involves amending and supplementing the risk register at every stage, in particular if new surveys have been carried out in an attempt to reduce uncertainties. *The risk mitigation in the construction phase* implies the choice of mitigation measures to be applied and of their entity, followed by the verification of mitigation measures on the basis of monitoring data gathered during the construction phases. If the threshold values are exceeded, modifications have to be introduced to take the risk level back below an acceptable threshold. In the mitigation phase, all the people interested have to be informed of the residual risk after the implementation of the necessary measures.

The risk management is an iterative process that must be repeated at the end of each design phase and carried on during the construction by means of survey activities, changes to the project, modifications of construction techniques etc. (Fig. 5.3).

5.2 Definitions and General Concepts

Traditionally, the geological risk is defined as the expected values of losses (generally considered in terms of casualties and economic losses, but the decrease in environmental quality can be included as well) due to particular natural phenomena or human interventions. The geological risk (*R*) defined in these terms, also called total risk R, can be evaluated as the product of three different factors:

$$R = H \cdot W \cdot V,$$

where *H* is the hazard, e.g. the probability that a potentially destructive phenomenon takes place in a certain time span and in a given area. The hazard is expressed in terms of probabilities. Therefore, this hazard is referred to a certain intensity *I* of the phenomenon (danger, magnitude) that can be expressed either by using a relative scale or in terms of one or more characteristic magnitudes of the phenomenon (velocity, volume, energy, etc.): $H = H(I)$

W is the exposition factor, e.g. the value of the elements at risk. It can be expressed in terms of number or quantity of exposed units (for example number of people, hectares of land) or in monetary terms. The value is a function of the type of element at risk *E* (population, properties, economic activities, public facilities and environmental goods in a given area at risk): $W = W(E)$.

V is the vulnerability (expected degree of loss for the elements at risk). It is expressed in a scale ranging from 0 (no loss) to 1 (total loss), and it is a function of the intensity of the phenomenon and the type of element at risk: $V = V(I, E)$.

It is also possible to define a specific risk (R_s), e.g. the expected level of loss as a consequence of a particular phenomenon of given intensity; it is expressed in terms of probabilities, for a given type of element at risk *E* and for a given intensity *I*:

$$R_s = R_s(I, E) = H \cdot V,$$

and a residual risk (R_r): $R_r = R - \Delta R$,

ΔR being the change in the risk level after the implementation of mitigation works connected to the hazardous phenomenon under exam. From the designing point of view, the residual risk is the value that has to be adequately limited. Unless other political and social issues are to be considered, the accepted residual risk is generally the one that minimizes the total cost due to the work realization and hazardous event. Otherwise, it can be assimilated to the risk accepted by insurance companies or the work financiers. Finally, it can correspond to the mortality rate due to natural causes of the population less at risk.

For more common collapse situations, the accepted occurrence probability is verified by means of a correct design using semi-probabilistic methods, e.g. with the application of partial safety coefficients for the material strengths and of amplification coefficients on design stresses. Nevertheless, observing the potential risk

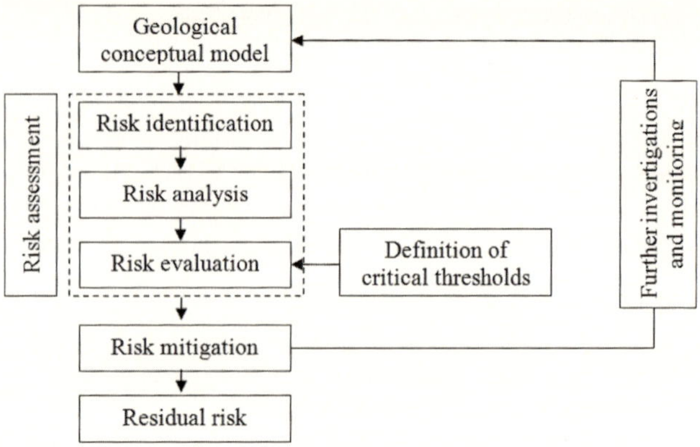

Fig. 5.3 Framework for geological risk management in underground construction

situations in Table 5.2 (the list is not thorough), it is clear that many mechanisms involved in the risk for underground works cannot be analysed with the most common semi-probabilistic methods adopted in the design of relatively simpler works and essentially aimed at the safety of the structure. This is due to the complexity of the phenomena, as well as to the number and variety of the features involved. All that makes it impossible to define the reductive or amplification coefficients.

5.3 Geological Risk Assessment for Underground Works

As stated above, the geological risk management methodology consists of three main phases (Fig. 5.3):

- The review of knowledge and uncertainties arising from the geological conceptual model;
- The geological risk assessment, which includes the identification, analysis and evaluation of the risk;
- The risk treatment through the definition of risk mitigation measures.

This procedure is an iterative process to be conducted throughout the study process and involves a continuum increase in the conceptual model knowledge.

The review of knowledge and uncertainties arising from the geological conceptual model allows, first of all, the identification of the potential hazards relating to tunnelling (Table 5.2) for a specific case study. After this phase of risk identification, the risk analysis and evaluation assessment has to be carried out for each of the hazards under consideration.

Risk analysis may be conducted to different levels of detail according to the risk, the purpose of the analysis as well as the information, data and sources available. This analysis may be qualitative, semi-quantitative or quantitative, depending on the circumstances.

Table 5.3 Qualitative or quantitative determination of likelihood in terms of probability

Matrix score	Likelihood scale	Indicative probability
4	Possible	20 %
3	Unlikely	5 %
2	Highly unlikely	2 %
1	Improbable	0.5 %

Table 5.4 Consequences of the hazard event expressed qualitatively

Risk matrix score	Scale of consequences	Delay expressed in terms of the overrun	Cost expressed in terms of the overrun
4	Very high	3 months	50 %
3	High	1–3 months	10–50 %
2	Medium	1–3 weeks	5–10 %
1	Low	< 1 week	< 5 %

Table 5.5 Matrix for the qualitative risk assessment

Risk matrix					
Likelihood	Possible	4	8	12	16
	Unlikely	3	6	9	12
	Highly unlikely	2	4	6	8
	Improbable	1	2	3	4
		Slight	Medium	Significant	Highly significant
		Consequences			

In general, the risk analysis includes the quantification of the likelihood (qualitative) or the probability of occurrence (quantitative) of the event identified as a hazard, and then the quantification of the consequences arising from the event.

5.3.1 Qualitative Methods for Risk Analysis

For a qualitative approach, a matrix can be used, showing likelihood and consequences expressed qualitatively (Tables 5.3, 5.4 and 5.5). Another example for a qualitative risk assessment is given by the rock engineering system (RES) method (Hudson 1992; Bernardos and Kaliampakos 2004).

5.3.2 Quantitative Methods for Risk Analysis: Safety Methods

When designing, the quantitative risk analysis is usually linked to the assessment of the system capability to respond to both functional and environmental requirements. The risk that a system may not perform its function is defined as collapse or failure probability p_{f}, referred to the life expectancy of the work in given operation conditions or, in a complementary way, as reliability of the system $r = 1 - p_{\mathrm{f}}$. It is, therefore, important to set a relationship between the system capability and the requirement, so that if they correspond, a limit state is identified. The safety factor and the safety margin are two examples of relationships used for this kind of assessment. More generally, it is possible to introduce the concept of the 'performance function' or limit state $g(X_{1,}X_{2}....X_{n}) = 0$ that is defined so that the safety conditions of the system are identified by that part of iperspace where $g(X_{1,}X_{2}....X_{n}) > 0$, whereas the complementary portion of iperspace identifies the combinations leading to the collapse of the system.

To that aim, the ground to be excavated is defined through the geotechnical parameters used to describe its behaviour using a constitutive model. The characterization is possible thanks to a number of tests and experimental observations providing precise information about its features. A more apt calculation model has to be chosen to describe the phenomenon to be simulated. It is clear that each one of these phases (choice of geotechnical parameters, choice of the constitutive model and choice of the calculation model) implicitly includes a certain degree of uncertainty. In particular, the first one is affected by a double degree of indeterminacy, both of knowledge (the single measure can be imprecise), and linked to the natural variability of geotechnical parameters (a geomechanical survey only describes the examined area) which are random variables.

The uncertainty linked to the determination of the properties characterizing the materials as well as great care in the choice of appropriate parameters to perform the calculations and the checks suggest the substitution of traditional deterministic analysis with probabilistic methods. In this view, the properties characterizing the rock mass have to be considered as random variables that can be represented by means of probability distribution.

A set of different measurements (events) can be considered; *S* represents the ensemble of events (e.g. all possible results of the measurements). The value of a mathematical function defined (in the set S) as *P*(*X*) (cumulative probability is associated to the possibility that an event *S* occurs, so that $0 < P(X) < 1$ is always verified.

There are two types of distributions:

- Discrete distributions (spectra that can be effectively represented through histograms)
- Continuous distributions (when represented on a graph, they look like curves)

Some characteristic parameters can be set for each distributions as: the average, the mode, the variance and standard deviation.

In any distribution of variables, it is possible to express the probability that a given value x is exceeded or not reached, according to the average value or standard deviation σ of the distribution. It is relevant to know the values x_p of the variable, characterized by a given occurrence probability p or P (cumulative); in that case, the following formula can be written: $p(x \leq x_p) = P = F(x_p)$, where $x_p = \bar{x} - u \cdot \sigma$

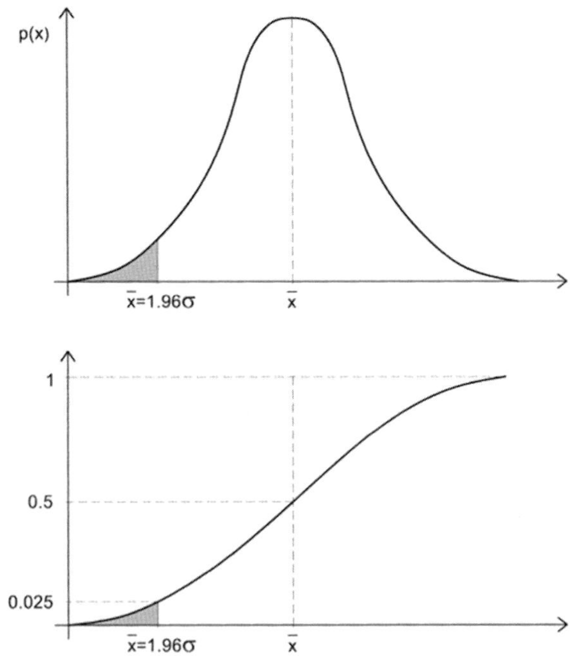

Fig. 5.4 Example of individuation of a fractile equal to 0.025

where $P(x_p)$, called fractile of order p, represents the cumulative probability P that the value x has a value lower than x_p. For each chosen fractile $P(x_p)$ it is possible to identify the corresponding value x_p when the parameter u changes. In other terms, a low fractile can be imposed varying the parameter u. For example, for a Gaussian distribution, $u = 2.58$ has to be imposed for a fractile equal to 0.005; $u = 1.96$ for a fractile equal to 0.025; $u = 1.64$ for a fractile equal to 0.05 (5 %) (Fig. 5.4).

As already stated, the risk for an underground work is defined as the probability to exceed predetermined damage thresholds. It is represented graphically on a plane F (probability to exceed the threshold)—N (number of events) through the complementary cumulative distribution. In the Eurocodes it is reported that the threshold probabilities for SLS (service limit states) and for SLU (ultimate limit states) of normal structures are the probability 1×10^{-3} and 1×10^{-6}, respectively.

In order to keep the probability P_r to reach the limit state lower or equal to the threshold value, two random variables S (stress) and R (strength) representative of a given state have to be considered. Three different methods can be used to measure the safety with regards to that state.

1. Exact method: the statistic distribution of the safety factor $F_s = R/S$ or of the safety margin $\mu = R - S$ can be made using the following relations (Fig. 5.5):

$$P_r = P\{R/S \leq 1\} = F_Y(1)$$
$$P_r = P\{R - S \leq 0\} = F_\mu(0)$$

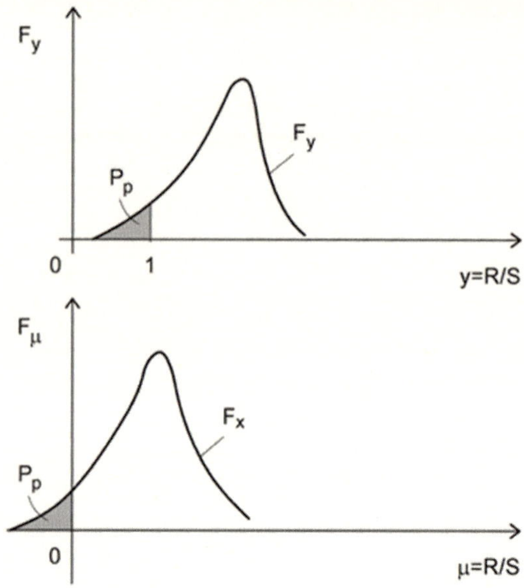

Fig. 5.5 Example of application of the exact method

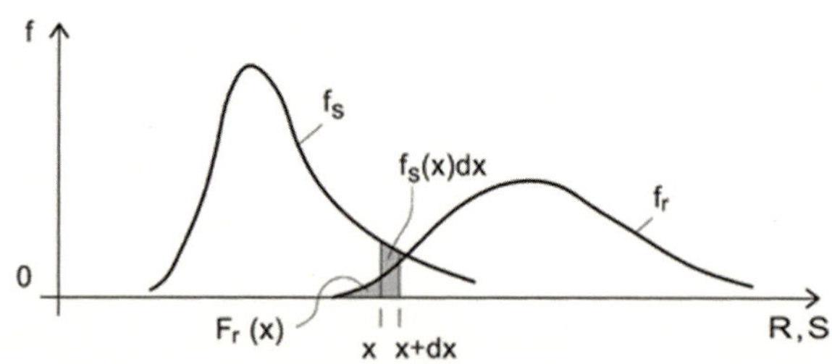

Fig. 5.6 Example of application of the method of the extreme functions

2. Method of the extreme functions: the statistic distribution of R and S are made separately and P_r is calculated as follows (Fig. 5.6):

$$P_r = P\{R \leq S\} = \int FR(x)fs(x)dx$$

3. Method of the extreme values: suppose R and S independently calculate the two values R_d (of design) and S_k (characteristic) needed to obtain the value of the safety factor $F_s = R_d/S_k$ or the safety margin $m = R_d - S_k$, following the criteria of decreasing the properties' values from which R depends and increasing those from which S depends (Fig. 5.7).

Figure 5.8 shows two sets of hypothetical distribution curves representing the uncertainty degree of strength R and of stress S in three different designing phases. During the preliminary design phase, the whole information available is usually very limited. Therefore, the distribution curves of the two variables considered are so large in the illustrative graph. If the margin for the safety factor is too low, there is a significant

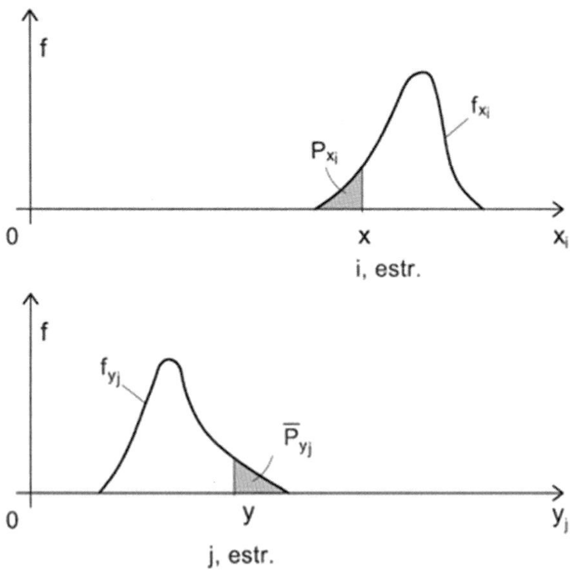

Fig. 5.7 Example of application of the method of the extreme values

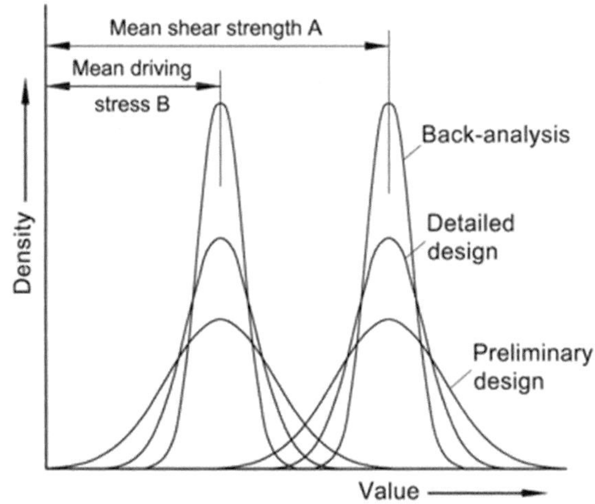

Fig. 5.8 Example of distribution curves of two variables (shear strength and disturbance-induced stress) linked to three different designing phases

collapse probability (overlapping area between the two distributions). To reduce to the minimum the collapse probability, a higher safety coefficient is normally used in the designing phase. During the constructive design phase, the information available is more detailed, therefore the collapse probability is lower (with the same ratio *R*/*S*): the overlapping between the two distributions is considerably smaller. In case a careful back analysis has been carried out on a collapse, the variable distributions show a reduced width and a negligible overlapping.

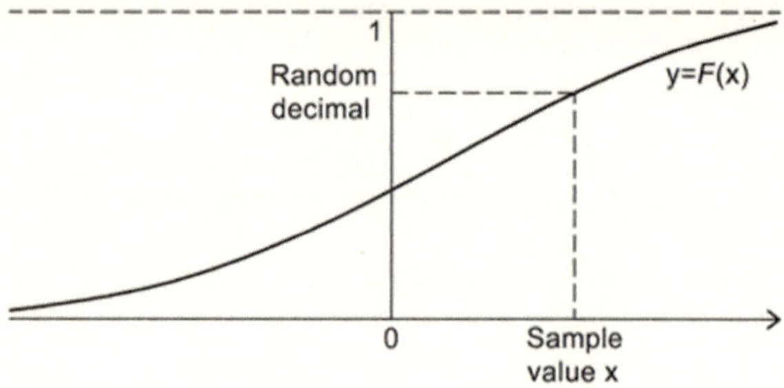

Fig. 5.9 Distribution curve of the probabilities of the output variable of a stochastic process

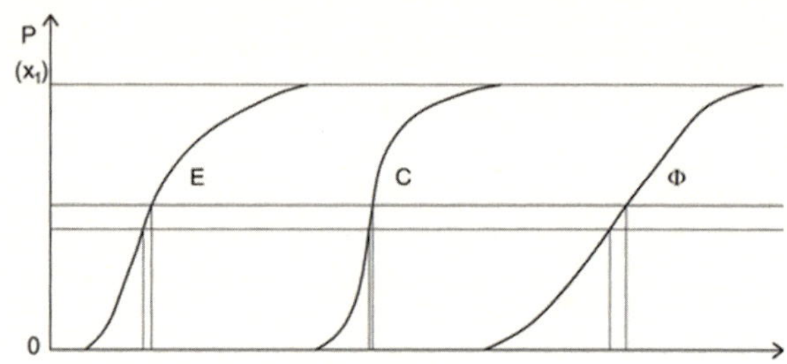

Fig. 5.10 Examples of cumulative distribution functions of different parameters: *E* elastic modulus, *C* cohesion, ϕ friction angle

5.3.3 *Monte Carlo Method for Quantitative Risk Analysis*

Situations are often faced where it is important to know the likelihood that a certain event might occur, but it is impossible to perform the analytic calculations due to the high number of variables affecting it. Simulated sampling methods are used in those situations, e.g. the situation in which the probability that a specific event might occur is simulated. The Monte Carlo method consists in producing a sufficiently high number *N* of possible values combinations the input variables can take and in calculating the output on the base of the model equation. In order to create each one of the *N* combinations, a value is randomly generated (e.g. 'extracted') for each input variable, according to the probability distribution specified and respecting the correlations among variables.

By repeating this procedure *N* times (*N* has to be high enough to allow statistically reliable results), *N* independent values of output variables will be obtained. They will represent a sample of the possible values of the output; then, the sample can be analysed using statistical techniques to estimate the descriptive parameters, reproduce histograms of the frequencies and obtain numerically the trend of the output distribution functions (Fig. 5.9).

For the random choice of an element from a universe described by the density function $f(x)$, follow the instructions below (Figs. 5.10 and 5.11):

1. Trace the graph of the cumulative function (u being the input variable)

$$y = F(x) = \int_{-\infty}^{x} f(u)\,du.$$

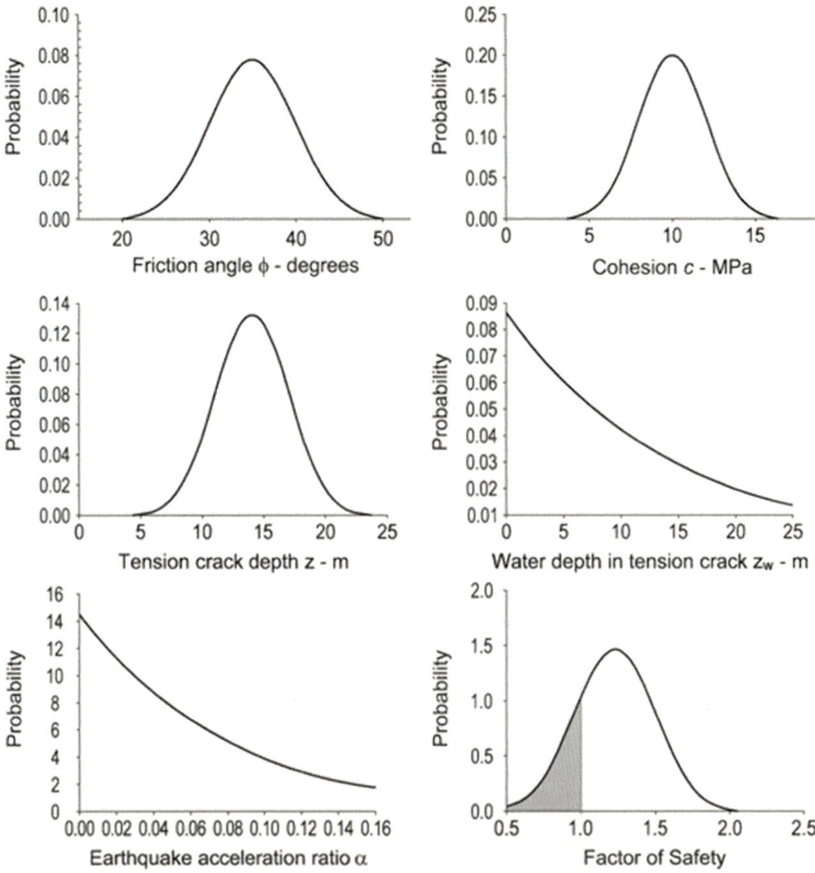

Fig. 5.11 Example of probability distribution of the parameters typical of the risk analysis and of the respective safety factor

2. Generate a random number ranging from 0 to 1.
3. Place the number thus found on the y-axis and project it horizontally on the curve $y = F(x)$.
4. The x value thus found on the curve is taken as one of the values of the sample x.

5.3.4 Risk Evaluation

The probabilistic approach to the risk evaluation always implies the identification of an acceptable collapse probability, a balanced compromise between the cost of risk reduction and the benefit it would bring along. It is important to notice that when the risk is kept 'as low as reasonably possible' (ALARP principle) taking into account the practical limits, and when the cost to further control or completely cancel the

Table 5.6 Example of definition and qualification of the level of risk. The colours correspond to the ones in risk matrix (Table 5.5)

Indicative qualification of the level of risk to be adjusted according to each project	
Negligible/minor risk	No action required, the risk factors must be subjected to specific monitoring by means of procedures
Significant risk (but in principle acceptable)	Construction work may commence; risk factors must be subjected to specific monitoring by means of procedures and the project may possibly be supplemented by a series of predefined measures which may undergo adjustement during the execution
Major risk (to be monitored)	Construction work may not commence until the risk has been reduced or removed. Solutions are possible without major changes to the project
Unacceptable risk	Construction work may not commence until the risk has been reduced or removed. If the risk cannot be controlled, the project may be abandoned or altered.

risk is excessive compared to the overall cost of work, the risk has to be considered residual and, as such, it must be acceptable and manageable. In other words, when designing it is important to learn to live with a risk level that is not null. As it is impossible to eliminate it completely, the effort has to be directed to the adequate limitation of that risk.

As a consequence, based on the results of the risk analysis, the risk evaluation helps decision makers to determine the risks requiring treatment and their priority. In particular, risk evaluation consists of comparing the level of risk determined during the analysis process with the risk criteria established by the project owner (Table 5.6). These criteria and the given threshold values may be different depending on the expected objectives (costs, time, environment, etc.).

On the basis of this comparison, the project owner can:

– Refuse the risk and request that the designer either revises the project eliminating the risk source (e.g. modifying the alignment) and/or carries out more investigation works in order to determine the level of risk more accurately
– Accept the risk, with or without treatment; the first case involves the evaluation of the residual risk

As a null risk is not possible, it is essential to learn to live with a certain level of risk as well as to limit that risk through different limit levels that are considered acceptable. An example of that concept can be seen in Figure 5.12 that represents risk levels considered acceptable by some government organizations of different countries. The point identified as 'Proposed BC Hydro individual risk' equal to 10^{-4} (1 on 10,000) is based on the concept that the risk linked to an engineering work cannot exceed the level of individual risk of 'natural death' that characterizes the safer part of the population (teenagers from 10 to 14). That value is also acknowledged as the limit between voluntary and involuntary risk (Nielsen et al. 1994).

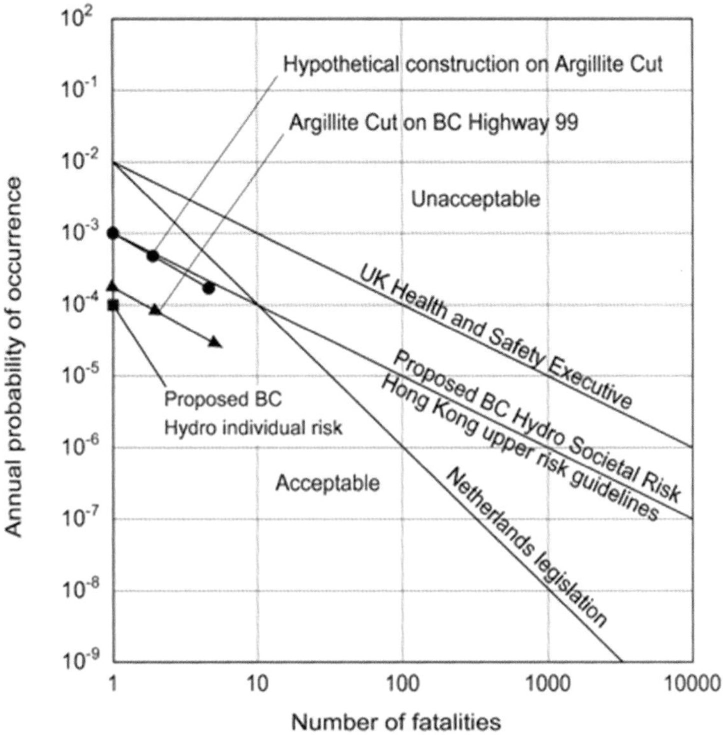

Fig. 5.12 Example of graph to determine the risk acceptability according to the probability of yearly occurrence. (Nielsen et al. 1994)

5.4 Applicative Example: The Decision Aid in Tunnelling (DAT)

When the choice criterion is not univocal and a certain weight has to be given to the different alternatives, the so-called DAT, 'Decision Aid in Tunneling' (Fig. 5.13; Einstein et al. 1999) has proved to be a valid system.

This method developed by MIT and EPFL can simulate the realization of a tunnel starting from a geological profile constructed considering the statistical distributions of the possible uncertain parameters (for example the length of the geological and geotechnical units and their features).

The results can be used for a risk analysis of different scenarios or design modifications (exceeding of certain times and/or costs). The analysis can be repeated in the different design phases to define more apt strategies. The tunnel excavation process is simulated considering the sequences of the construction cycles, starting from a programme of general activities and a programme of specific activities. The geological and the constructive modules of the programme are coupled. The programme selects a value of time and cost for each cycle of a constructive model associated to given simulated geological and geotechnical characteristics. Each simulation generates a

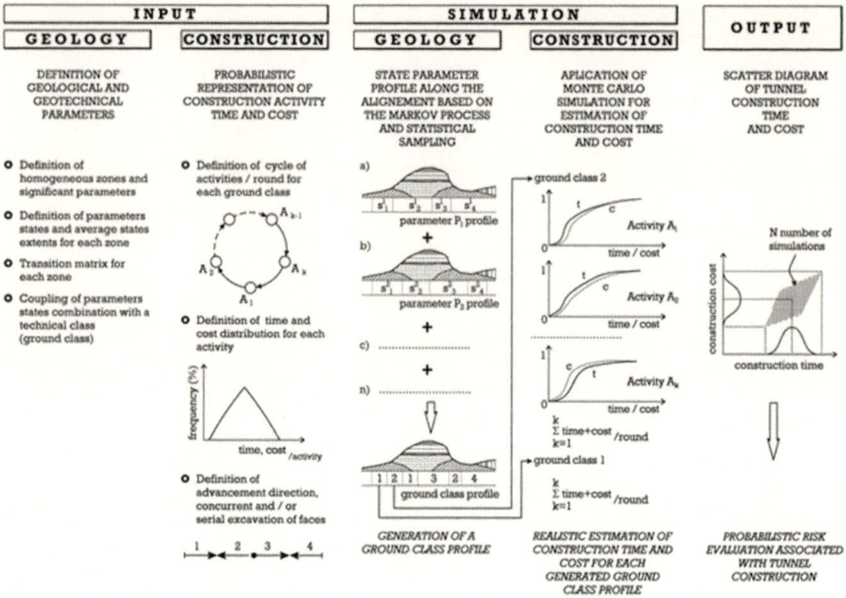

Fig. 5.13 Scheme of the DAT process. (Collomb et al. 2001)

point in a time-costs diagram. Performing a statistically significant number of simulations, the programme generates a cloud of points and a statistic of the construction times and costs. Thus, different realization scenarios can be compared considering the geological uncertainties linked to the realization of the underground work.

The main output data are the geological model and the construction model.

The geological model consists of:

- The geological geotechnical and hydrogeological profiles divided in homogeneous stretches associated to parameters and expected conditions defined in probabilistic terms
- The definition of potential risk situations, of their likelihood of occurrence and distribution along the tunnel layout.

The construction model consists of:

- The definition of the constructive methods (type section) associated to the potential risk situations and the times and costs for their application along the layout
- The evaluation of the risk associated to deviations from the expected conditions and of the resulting deviations of the performance (times and costs). For example, probability of collapse or of relevant water flow and costs and times associated to it. If there are more critical events, more than proportional increase in times and costs can occur
- The programme of general activities with the definition of the construction activities for each of the scenarios to be studied, the connections among activities and the time restrictions.

5.5 From Risk Assessment to Risk Mitigation

The described analysis process involves reducing the level of risk, or even eliminating it altogether, by using the following types of means: reducing uncertainty by carrying out additional investigations, reducing hazard and/or consequences by modifying tunnel axis, layout, profile, excavation methods or by using improvement methods. These methods to minimize the risk are described in the following chapter.

Once these measures have been applied, the level of risk is re-evaluated and compared to the acceptability criteria.

References

AFTES (2012) Recommendation on the characterisation of geological, hydrogeological and geotechnical uncertainties and risk, AFTES Recommendation n° GT32.R2A1

Bernardos AG, Kaliampakos DC (2004) A methodology for assessing geotechnical hazards for TBM tunnelling—illustrated by the Athens Metro, Greece. Int J Rock Mech Min Sci 41:987–999

Collomb D, Kalamaras GS, Vignat P, Bochon A (2001) LGV Lyon-Turin—Modélisation probabiliste de l'incertitude sur les couts et la durée de réalisation du tunnel de base Maurienne-Ambin. Progress In Tunneling After 2000, Milano. June 10–13:2001

Einstein HH, Indermitte CA, Sinfield JV, Descoeudres F, Dudt JP (1999) Decision aids for tunneling. J Transp Res Board 1656:6–12

Hudson JA (1992) Rock engineering systems: theory and practice. Ellis Horwood, New York

Gattinoni P, Scesi L (2006) Analisi del rischio idrogeologico nelle gallerie in roccia a media profondità. Gallerie, Patron Editore (Bologna), n. 79, pp 69–79

Nielsen NM, Hartford DND, MacDonald JJ (1994) Selection of tolerable risk criteria for dam safety decision making. Proc. 1994 Canadian Dam Safety Conference, Winnipeg, Manitoba. BiTech, Vancouver, pp 355–369.

Chapter 6
Risk Mitigation and Control

6.1 Introduction

Generally speaking, it can be said that the correct static design of an underground work consists of forecasting in a detailed way the potential risks and in designing the consequent mitigation measures. Therefore, the risk mitigation and control techniques can be selected according to the risk considered (Table 6.1).

Considering the variety and the contemporary presence of the different hazard factors during the construction of an underground work, the techniques described in the following pages are often combined to face more complex situations.

The measures that allow to mitigate and control the risks arising from the construction of an underground work can be classified according to their effects: improvement of the ground around the cavity or application of a confining pressure on the boundary of the cavity (Table 6.2).

The methods that are mainly aimed at creating a wide zone of material with improved mechanical features act around the cavity. Those providing a confinement pressure act on the excavation perimeter, creating a resistant and/or impermeable shell.

There is a third methodology, where the supports are concentrated right outside the excavation: it is often referred to as reinforced umbrella protective method (RPUM), and it is characterized by a truncated cone-shaped geometry (jet-grouting vaults, forepoling, precutting, pretunnel etc.).

6.2 Excavation Methods

The risk mitigation and control techniques to be adopted are strongly influenced by the advancing method, that is to say by the type of activities carried out for the excavation.

Excavation methods (Fig. 6.1) can be roughly divided into traditional and mechanical methods. Historically, traditional methods required many workers (Fig. 6.2); nowadays, also traditional methods make use of machines (Fig. 6.3) that have

P. Gattinoni et al., *Engineering Geology for Underground Works*,
DOI 10.1007/978-94-007-7850-4_6,

Table 6.1 Main hazard situations during the excavation phase and the respective mitigating measures

Main hazard situations during the excavation phase	Main mitigating elements
Face instability: instability of the core in loose soils or little cohesive soils; sliding or release in fractured rock; extrusion due to excessive loads	To be applied on the core during the advancing phase: Cemented or injected bolts (GRP bolts, self-drilling bolts) Jet-grouting columns Injections To be applied at the face: Fibre-reinforced shotcrete
Cavity instability, large deformations (plastic or viscous-plastic behaviour), associated with collapses and/or excessive deformation of confinement structures	First-phase lining with the invert with circular or sub-circular shape as close as possible to the face Radial bolting First-phase lining with yielding steel ribs and deformable elements
Walls instability: collapse of the confining structures due to the loss of load-bearing capability, loss of the equilibrium and weakness while excavating the bench in subsequent phases, loss of structural continuity between the different parts of the structure	First-phase lining with steel ribs having large base or closing of the crown arch with invert Reinforcing on the base of the first-phase lining with jet-grouting or micropiles
Very asymmetric stress state: asymmetry of both deformation and load	First-phase lining reinforcing (steel ribs, steel reinforcement bars, increasing of the resistant section) Stress-strain constrain through radial bolts
Sections with different lithotypes: behaviour dishomogeneity (both mechanical and hydraulic), with possible water or gas inflow	First-phase lining reinforcing (steel ribs, steel reinforcement bars, increasing of the resistant section) Stress-strain constrain through radial bolts Draining in advance Spilling
Squeezing	First-phase lining including invert placed as close as possible to the face with circular or sub-circular geometry
Overburden collapse	Forepoling Jet-grouting vault To be applied on the core during the advancing phase: Cemented or injected bolts (GRP bolts, self-drilling bolts) Jet-grouting columns Injections To be applied at the face: Fibre-reinforced shotcrete First-phase lining to be placed as close as possible to the face

Table 6.1 (continued)

Main hazard situations during the excavation phase	Main mitigating elements
Spalling and spitting: fall or sliding of rock blocks, flexural instabilities	Radial bolts or nails
Rock burst, strain burst, spalling, spitting, slabbing	First-phase lining reinforced with steel ribs as close as possible to the face Radial bolts or nails Meshes having high adsorbing energy characteristics First-phase lining having high adsorbing energy characteristics as close as possible to the face
Faults and overthrusts: falls, instabilities, water and gas inflow	Fault zone drainage Forepoling Spilling GRP bolts in advance Injections
Water inflow and pressures: sudden inflows even with high gradient and possible ground instabilities	Drainage in advance Waterproofing injections Freezing
Erosion and transport: changes in the physical characteristics of the ground as a consequence of changes in water content or transport of the finest particles	Immediate first-phase lining Drainage in advance Waterproofing injections Freezing
Interference with neighbouring cavities	Rigid first-phase lining with invert Bolts, nails and cables on the pillar between the two cavities Injections
Interferences with surface infrastructures, surface settlements, even arising from groundwater table drawdown	Rigid first-phase lining with invert Injections Freezing Pressurised TBM with pre-cast waterproofing concrete lining
Slope interference: high deformations of surface slope, rock slope instabilities	Rigid first-phase lining with invert as close as possible to the face Forepoling in advance Jet-grouting vault To be applied on the core during the advancing phase: Cemented or injected bolts (GRP bolts, self-drilling bolts) Jet-grouting columns Injections To be applied at the face: Fibre-reinforced shotcrete

Table 6.2 Classification of the measures according to their effect and position

Risk mitigation method	Effect	Position in relation to the cavity
Injections	Ground improvement Waterproofing	Around the cavity
Freezing	Ground improvement Waterproofing	Around the cavity
Anchors	Ground improvement Confining pressure	Around the cavity
Drainage	Ground improvement Groundwater control	Around the cavity
Forepoling	Confining pressure	External perimeter of the cavity
Jet-grouting vaults	Ground improvement Confining pressure	External perimeter of the cavity
Precutting/pretunnel	Confining pressure	External perimeter of the cavity
Linings	Confining pressure Waterproofing	Internal perimeter of the cavity

drastically reduced the presence of workers underground. Still, in traditional excavation methods, the risk mitigation measures are mainly applied by workers who often use the same machines that are already being used to advance.

A third type of methods can also be included: the so-called semi-mechanized or industrial methods. These makes use of traditional excavation methods for advancing and of machines for installing risk mitigation measures (positioners, jumbo, precutting, pretunnel etc.).

Mechanized excavations are carried out by means of different types of tunnel-boring machines (TBM) according to the soil to bore and/or the risks to mitigate. In this case, workers are required only to drive, control and manage the TBM. According to the type of machine, risk mitigation measures are installed in a more or less mechanized way, in the extreme case of pressurized TBM with installation of prefabricated lining, where direct contact with the surrounding ground is totally avoided.

6.2.1 Shielded and Pressurized TBM

Generally speaking, mechanized excavation methods can be divided in three large groups:

a. Machines not providing immediate support
   - Boom-type tunnelling machine
   - Main-beam TBM
   - Tunnel-reaming machine
   - Non-conventional TBM

b. Machines providing immediate support peripherally
   - Shielded TBM
   - Double-shield TBM

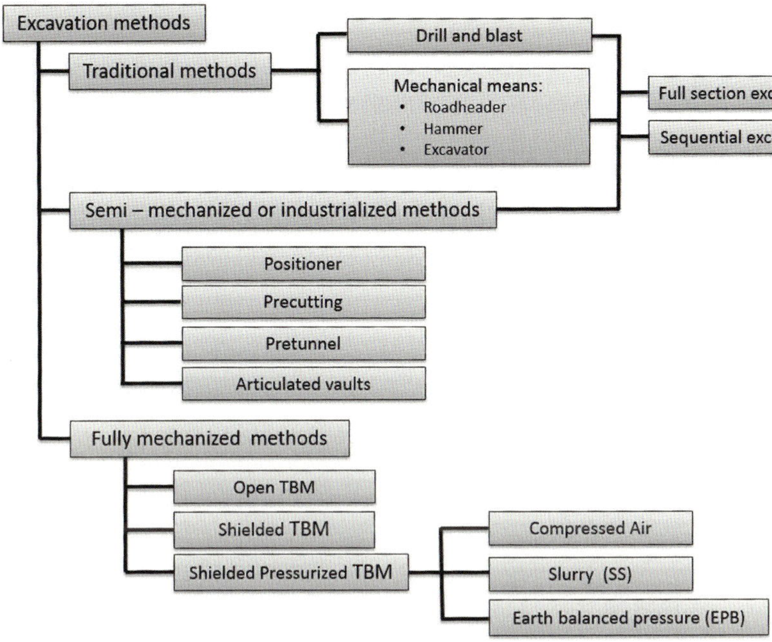

Fig. 6.1 Excavation techniques

Fig. 6.2 South ramp of the Lötschberg tunnel: execution of the portal in loose soil, about 1910. (Kovari and Fechtig 1996)

c. Machines providing immediate peripheral and frontal support simultaneously
   - Mechanical-support shield TBM
   - Compressed-air shield TBM
   - Slurry shield TBM
   - Earth pressure balance shield TBM
   - Mixed-face shield TBM

Fig. 6.3 Examples of Machines. (By Pizzarotti)

The first group (a) only constitutes a totally mechanized excavation methodology (other works are done manually).

The second group (b) group also provides immediate support in the shield stretch, to be substituted with a first-phase lining or precast lining as the machine advances.

Thanks to the application of a considerable pressure at the face and the immediate installation of a precast lining, TBM in the third group (c) can be seen as real risk mitigation equipment for surface settlements, sliding or excessive deformations.

a. Unshielded TBM

Open TBMs (Fig. 6.4) are suitable for boring in rock masses with characteristics ranging from excellent to discrete and with a self-supporting time from middle to high. The cutter head rotating along the tunnel axis is pushed against the excavation face, the cutters placed in the boring head shatter the rock and the obtained debris are cleared away from the face through openings on the boring head and a conveying system. The working cycle of an open TBM is discontinuous and consists of two phases cyclically following one another: excavation for a length equal to the working run, repositioning of the machine.

b. Shielded TBM

Shielded TBMs (Fig. 6.5) are suitable for the boring in rock masses with characteristics ranging from discrete to poor. They are characterized by a protective shield that can be integral with the cutter head or divided in two parts, an integral and a

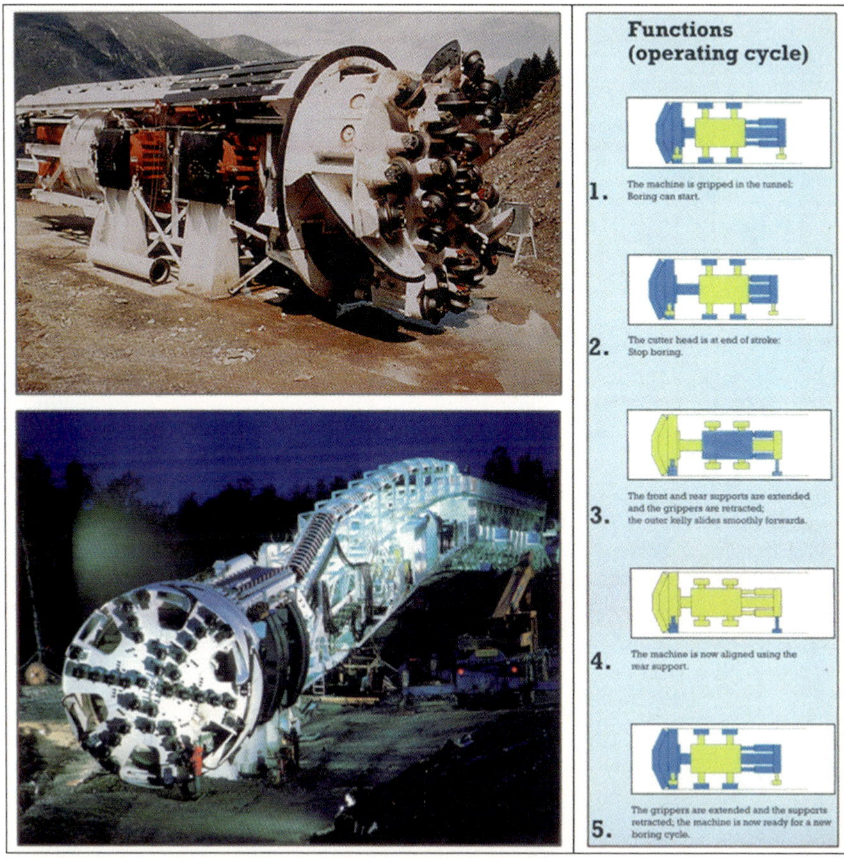

Fig. 6.4 Main beam TBM, opened TBM. Operating cycle of an open TMB. (By The Robbins Company USA and by Aker Wirth, Germany)

telescopic part (double shield). In case of shielded TBM, the excavation occurs by means of the thrust of longitudinal jacks placed inside the shield, finding the contrast on the precast segments positioned behind the machine. Advancing with double-shield TBM can be similar to advancing with unshielded or shielded TBM.

Double-shield TBM are characterized by a double-thrust system (grippers and hydraulic thrust cylinders) allowing a steady advance (Fig. 6.6).

c. *Shielded-Pressurized TBM*

Compressed-Air TBMs are suitable to excavate in soil characterized by a high water content and medium–low permeability ($k < 10^{-4}$m/s); if permeability is locally higher, it is possible to pump bentonite slurry in the excavation chamber. The excavation face is supported by a sufficiently high air pressure that can balance the hydrostatic pressure and the soil pressure. An impermeable membrane divides the

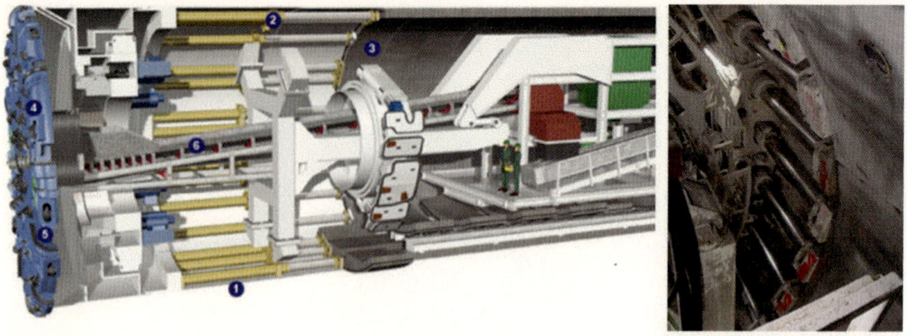

a b

Fig. 6.5 **a** Single-shield TBM scheme (Herrenknecht website): *1* shield, *2* hydraulic thrust cylinders, *3* last segment ring, *4* cutting wheel, *5* muck bucket lips, *6* muck conveyers. **b** Hydraulic thrust cylinders detail

Fig. 6.6 Shielded TBM boring cycle. (By The Robbins Company, USA)

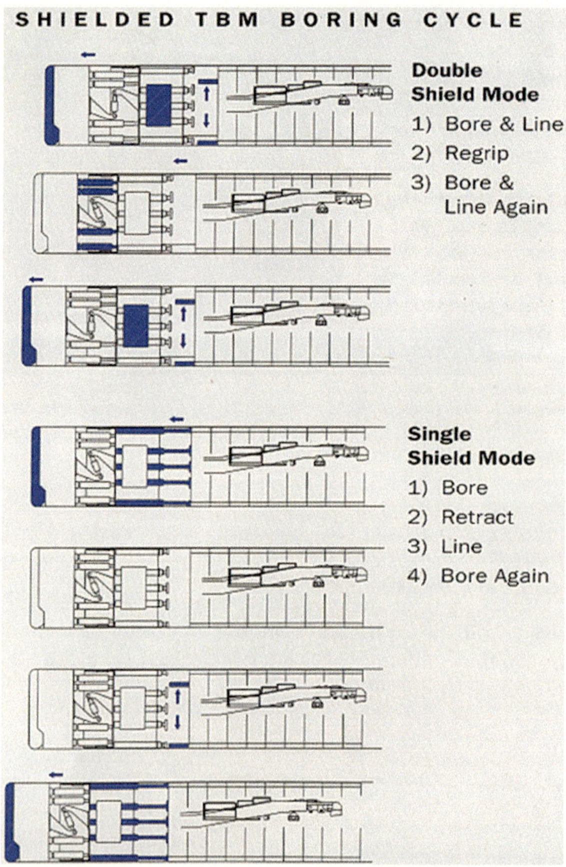

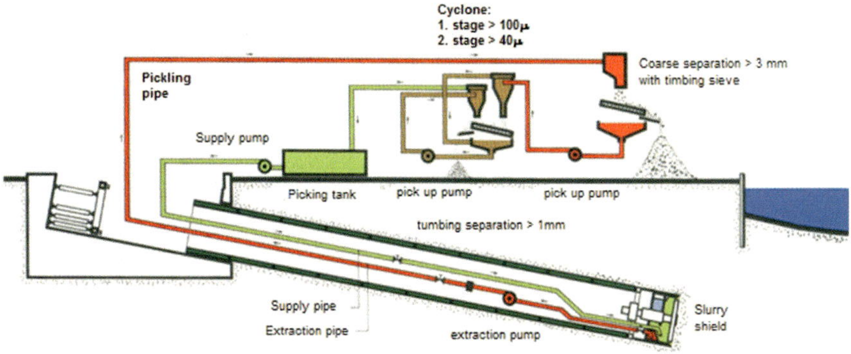

Fig. 6.7 How an SS-TBM works. (By Hochtief, Germany, modified)

shield in a pressurized frontal part (excavation chamber) and a second, not pressurized part. A "ball valve-type rotary hopper" removes the muck from the face allowing to keep the pressure level at the face.

Slurry Shield TBMs are suitable for excavation in sands and gravels with silts (Figs. 6.7 and 6.8) under the water table. The TBM head can have cutting disc tools that allow to bore in rocks. There is a hydraulic muck conveying system and the blocks that would not pass through that system are broken apart by a crusher placed in the excavation chamber. The support of the face is assured by the pressure of the slurry, water and bentonite (or clay), which is pumped in the excavation chamber. When touching the soil, it creates a soaked impermeable zone (cake) that allows to transfer the pressure of the slurry to the soil. During the advancing process, the muck gets mixed to the slurry and a hydraulic plant removes the debris from the face and conveys them to the separation plant. After the separation, the slurry can be re-used.

Hydroshield TBM When slurry shield (SS) TBMs are used, the pressure of the mixture of slurry and excavated material ensures the face stability; this pressure must be as steady as possible during the advancing phase because an abrupt reduction would cause the collapse of the face. To that aim, during the excavation, the density imbalance of the material inside the excavation chamber caused by the continuous mixing between injected fluid and extracted material is counterbalanced by continuous adaptations of the pressure of the injected suspension. In the slurry shields, this happens by means of pumps and compensation valves between the fluid input line and the exit line of the excavated material. *Hydroshield TBM* (Fig. 6.9) works in a different way: in the excavation chamber, there is a chamber with pressurized air that acts as a buffer in case of possible pressure falls.

Earth Pressure Balance Shields (EPBS) TBM are suitable for excavation in silt and clay with sand (Figs. 6.10–6.12), which are not self-supporting in the presence of water. The use of additives and foams allows the excavation by means of EPBS also in sandy-gravelly soils. The support of the face is ensured by the pressure of

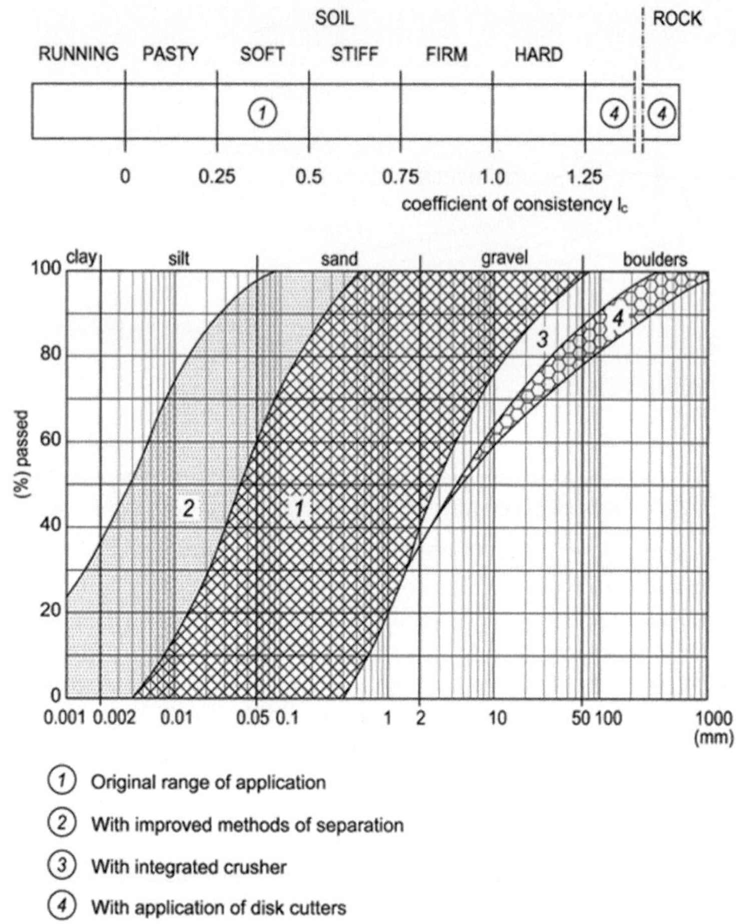

Fig. 6.8 Application range of SS-TBM (By Herrenknecht, modified)

the excavated soil that is compressed by a diaphragm that separates the excavation chamber and the remaining part of the shield. The diaphragm is pushed toward the face by jacks. During the advancing phase, the excavated material is removed from the face by a conveyor, whose rotating speed rules the soil pressure at the face.

6.3 Injections

6.3.1 Injections via Impregnation and Fracturing

This technique consists of injecting a fluid mixture in the soil that tends to solidify in time. Injections are suitable to treat weak and/or fractured rocks, fault zones and cohesionless soils. Their effectiveness depends on the ease with which the mix

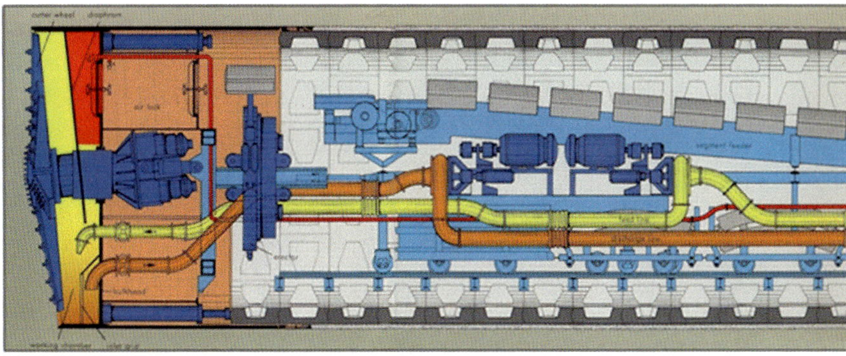

Fig. 6.9 How a Hydroschield-TBM works. (By Wayss & Freytag Ingenieurbau, Germany)

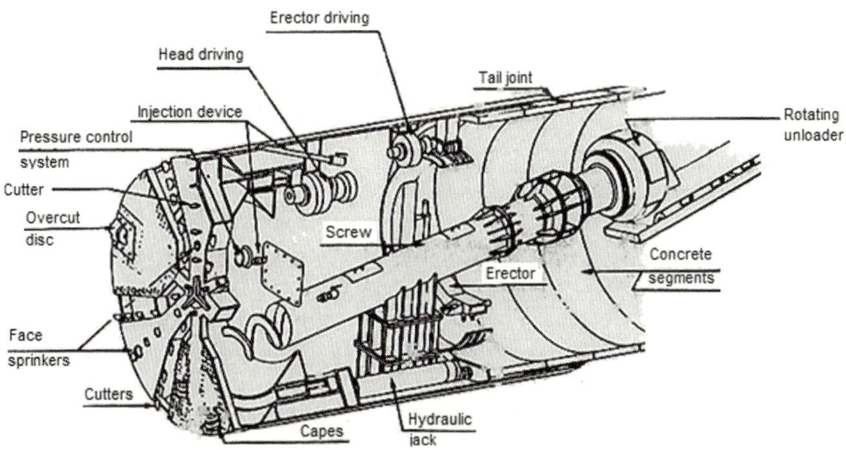

Fig. 6.10 How an EPBS-TBM works

penetrates the treated soil, according to the applied pressure. The main aim is to fill the cavities in the rocks (injections in the fracture network) or in soils (injections in pores) and to integrate the structure of the soil with a resulting increase in its strength and/or reduction of its deformability and permeability. Mixtures used with that aim can have different origins: cement, organic chemical, inorganic chemical, synthetic resins, etc. An adequate choice of the type of mixture has to take into account the intrinsic features of the soil to be treated (effective porosity and permeability) as well as the application field of the injecting mixtures (according to their stability, viscosity, gelling time and setting time, unit dimension of the components, etc.).

In granular soils (with permeability $k > 10^{-4}$cm/s), injections are normally carried out using PVC pipes with valves, installed in the boreholes. Before the injection, the interspace between borehole and pipe is filled with a low strength mixture (lesser or equal strength to that of the surrounding soil) to prevent the injected material to go back up along the pipe. Moreover, packers are placed to allow the selective

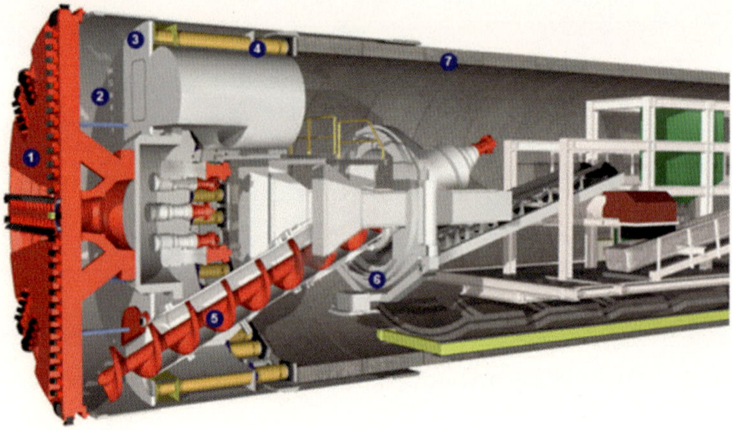

Fig. 6.11 How an EPBS-TBM works (Herrenknecht website): *1* cutting wheel, *2* excavation chamber, *3* pressure bulkhead, *4* hydraulic thrust cylinders, *5* auger conveyor, *6* erectors, *7* lining segments

treatment of the soil section. Injections are carried out using cement-based mixes (stable ternary mixtures: water, cement and colloid additive—bentonite, with ratio water/cement (W/C) > 1—using microfine cements or silica fumes if necessary to treat less permeable layers) and inorganic chemical (silicatic). Those treatments make it possible to obtain artificial cohesions of a few tenths of MPa, deformation module increases equal to maximum 2–3 times of the natural soil and a drastic reduction in permeability can be obtained.

In weak and/or fractured rocks, injections can be made with pipes equipped with valves using the Multiple Packer Sleeved Pipe system that allows not injecting the grout between the pipe and the borehole beforehand. This methodology consists in applying bag packers on the injection pipes that are filled with cement-based mixture before the real injection, in every second valve (Fig. 6.13).

More traditionally, injections in rocks are carried out directly while advancing or coming up, through an injection pipe equipped with a packer. The mixtures used in rocks are binary cement (water–cement) stabilized with a low water/cement ratio (W/C < 1). Superfluidifier additives and microfine cements are used to increase the penetration capability.

Both in rocks and in soils, chemical (poliurethane, acrylic, organic-mineral etc.) mixtures are used in very particular cases (very low permeability or, vice versa, important inflows) and in a very focused way due to their high cost. Usually, these mixtures feature low viscosity and very small unit dimensions of the components, therefore they can penetrate better with respect to cement-based mixtures and have variable properties according to the composition: high strength, short hardening times, relevant volume expansion etc.

Injections of synthetic resins in rocks are carried out directly, as described above, while advancing or coming up, through an injection pipe equipped with a packer.

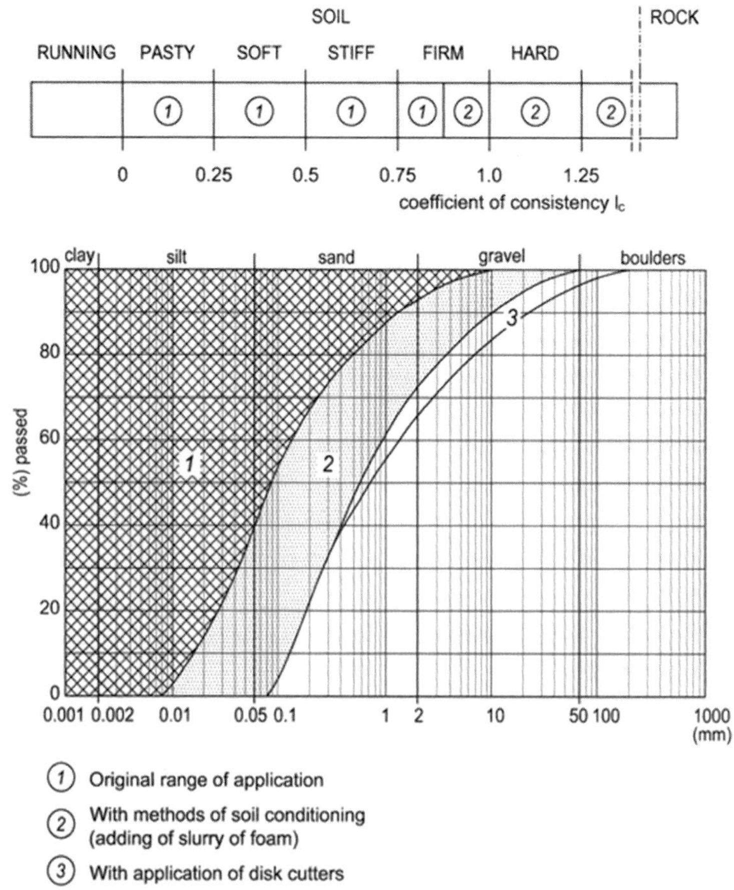

Fig. 6.12 EPBS-TBM range of application. (By Herrenknecht, modified)

Injection pressures vary according to the characteristics of the medium, the injecting mixture, the expected action range, the depth and the boundary conditions.

Generally, in porous media and close to the surface or to the cavity walls, injections are carried out at limited pressures to prevent hydrofracturing with its undesirable effects: mixture dispersion, ground deformation phenomena, overpressures etc. The maximum pressure reaches a few tenths of MPa, whereas the maximum volume of the injected mixture has to be set according to the soil porosity as a percentage of the overall volume to be treated (usually 15–30 %). As a consequence, the distance of the injection points, both among different drills and in the same drill, has to be quite limited to perform a homogeneous treatment. For example, in soils, the distance of the valves in the injection pipes ranges from 0.25 to 0.5 m, and the distance among the different injection pipes reaches 2 m at maximum.

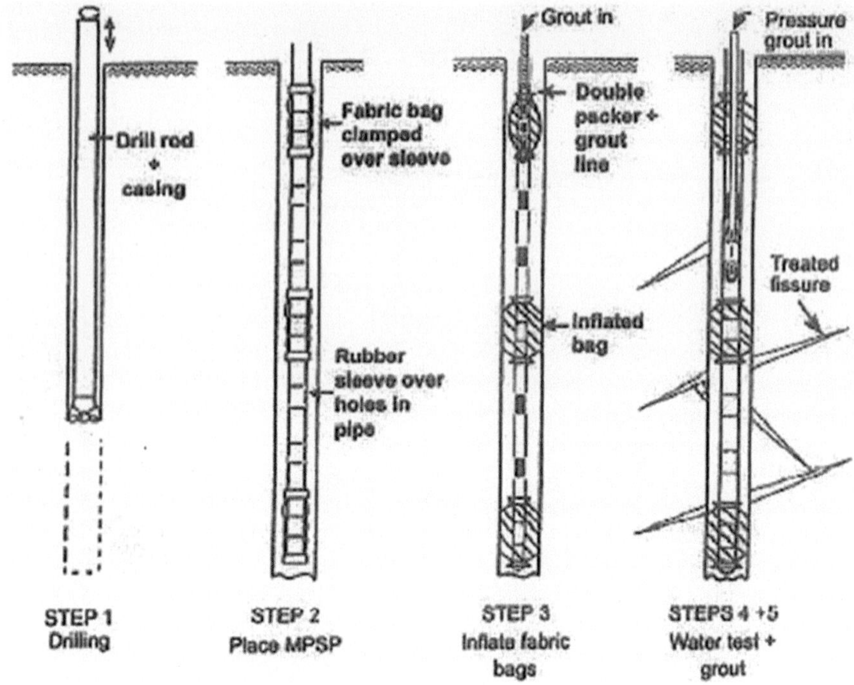

Fig. 6.13 MPSP injections

For injections in fractured rocks, quite high pressures (up to 50 bar) can also be used, in particular, if injections are carried out at a depth that will not cause any surface uplift. Usually, injection volumes are modest (maximum 15 % of the volume to be treated) considering the reduced volume of the voids to fill; the distance of the injection points is higher both among different drills and in the same one: 1 m in the same drill and up to 4 m among different drills.

When treating porous media, limit pressures and volumes and distances among injection points are decided beforehand according to the soil characteristics or the results of specific tests. Moreover, the injection flow rate can vary according to the effective performance (excessive flow rates lead to reaching the limiting pressure without reaching the maximum forecasted volume).

Injections in rock are carried out setting beforehand a constant value of the product $P \times V$ (injection pressure $\times$ injected volume from the single injection point—Grout Intensity Number (GIN) method—Lombardi and Deere 1993), the limit values of pressure and volume, and the injection flow rates. The distances of the drills decrease progressively: primary injections are carried out very far from each other (for example with 4 m spacing), secondary injections at an intermediate distance and, in case, tertiary ones at a further reduced distance, till reaching the limit pressures at a negligible injected volume.

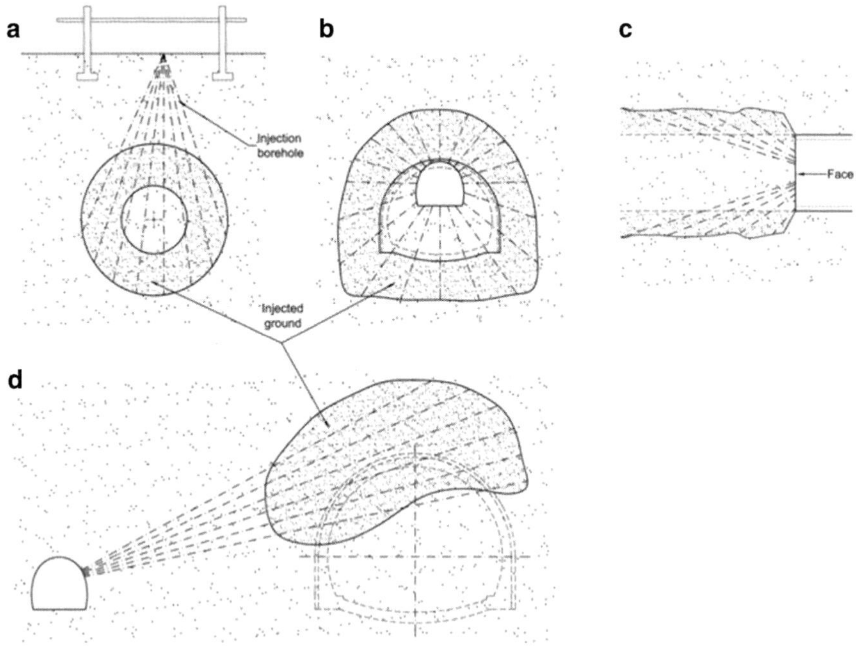

Fig. 6.14 Injection modalities. **a** Injection from the surface for shallow tunnel. **b** Injections from the pilot tunnel. **c** Injections from the face, while advancing. **d** Injection from a lateral tunnel. (From https://sites.google.com/site/eros84via/in-the-news/arguments/infrastrutture-delle-grandi-opere/vol-3/cap-8-gallerie-e-opere-in-sotterraneo-parte-1)

The maximum drilling-run length for the treatment with injections reaches indicatively 30 m. Injections can be carried out from the surface, from an exploratory tunnel or a nearby tunnel to that whose boundary has to be consolidated, or while advancing (Fig. 6.14). The smaller equipment allows to work in confined spaces of about 3 m of span (Fig. 6.15).

Generally, the zones treated with injections around an underground cavity aren't wider than the excavation radius (Fig. 6.16).

6.3.2 Jet-Grouting

Jet-grouting consists of an injection of cement suspension (usually water–cement with a ratio W/C close to 1) at very high pressure (300–600 bar). From the operational point of view, a boring is drilled up to the pre-set depth and then the boring rod is withdrawn and rotated performing simultaneously the injection. In the meantime, the soil is fractured and mixed with the cement suspension to create a column of consolidated soil whose shape and dimension can be controlled acting on the execution parameters (injection pressure, return and rotation speed of the boring rod; number,

Fig. 6.15 Radial injections from the pilot tunnel using PVC-valved pipes. (By Pizzarotti)

diameter and tilt angle of the nozzles). Jet-grouting is suitable to treat cohesionless soils and, in case, cohesive soils if their disgregation is possible within the range of the applicable pressures in the normal praxis.

The jet-grouting technique can be carried out by three different approaches, according to the soil disaggregation modality:

- Single-fluid method: the cement suspension has a double function of disgregating and stabilizing the treated soil.
- Double-fluid method: a higher disgregation and, therefore, a wider range of action is obtained by pumping water in the soil by means of a suitable nozzle placed over the cement suspension nozzle.
- Triple-fluid: in this case, the disgregating fluid is a mixture of air and water that is pumped at high pressure by a coaxial nozzle breaking up the surrounding soil, whereas the cement suspension injected through the nozzles placed underneath is only aimed at stabilizing the disgregated volume.

The action radius varies indicatively up to 0.5 m for single-fluid jet-grouting, up to 1 m for double-fluid jet-grouting and up to 2 m for triple-fluid jet-grouting.

With respect to low pressure injections, the jet-grouting technology provides a better penetration of the suspension in the surrounding ground, whereas from the mechanical point of view, the jet-grouting treatment sensibly increases the shear strength of the soil. Side effects can be the uncontrolled swelling of the ground and the suspension dispersion in areas far from the injection and, especially in case of double-fluid and triple-fluid jet-grouting under the groundwater table, the generation of overpressure. Moreover, triple-fluid jet-grouting can generate dangerous subsidence due to the soil collapse following the disgregation.

The maximum drilling-run length for the jet-grouting treatment reaches 30 m if drilled from the surface, and 24 m if drilled while advancing (see the following chapters on RPUM). In confined spaces the smaller equipment allows maximum column length of about 4.5 m in vertical or subvertical direction and 6–12 m in

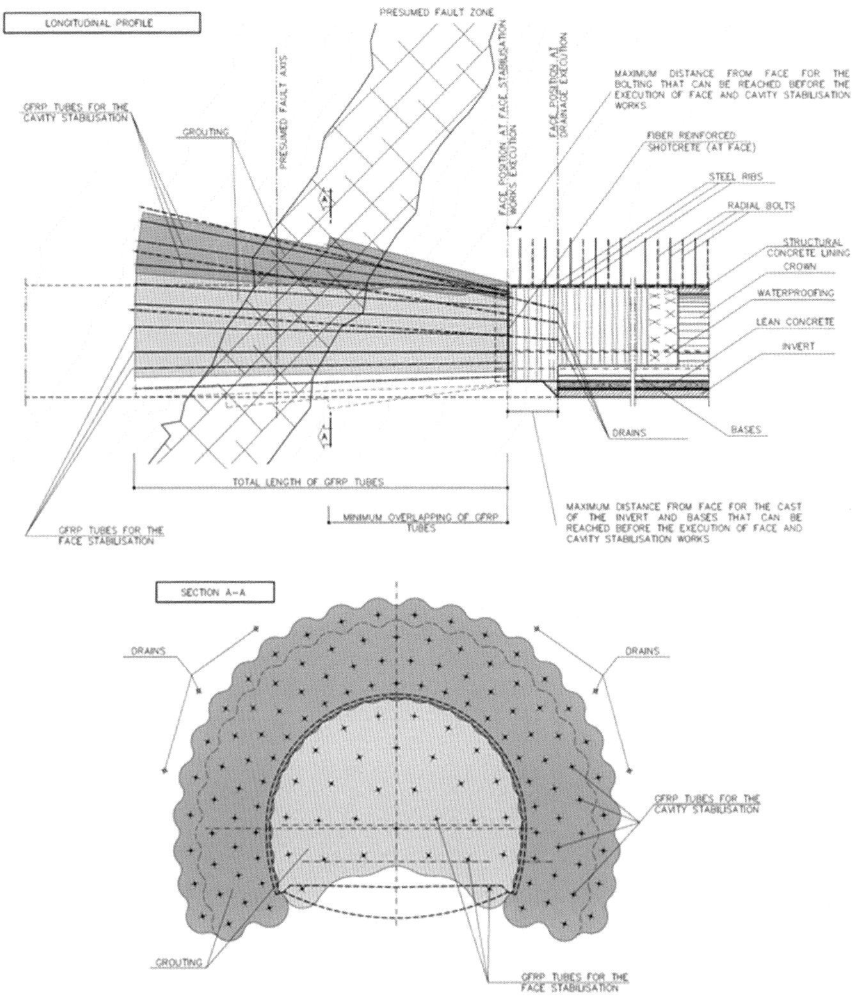

Fig. 6.16 Injections while advancing to cross a fault (*longitudinal profile*, *cross section*)

horizontal or subhorizontal direction. Double-fluid and triple-fluid jet-groutings are carried out almost exclusively from the surface and vertically due to the fragility of the drilling rods with concentric pipes (Fig. 6.17).

The treatment widths around the cavity are smaller than those of impregnation injections, as the strengths (5–20 MPa) and deformation modules (5–20 GPa) of the treated soil are much higher; with the RPUM, a crown of columns is obtained that is composed by 1–3 overlapping layers.

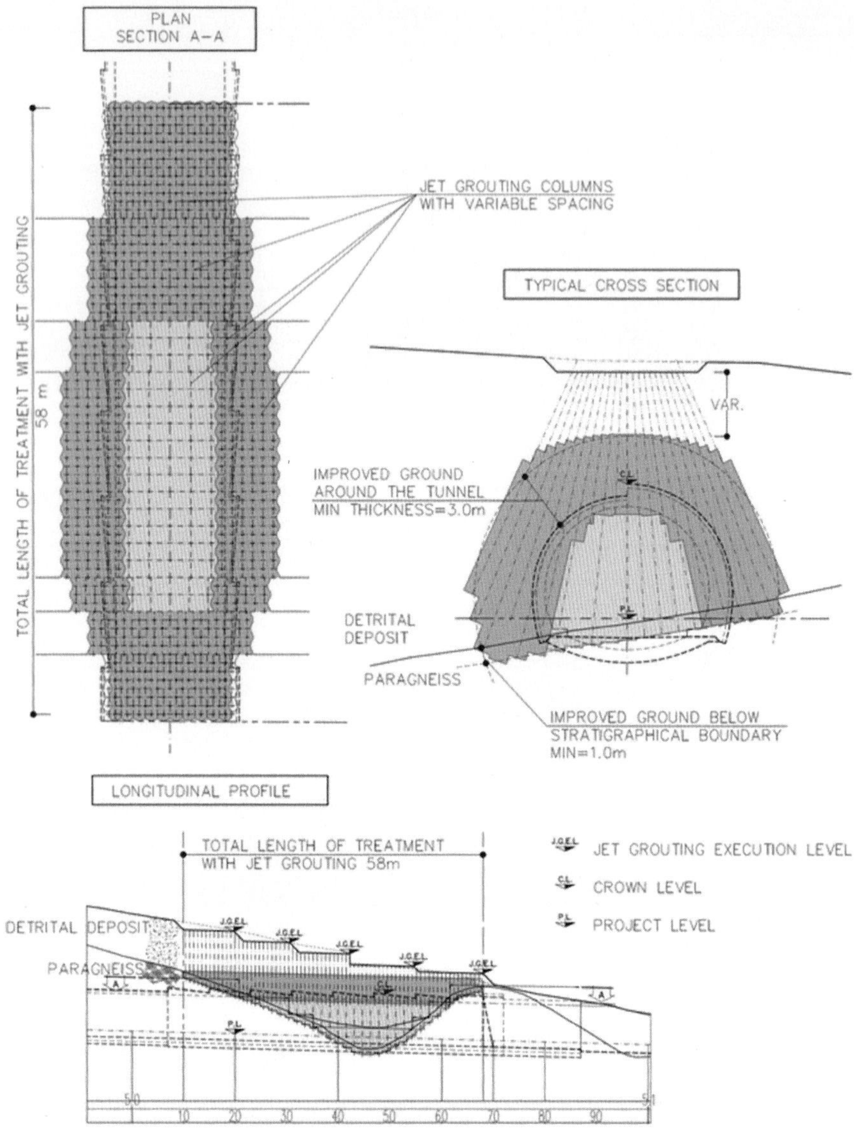

Fig. 6.17 Jet-grouting soil improvement at the portal with low overburden. Scheme of the positioning of the treatment, cross section, layout of the portal area, and longitudinal profile

6.4 Freezing

The freezing technique can be used when the soil to be stabilized or impermeabilized is characterized by a high water content (in particular, when it is under the water table). The technique, applied to granular and cohesive soils, consists of the drilling

Fig. 6.18 Ground freezing. (By Rodio)

of boreholes in which pipes are placed to let freezing fluid circulate (Fig. 6.18). The boreholes can be drilled starting from the surface, from a nearby cavity, from the cavity itself, while advancing (but with the limitation that drills must have such a geometry that the following excavation will not affect them). The freezing procedure allows the creation of an "ice wall", e.g. a belt of very resistant and impermeable soil that usually is 1–2 m thick.

Freezing fluids can be of two different types: brine (calcium chloride) or liquid nitrogen/hydrogen. When using brine, the freezing system is made by a close circuit of pipes, tanks and cooling pump (indirect method), whereas when using nitrogen/hydrogen the system is open and, at the end of the circuit, the fluid that has become gas after transferring frigories to the soil is released in the atmosphere (direct method). The two types of freezing plants are characterized by some peculiarities that make them more or less apt to different aims. Freezing the soil with the indirect method (Fig. 6.19) requires a more complex plant, the setting times of the "ice wall" are longer (a few weeks), but this method is cheaper than using liquid nitrogen/hydrogen. Therefore, it is suitable for treatments that extend over a long time. The direct plant, on the other hand, has the advantage of speeding up the working process (easier plant, the creation of the "ice wall" takes only a few days), but the use of nitrogen/hydrogen implies higher costs, whereas the releasing of the gas in the atmosphere requires a steady monitoring of the air quality in the working area. It is suitable for short-time treatments.

Freezing borings are carried out together with other boreholes in which thermometric probes are placed to control that the desired temperatures are reached and maintained.

After the freezing treatment, in particular in fine soils, a recompression is often carried out with cement injections to avoid deformations and settlements due to the destructuring caused by the ground expansion during the freezing phase (Fig. 6.19).

6.5 Cutter Soil Mix (CSM)

The Cutter Soil Mix (CSM) is an improvement technique operating from the ground surface. Through the execution of 0.5–1 m thick panels reaching a maximum depth of 30 m, it performs the improvement of both the face and the tunnel contour. This technique is an alternative to the soil treatment by means of jet-grouting or injections from the surface.

The CSM cutter head (Fig. 6.20) is made of two wheels moving on the horizontal axis and is linked to a system of rigid beams or ropes, according to the depth of the panel to realize. For a maximum panel length of 20 m, the cutter head is assembled on rigid beams, but ropes are used when reaching deeper depths.

During the downgoing of the cutter head, the pronged wheels break up and disaggregate the soil while a nozzle placed between the heads releases the cement mixture at low pressure. After reaching the depth set by the project, the cutting head is withdrawn while the release of the cement mixture goes on and the wheels go up mixing continuously the previously fractured soil (Figs. 6.21 and 6.22).

The CSM technology can be applied in soils (cohesive and not cohesive) and in weak rocks, where the effectiveness of the cutter wheels is guaranteed.

6.6 Anchors

There are different kinds of anchors, according to their realization and the way they work.

- *Nails*: They adhere completely to the surrounding rock mass (by cementation, through resins, by expansion of a tubular section bar or through other forms of mechanical adherence). They are passive elements, i.e. the stress in the nail is due to the deformation of the rock mass.

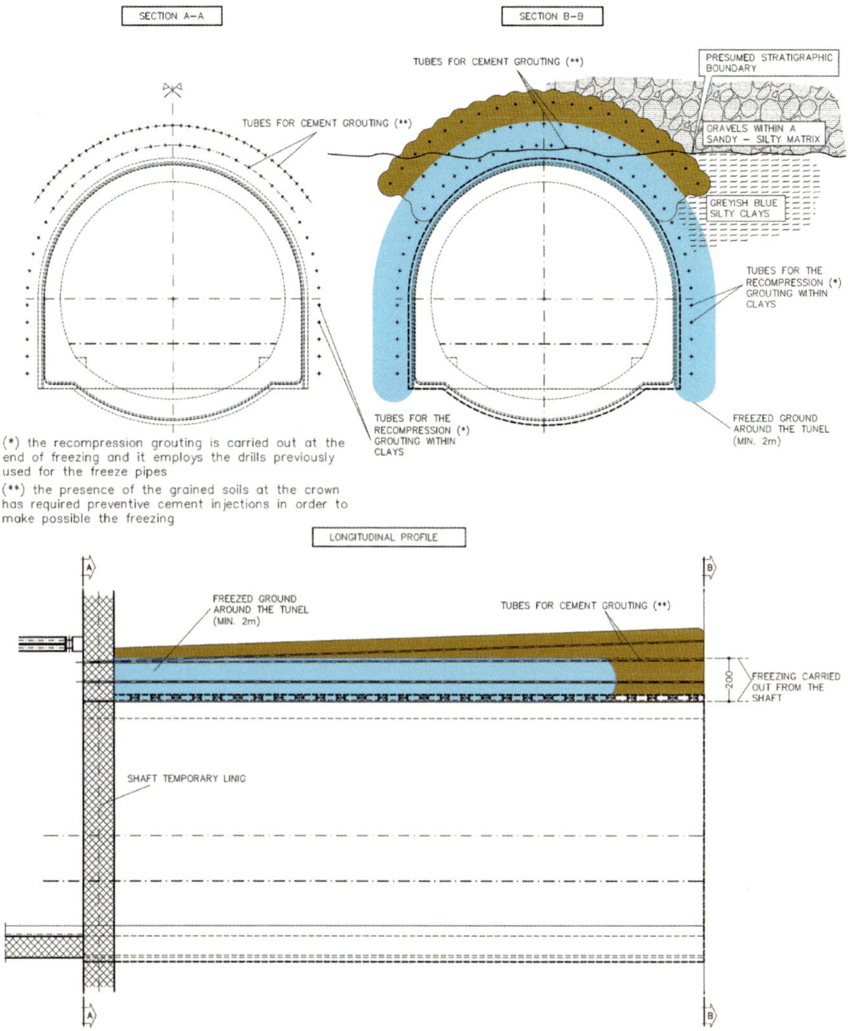

Fig. 6.19 Consolidation on the cavity contour carried out by freezing with the indirect method combined with injections. The carrying out of a preventive treatment of the materials by means of injections of consolidating and waterproofing cement-based grout was necessary in order to provide the soil with an adequate uniformity where the crown excavation was carried out and, above all, to prevent the few movements of the groundwater within the ground volume to be treated by freezing. Treatments have been carried out from a previously drilled access shaft. The freezing probes were distributed outside the cavity and did not interfere with the excavation. The minimum width of the frozen stripe on the tunnel contour is 2 m. At the end of the freezing phase, recompression injections were carried out. The freezing treatment has been maintained for the whole period of the tunnel excavation till the completion of the final lining and at least till the reaching of the concrete strength foreseen in the project

Fig. 6.20 Cutter head used for Cutter Soil Mix. (by Bauer)

- *Bolts*: Elements mechanically anchored to the ground by means of an expansion head. They are active elements, installed with a pre-load transmitted by tightening a nut on a contrast plate that also allows the opening and gripping of the expansion head.

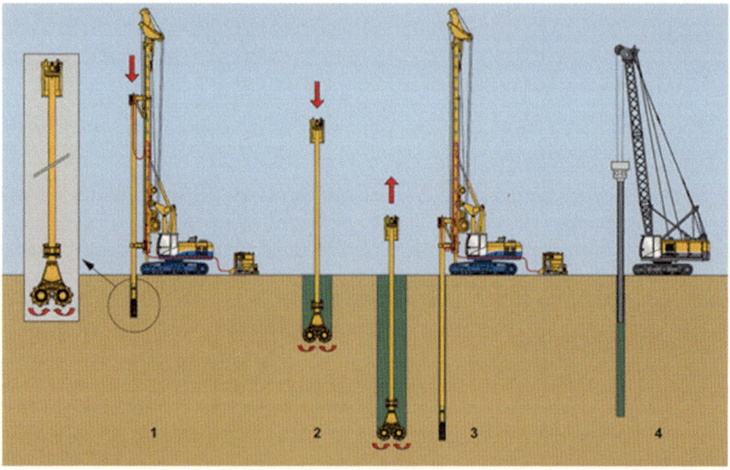

Fig. 6.21 Execution phases of a CSM panel, *1* positioning of the cutter head, *2* downgoing phase, *3* upcoming phase, *4* reinforcing elements can be inserted into the wall. (by Bauer)

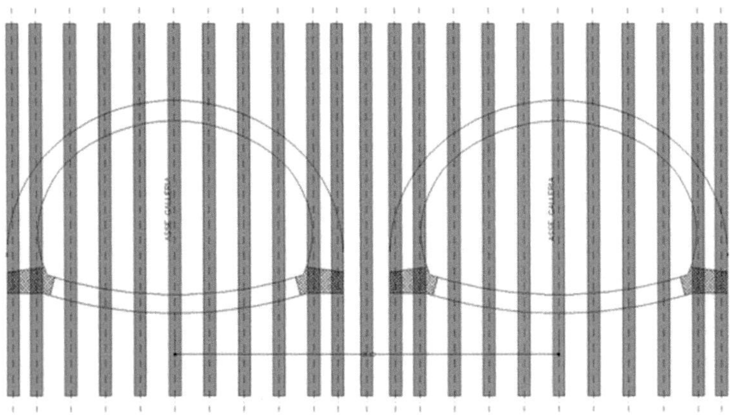

Fig. 6.22 Soil improvement around shallow tunnels in difficult soils using the CSM technology

- *Tiebacks*: Elements anchored to the ground by a bond length, installed with a heavy preload applied on a contrast device (beam or load distribution plate) by means of jacks.

6.6.1 Nails

Nails are elements, normally placed radially with regards to the excavation axis (Fig. 6.23), but also orthogonally to the core, on the face or around the cavity in advancing (spiles).

They are made of steel bars or pipes, GRP or carbon fibres; usually, they are fitted with a load distribution plate at the end paced on inner surface of the cavity. They sew the existing cracks or the surfaces affected by failure following the excavation along the shearing discontinuities, with modalities that may differ according to the values of the angles α and ϕ (where α is the angle between the nail and the discontinuity surface and ϕ is the shear strength angle). In particular:

- By direct shear strength T_d (dowel effect, cases *e* and *f* in Figure 6.24) for $\alpha >> \phi$ and close or bigger than 90°. This contribution T_d reaches the limit value equal to the nail tensile strength (N) for big deformations following the reaching of the local rock strength in correspondence to the sliding surfaces.
- Thanks to a combined effect of sliding strength and the shear strength increment (case d in Figure 6.24), by transferring a pressure to the surface. This contribution equals to $T = N\ (\cos\alpha + \text{sen}\alpha\ \text{tg}\phi)$ and reaches its maximum when $\alpha = \phi$.

Usually, the lengths of radial nailing range from the radius to the diameter of the cavity (Fig. 6.25). Nailing while advancing, on the contrary, are effective for depths equal to about the excavation radius, but they have to be much longer (at least

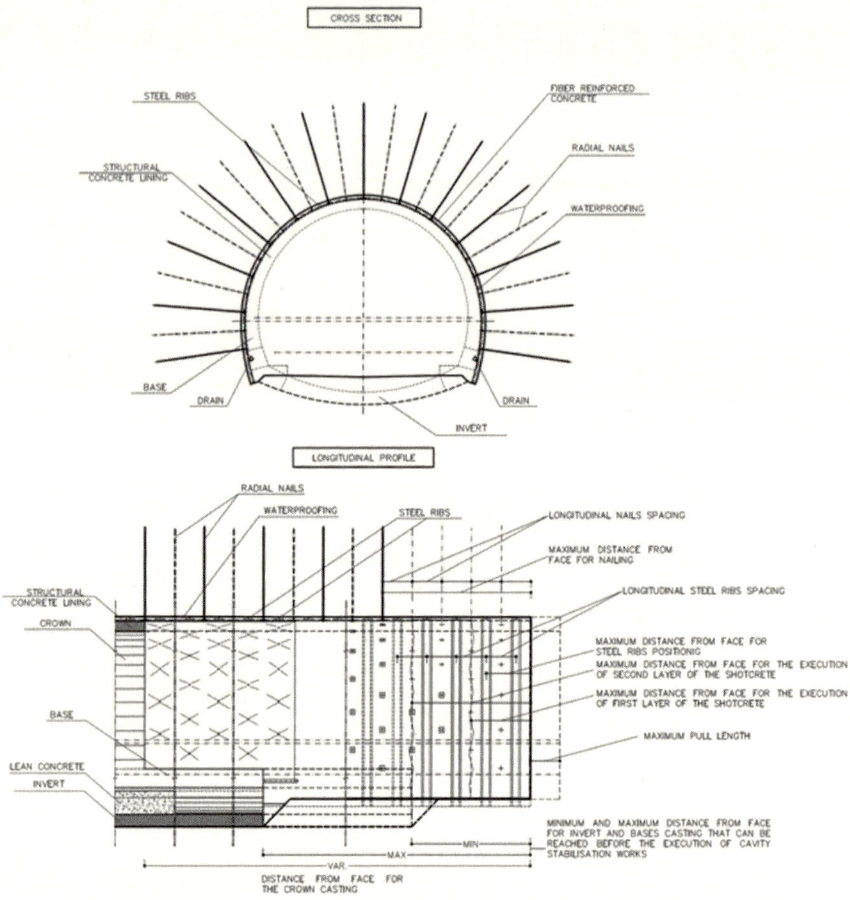

Fig. 6.23 Radial nailing: cross section and layout

three times the excavation radius) to allow an adequate anchoring and an effective overlapping of the following measures.

Nailing is useful to reinforce both homogeneous rock masses with low mechanical strength and rock masses with high matrix strength, but is characterized by extended fracturing.

Nailing is aimed at:

- Sewing discontinuities and weak surfaces, with a subsequent decrease of the hazard of detachment and/or relative sliding of rock blocks
- Increasing the strength and decreasing the deformability of the ground interested by this measure

Nailing does not necessarily require very resistant elements (maximum a few hundred kN), because of three reasons:

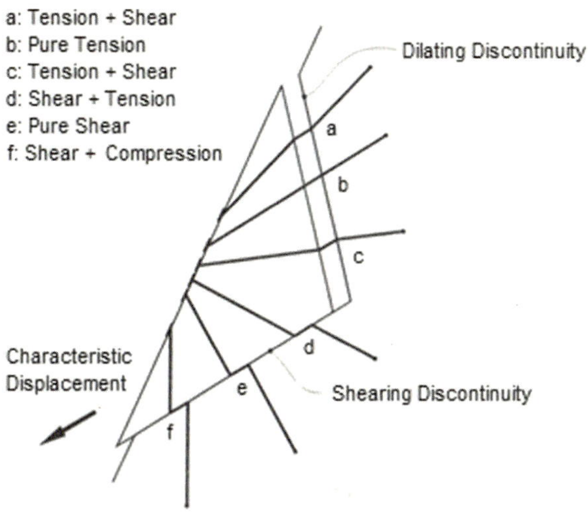

Fig. 6.24 Strength mechanism of a nail versus its orientation

Fig. 6.25 Tunnel realized with crown nailing. (By Pizzarotti)

- Nails are mainly subjected to axial stresses; therefore, their tensile stress is very well exploited.
- The global effect of a more widespread reinforcement is better on the mass strength because it distributes the confinement pressure.
- The higher is the strength of the element, the longer must be the anchored length and therefore its total length.

A traditional nail can be positioned by a suitable drilling rig or jumbo drill, the one used to bore drill holes in mines.

The drilling rig (Figs. 6.26 and 6.27) can be equipped with a feed system in order to install nails. It can bore a hole and mechanically position a nail using the same device. In this way, it is possible to insert the nail with a push that overcomes the possible obstruction of the borehole walls.

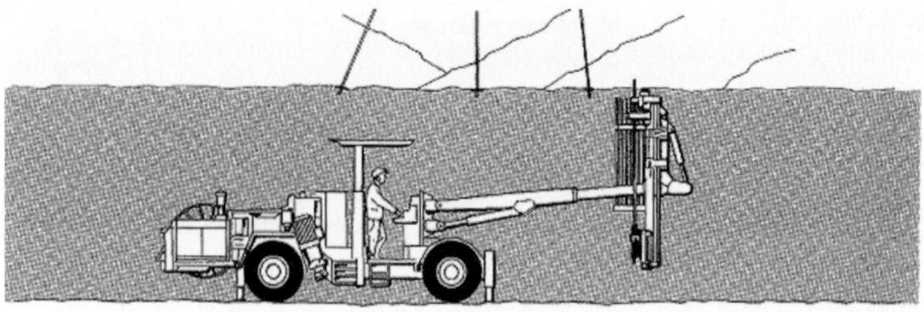

Fig. 6.26 Mechanised nailing with a drilling rig. (By Atlas Copco)

Fig. 6.27 Nail installation on the crown using a drilling rig. (By Atlas Copco)

In certain situations, a more common jumbo is preferred to the drilling rig. Usually, a jumbo (Figs. 6.28 and 6.29) is not fitted with mechanical devices to install nails, but it simply drills the holes using slides similar to those of a drilling rig and therefore it has the same length limitation (maximum 5–6 m).

In case of jumbo drill, the installation of a nail is performed by a couple of workers on a platform (Fig. 6.30). In these conditions, the weight and the flexibility of the nail make the installation difficult and the smallest blockage of the hole can jeopardize the whole operation.

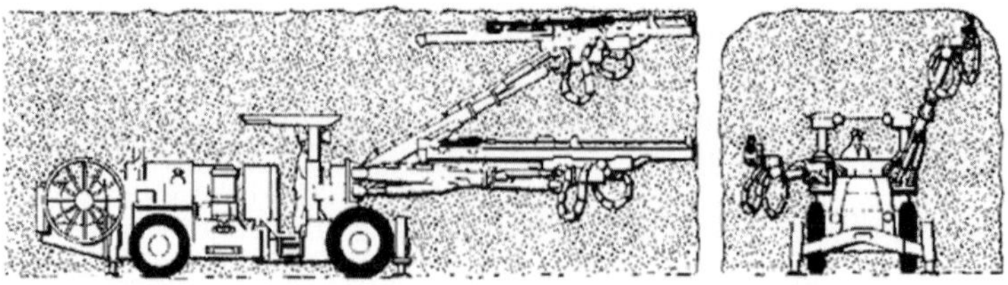

Fig. 6.28 Jumbo drill. (By Atlas Copco)

Fig. 6.29 Drilling with jumbo to install nails in the crown. (By Atlas Copco)

Fig. 6.30 Manual installation of nails performed from a platform. (By Pizzarotti)

Nevertheless, it is often convenient to choose a jumbo, in particular in tunnels where only occasional measures and not systematic nailing is required or in those cases where it is not economically convenient to invest in a drill rig equipped with a feed system and a jumbo at the same time.

Moreover, it has to be taken into account that a jumbo allows a reduction of the boring time as it can work with more drills simultaneously.

Anyway, nailing becomes difficult when it overcomes the maximum length allowed by usual boring devices. In this case, boring rods and jointed nails must be used. Drill rig can be equipped to joint mechanically the rods (rod adding system), but operations become quite long. The rod has to be installed progressively during the drilling, disassembled during the removal phase from the hole, and then the nail has to be assembled manually during its installation.

Inconveniences and limitations of traditional nailing methods (hole instability, maximum length of the boring devices) can be overcome using self-drilling bars and/or techniques that allow the installation of tubular elements while drilling (Fig. 6.31) that reduce the installation time.

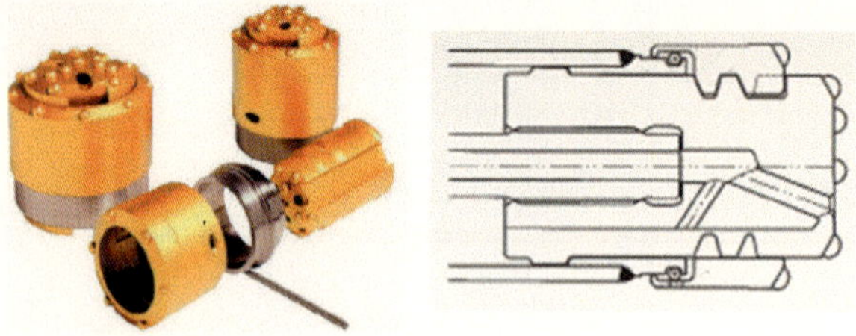

Fig. 6.31 The driving head is linked to the sleeve with a bayonet twist lock. Both rotate clockwise boring with a diameter wide enough to allow the shoe to tug the lining pipe constituting the reinforcing element. (By Atlas Copco)

In particular, the use of self-boring nails eliminates the disassembling time of the rods during the extraction phases and assembling and positioning times of reinforcement bars, independently from the use of jumbos or drill rigs. Moreover, the problem linked to the instability of the borehole walls, which is particularly frequent in much fractured rocks and in unstable soils, is definitely solved.

Spilling (Fig. 6.32) is a technique that, according to the nail inclination, may perform the function of forepoling (Sect. 6.8.1) and radial nailing (Sect. 6.6.1), at the same time. It is used in rocks requiring the improvement of general mechanical

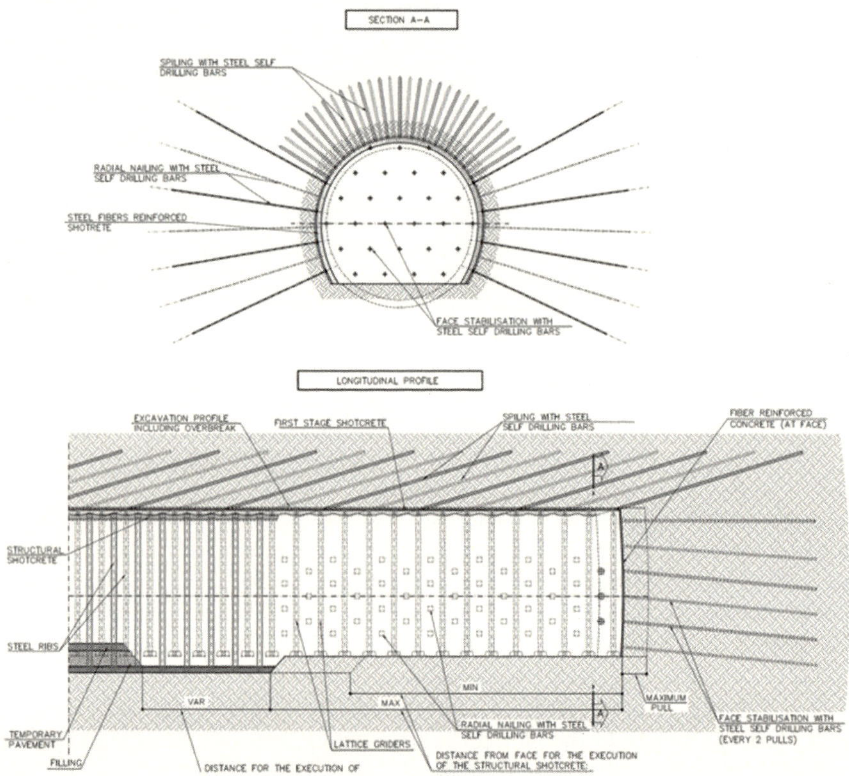

Fig. 6.32 Spiling: cross section and longitudinal profile

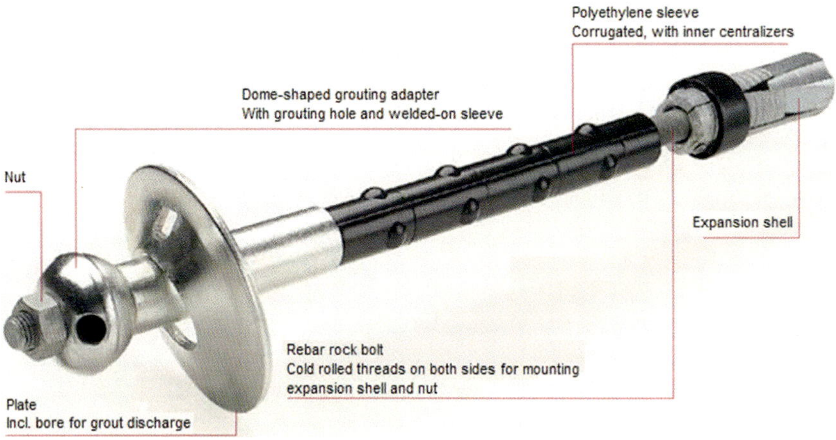

Fig. 6.33 CT-Bolt system components (by DSI, modified). The CT-Bolt system allows the installation of the bolt and its secondary cementation through a PVC sheath that provides an excellent protection against corrosion and guarantees the element durability

properties that also present a fracturing degree that might lead to the detachment of rock volumes from the ceiling close to the face.

The improvement measure is realized while advancing with respect to the excavation in the ground above the face and inserting different layers of nails having a dip of $i < 45°$, at a distance equal to a multiple of the steel ribs pitch. The smaller is the angle on the horizontal surface used to insert the nails, the lesser is the nailing effect on the mass; still, the protection offered on the advancing area is higher.

The use of nailing while advancing with respect to the excavation face is strictly linked to the mechanical qualities of the rock and to the ratio between the orientation of the discontinuities present in the rock mass and the excavation geometry. Similar to what happens with an ordinary radial nailing, the first effect obtained by this measure is the improvement of the mechanical properties of the ground around the cavity. The dip given to the nails actually limits the thickness of the ground interested by nailing, but nails placed in this way secure the ground that will constitute the crown of the future advancing tunnel. This technique is particularly effective when used on rock masses with intense and pervading fracturing. In this context, it can allow the securing of blocks isolated by discontinuities that otherwise would be unstable.

The features of the elements used for nailing while advancing must be carefully studied in relationship to the main function that they have to perform, to the tunnel dimension, the mutual distance and the general qualities of the rock. Wider diameters are used when important falls of material from the ceiling are feared and therefore the improvement of the contour of the excavation while advancing is preferred; it is better to use a higher number of nails of smaller diameter to diffuse the reinforcement if the main desired effect is the one obtained with radial nailing.

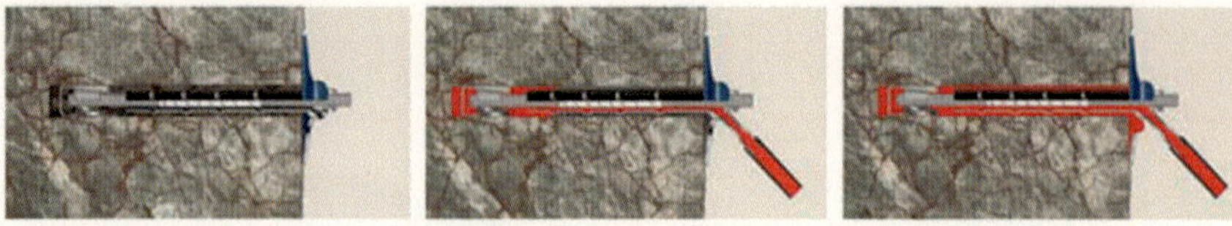

Fig. 6.34 CT-Bolt-installation procedure. (by DSI)

6.6.2 Bolts

Bolts (steel or GRP bars, more rarely, pipes) are particularly suitable in the cases of possible detachment of rock blocks, failure of thin layers or bedding planes of rock and rock bursts, because they can provide a preventive confinement pressure. Usually, their length is similar to the cavity radius. After the installation, they can be completely connected to the rock mass by cementation (Figs. 6.33 and 6.34). The anchor head is often used to facilitate the installation of the element (in particular, if it is oriented upward) before cementation, even when pre-tensioning would not be strictly necessary.

Similar to the nails, they act in the shear-traction domain, but their characteristics are different in terms of action and anchorage mechanism. In particular:

- Bolts are active elements. Tensile stress is the result of the initial tightening;
- Bolts have anchor points, they adhere to the walls of the hole for a limited stretch with respect to their length (expansion head);
- The dowel effects are forwarded in case of not cemented bolts, when the walls of the hole get in contact with the anchor bar after sliding.

6.6.3 Tiebacks

The installation of tiebacks (bars, cables, steel strands, etc. having high yield limit or synthetic fibres) in underground works is limited to few particular cases where the application of a strong action preventive to any deformation (large cavities as, e.g. underground plants) is required. They are quite long because the anchor bond length, which is already considerably long, has to be positioned outside the area disturbed by excavation. Their use is concentrated mainly outside the tunnel (e.g. in the portal area to stabilize the slope).

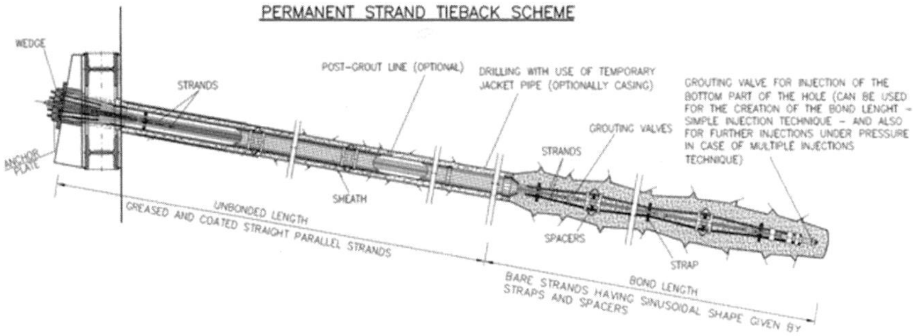

Fig. 6.35 Scheme of a permanent strand tieback

Anchors are made as follows (Fig. 6.35):

- Anchor head: it is made by a perforated steel plate fitted with a strand blocking system that is usually connected to a load distribution structure in reinforced concrete or in steel.
- Reinforcement: it is characterized by an anchored stretch-bond length (along which the adherence between steel and cement injection is developed) and a free stretch-free length (between the head and the anchoring bulb) where strands can extend freely.
- Bond length: it is accomplished through cement mixture injections; it is aimed at blocking the anchor in the surrounding ground, the anchoring action is guaranteed thanks to the cohesion developed between bond length and ground; the length of the anchored stretch depends on the strength to be transmitted and the characteristics of the surrounding ground.

6.7 Drainage

Water inflows in tunnel always represent a hazard element and may cause damage on the ground surface. Hazard situation linked to the presence of water during the execution of underground works, already discussed, may be:

1. Potential risk for the environment

- Settlements due to the water table drawdown
- Impacts on superficial and underground waters: drawdown of the original aquifer
- Pollution

2. Potential risk for the tunnel

- Infiltration: sudden important inflows
- Interstitial pressure: development of gradients and possible incoming of material or instability of the face and the walls

Fig. 6.36 Working problems due to the presence of water. (By Co.ge.fa. S.pa. Italy)

- Dissolution: change of the physical characteristics of the material due to the water content
- Transportation: change in the physical characteristics of the material due to the transportation of fine particles
- Difficult working conditions in presence of important inflows (Fig. 6.36)

It is clear that the correct forecast of groundwater, its regulation through a suitable draining system and the management of its drainage play a very important role in the mitigation of risks linked to the presence of water during construction.

Drainages are elements that allow to eliminate, even if only partially, groundwater from ground, conveying it toward the cavity in a controlled way. Drainage consists in creating artificial preferential flow paths. They can be: radial or in advance with respect to the tunnel face holes, usually fitted with drainage pipes (slotted or microfissured) in a geotextile sleeve or equipped with a filter to avoid or reduce the transportation of solid particles or drainage tunnels if the flow rate is very high. The drainage reduces interstitial pressures and, as a consequence, it improves the mechanical characteristics of the soil and reduces the water thrust on the lining (Figs. 6.37 – 6.39).

In case of high hydraulic pressures, the boring to install draining pipes has to be performed with appropriate devices called preventers. A preventer (Fig. 6.38) is a safety hydraulic equipment used during boring in presence of high pressure water, aimed at preventing the uncontrolled flow from the ground.

6.8 Reinforced Protective Umbrella Methods (RPUM)

In a poor quality rock mass (risk of face and cavity instability also in the short term, even for low stresses), the excavation can be carried out safely proceeding with the systematic execution of forward treatments (forepoling, jet-grouting, precutting,

Fig. 6.37 Draining pipes

pretunnelling) on the excavation boundary which are normally coupled with face stabilization by jet-grouting or VTR nailing, for consecutive fields, that is with cone-shaped section (umbrella), the so-called RPUM.

In general, the implementation of RPUM methods consists in advancing by consecutive cone-shaped fields. In particular, following working steps are repeated cyclically:

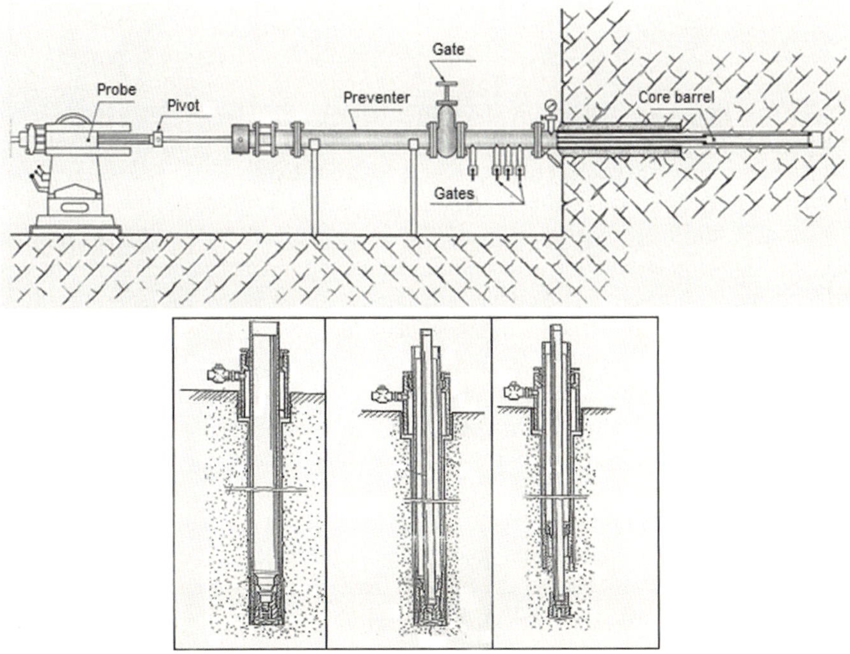

Fig. 6.38 Drainage installation system with pipe, sleeve and preventer

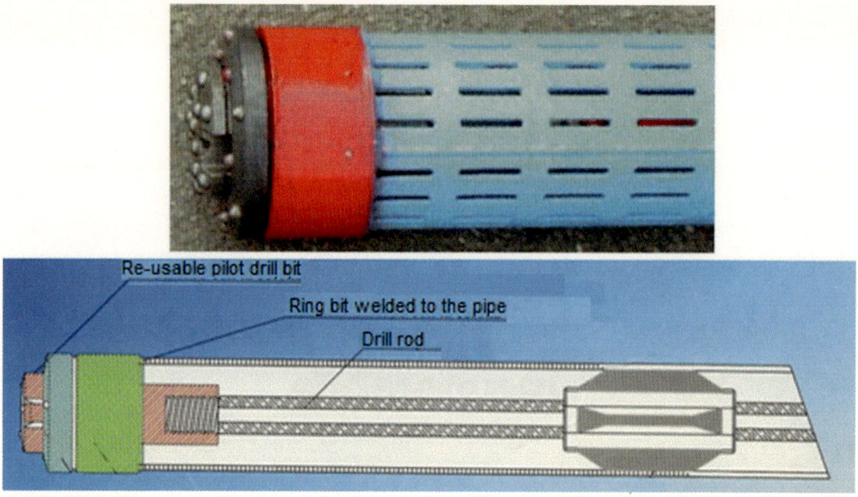

Fig. 6.39 Drainage installation system during boring with partially non-retrievable drilling head

- Implementation of the measures while advancing
- Excavation and quick positioning of a first-phase lining on a stretch shorter than the length of the forward treatments

6.8.1 Forepoling

Forepoling the excavation face is a technique allowing the installation of metal tube-shaped elements with longitudinal strike on the outside of the excavation perimeter, while advancing with respect to the excavation, aimed at protecting the cavity from material falling from the ceiling before lining installation.

Usually, in soils with poor mechanical qualities, forepoling requires the drilling and the following installation of cemented (grouted) pipes in the boreholes by means of a positioner. Sometimes, the pipes are equipped with valves for later injections. Normally, this kind of forepoling reaches a length up to 15 m (Fig. 6.40). In this way, the first-phase lining can be placed under the pipes allowing to proceed with the next forepoling field. The big dimension of the more common positioners precludes their use in small tunnels (minimum diameter 6–7 m). Usually, smaller boring machines with pusher leg are used that can bore holes suitable to host poles. In this case, the poles are shorter.

The protective umbrella is made by a variable number of elements installed at a distance of some decimetres from each other. If no injections are carried out, the stability of the mass portions between the different elements depends on the shear strength of the material (arch effect, Fig. 6.41), thus limiting the use of forepoling in materials with at least a small cohesion degree. Combined to injections or jet-grouting, the use of forepoling is effective also in cohesionless soils.

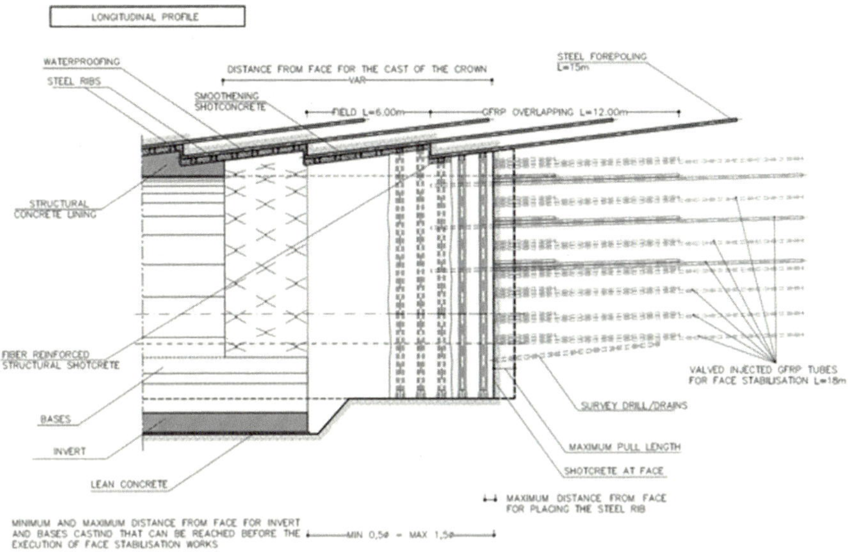

Fig. 6.40 Forepoling

Fig. 6.41 The arch effect in the transverse cross section (between reinforcement pipes) and the longitudinal one (between steel ribs) before the realization of shotcrete

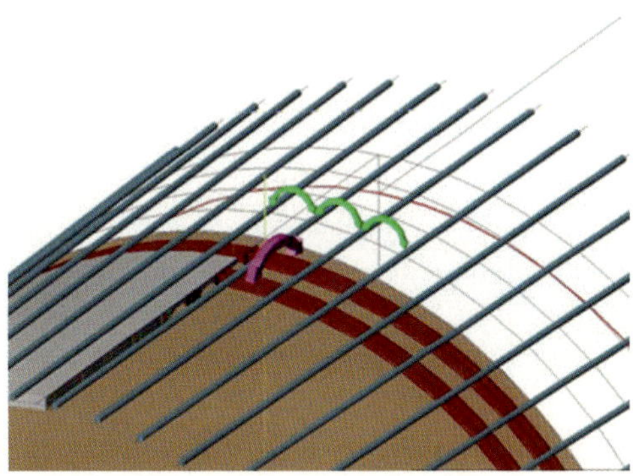

As already pointed out for nailing, the installation speed of this type of elements depends on the working steps (boring, insertion of the reinforcement, grouting or injection). Some techniques can be used for forepoling, very similar to those discussed for the nails that allow the improvement of productivity (e.g. use of self-boring elements or boring with partially non-retreivable drilling-with-casing systems) (Fig. 6.42).

6.8.2 *Jet-grouting Vaults*

The procedure to build a jet-grouting vault is similar to the one already described for forepoling: subhorizontal columns are realized in the contour of the excavation

Fig. 6.42 Forepoling (from *top*). **a** portal area with forepoles on the contour and jet-grouting columns at the face. **b** detail of the forepoles on the steel ribs. **c** implementation of forepoles on the tunnel contour. (By Pizzarotti)

according to a truncated cone-shaped geometry. Their length can reach 24 m. Jet-grouting columns can be disposed along one or more lines. If the ground features require it, jet-grouting columns can be reinforced with steel pipes or bars introduced in the columns before the mixture begins to set or under the columns by means of forepoles. As in other RUMP technologies, the advancing field is shorter than the columns in order to ensure the creation of an overlapping area.

An example of implementation of this technology is shown in Figure 6.43.

6.8.3 Precutting

This technique consists in the realization of a cut around the tunnel shape while advancing (Figs. 6.44 and 6.45). Then, the cut is filled with fibre-reinforced shotcrete to create a pre-ceiling with lining function. This technique is suitable for the excavation in weak rocks and in clayey soils that can be quite easily cut and maintain a sufficient stability (Fig. 6.46). The materials have to be self-supporting for a time span that allows to perform the cut and the subsequent filling with shotcrete.

The lining ceiling thus created while advancing has a width ranging from 15 and 25 cm and a length up to 3.5–4 m. Each single stretch of the protective umbrella is

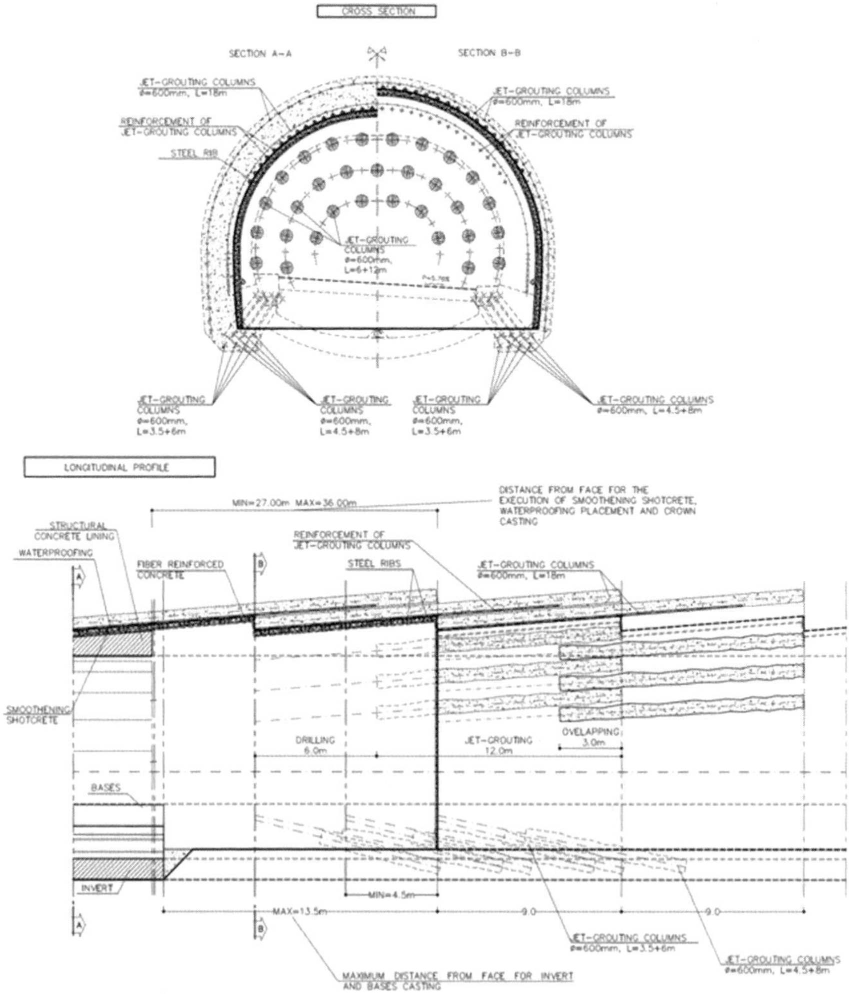

Fig. 6.43 Jet-grouting vaults reinforced underneath, with total overlapping: cross-sectional and longitudinal profile. In this case, the consolidation of the core with jet-grouting columns was foreseen as well

cone shaped in order to allow the realization of the subsequent treatments. A 0.5 m overlapping ensures the continuity of the improvement measure. A few hours have to be waited between the completion of the vault and the beginning of the next field, depending on the time required by the shotcrete to reach the suitable strength characteristics. In general, while the excavation advances, steel ribs are mounted and covered by a supplementary layer of shotcrete to allow the subsequent coat of the final waterproofing lining.

The precutting method provides other advantages, as:

- Avoiding being out of shape thanks to the regular cut carried out with a band saw
- Lower incidence of steel ribs and shotcrete for the first-phase lining
- Increased safety for the workers

6.8.4 Pretunnel

Precutting allows the realization of a first-phase lining in advancing, to be integrated during the excavation phase and completed with the final lining. The pretunnel technology may allow the realization of a final completely supporting structural lining of the tunnel before excavation (even if it is usually integrated with a waterproof inner lining). The operations below are typical of this methodology and are cyclically repeated during the tunnel construction (Fig. 6.47):

- Realization of a structural lining while advancing (segment width 0.9 m, segment length 12 m)
- Excavation of the previously lined stretch with suitable tools (hammer, roadheader, excavator etc.)

The lining is implemented using a cutting module with a maximum length of 12 m that is moved crosswise within the ground along the section to be excavated. The cutting module is paired with a formwork that can follow the displacement of the cutting module and that separates the cutter head from the cavity that is immediately filled with concrete, fibre reinforced if required (Fig. 6.48).

The smaller is the cavity convergence when the lining is installed, the heavier will be the load on the lining. Therefore, it is evident that if the concrete is casted in advancing, as in pretunnel, the convergence is immediately limited leading to heavy loads on the lining. If no particular measures are adopted, this method can be used in shallow tunnels or for the portal area of deep tunnels.

The technology described can be used for both the execution of open structure (prevault) so that the invert is constructed in a second phase, and for closed structures with round or polycentric section (pretunnel).

6.9 Linings

As already said in Chap. 4, lining is any kind of measure that can provide a confining pressure aimed at guaranteeing the total stabilisation of the cavity. In a tunnel, the lining is loaded by possible falls of material, or as a consequence of the cavity deformations developing after construction (Sect. 7.8).

Generally, a difference is made between the lining realized during the excavation phase, usually after each excavation step—first-stage lining—and the lining realized

Fig. 6.44 Precutting equipment: The pronged blade is mounted on a structure with the shape of the tunnel to be excavated. (By Rodio)

Fig. 6.45 Precutting equipment: Detail of the pronged cutting blade. (By Rodio)

Fig. 6.46 Tunnel excavated with precutting method. (By Rodio)

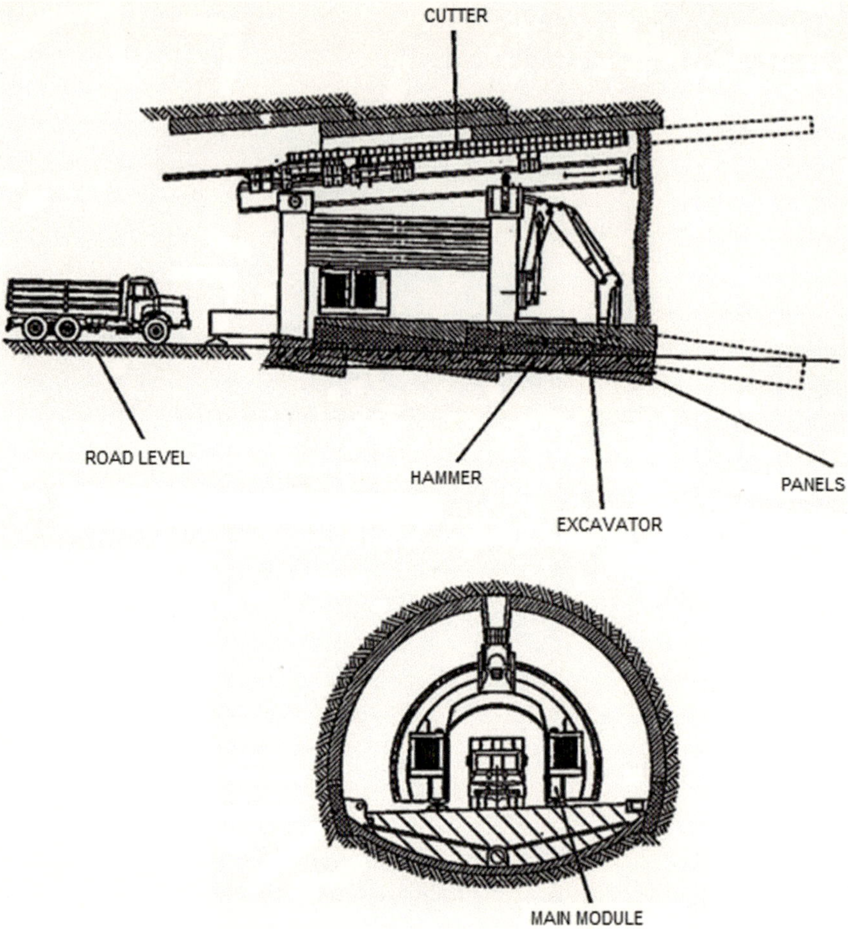

Fig. 6.47 Scheme of pretunnel technology: *side* and *frontal view*. (By Trevi)

at great distance from the face—final lining. The first is aimed at guaranteeing the immediate stabilization and deformation control just after excavation; the latter, usually fitted with a waterproofing system, either full round or limited to the ceiling area, has to guarantee the static and service functionality in the long term.

In particular cases, a single shell or monocoque lining system can be adopted. Also, this one is realized with subsequent layers and possible insertion of a waterproofing layer, and it combines the function of first stage and final lining.

First stage and single-shell lining are mainly realized with simple or reinforced shotcrete layers. Final linings are realized with simple or reinforced concrete.

Finally, reinforced precast concrete linings often combine the functions of first and final linings.

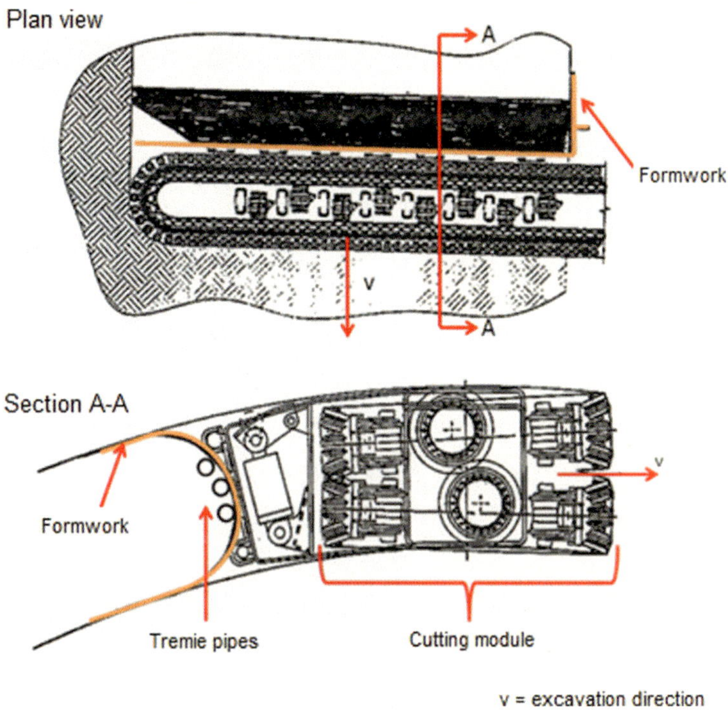

Fig. 6.48 Detail of the cutter head. (By Trevi)

6.9.1 First Stage Linings

6.9.1.1 Shotcrete

Sprayed concrete (shotcrete, Fig. 6.49) is aimed at providing an immediate protection against falls due to decompression of the ground; it also helps in realizing a structure that can offer a confinement pressure and creating an homogeneous surface for the subsequent coat of waterproofing.

Shotcrete is projected directly on the surface to be treated using a high pressure nozzle (final part of the transportation hose). Shotcrete can be projected either wet or dry. The latter methodology is not often used; water is added only at the nozzle; the components of the sand and cement mixture are suspended in the high pressure air flow (*diluted conveying*). If a wet mixture is projected, water is added while concrete is being prepared in a fix or truck mixer. The mixture is transported with the so-called *flow conveying*. It consists in the transportation of a continuous mixture flow by means of pneumatic pressure, pistons and archimedean screws, adding a small quantity of compressed air to melt the mixture.

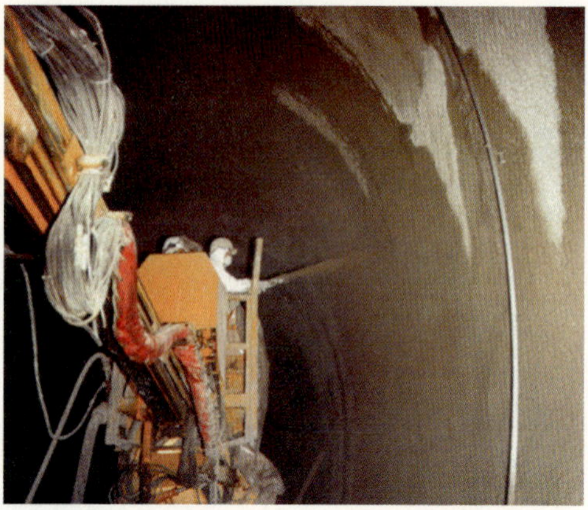

Fig. 6.49 Laying of shotcrete. (By Pizzarotti)

Fig. 6.50 Preparation of steel ribs and electrowelded mesh for the laying of shotcrete. (By Pizzarotti)

Fig. 6.51 Steel profile ribs

Fig. 6.52 Reinforcement in lattice girders. (By Pizzarotti).

Sprayed concrete must have following characteristics:

- Steady mechanical characteristic: homogeneity of the mechanical characteristics of the final product is required; those characteristics must comply with the design prescription.
- Steady ratio water/cement: this is important in order not to affect negatively the mixture with possible consequences on the strength.
- Low rebound (rebound is the percentage of sprayed material that does not adhere and fall and which increases the cost of the work).
- Dust reduction: it is aimed at protecting the health of the workers and the environment.
- High productive performance: the operation velocity influences the cost of the work.
- Good adherence to the rock wall.

These requirements are more easily satisfied with the projection of wet mix; this is the reason why, at present, the wet mix is almost the only one sprayed concrete adopted in tunnels.

Layers of shotcrete are usually 5–20 cm thick. From the reinforcement point of view, shotcrete can be:

- Not reinforced: used to regularize the excavation surface and prevent the contact between rock and atmosphere, usually with limited structural functions.
- Reinforced with different types of metal or synthetic fibres: the presence of fibres reduces the shrinkage and increases the shotcrete strength and deformability; moreover, it allows the substitution of the steel mesh thus accelerating the working process.
- Reinforced with steel mesh: the steel mesh is a reinforcement element of the section offering a higher strength.
- Reinforced with steel ribs or other metal reinforcement elements (described more in detail in the following paragraph, Fig. 6.50).

6.9.1.2 Steel Ribs

Steel ribs are very adaptable reinforcement devices for shotcrete because they adapt to the different excavation shapes. Steel ribs support high convergences (in particular deformable/yielding ribs) and are characterized by good strength characteristics both to compression and to bending; they also provide immediate protection for workers.

Steel Profile Ribs

Ribs can be made with structural steel with UPN, IPN and IPE sections (preferably paired with cross stiffening brackets) or HEA and HEB (Fig. 6.51). Generally, the rib shape is adapted to the excavation by the adequate calendaring of different portions of steel jointed during installation by means of bolted plates. The different steel ribs placed sequentially are connected to one another with steel bars. Their correct operation depends not only on the characteristics of the sectional steel used, but also on those of the accessories such as: flanges, bolts and tie bars, bases, reinforcing sheets etc.

Particular steel profiles with Y section can be used as well, with masses concentrated at the ends of the Y wings to provide moment and compression strength equal to that of traditional steel sections. They allow a better quality of the first-phase lining reducing the vacuum due to the shadow effects of the steel profile wings during the shotcrete projection; moreover, they are lighter than traditional steel beam profiles, without the disadvantage of a lower stiffness during the assembly. Still, they offer a lower cooperation between steel and shotcrete and a lower structure ductility with respect to lattice girders.

Reinforcement in Lattice Girders

Also, reinforcement in lattice girders with triangular or rectangular section, with strength moment and compression strength equal to those of steel profiles, can be used for the reinforcement of first-phase shotcrete lining (Fig. 6.52). With respect to a traditional rib in steel, the lattice girders allow a better interaction between steel and shotcrete, a better quality of first-phase lining reducing the vacuums due to the shadow effect of the steel beam wings during the shotcrete projection, a limitation of fissuring and a higher ductility of the structure. With respect to reinforcement in steel profile, lattice reinforcements are lighter but with the disadvantage of a lower stiffness during the assembly that generally is not welcome by workers who are not used to work with them.

Yielding Ribs

Sliding or yielding ribs are particularly suitable if big deformations are expected. They are made by structural steel with Ω section. Each rib is made by different

Fig. 6.53 Bolted jaws. (by Belloli)

Fig. 6.54 Yielding ribs. (by Belloli)

portions placed on a telescopic arch fitted with suitable joints, the different portions can slide with respect to the others under the action of a well-defined load (Fig. 6.55e). The initial configuration of the steel beams of the different parts of a yielding rib are overlapped for at least 50 cm and blocked by bolted clamps (Fig. 6.53). The driving torque, the dip of the bolts and the number of clamps for each joint define the friction strength between the steel profiles and therefore the minimum load that causes the mutual sliding between the portions of the rib. The sliding can be blocked after reaching a pre-set value placing, for example, a second series of fixed steel profiles paired to the sliding ones, but shorter. In any case, after a certain deformation, the joint tends to have an increasingly higher friction strength due to the change in bending.

Yielding ribs are preferably mounted with a complete round profile (Fig. 6.54), but also with open inverts, and can be coupled with radial nailing and/or a yielding first-phase lining made, for example, by arches of shotcrete fields spaced out by yielding HiDCon elements; these elements can withstand deformation in the measure of 50 % before reaching strength of the same order of those of shotcrete. Shotcrete projection has to be limited to the intervals between steel ribs; otherwise, big deformations would crack it preventing the underground working process.

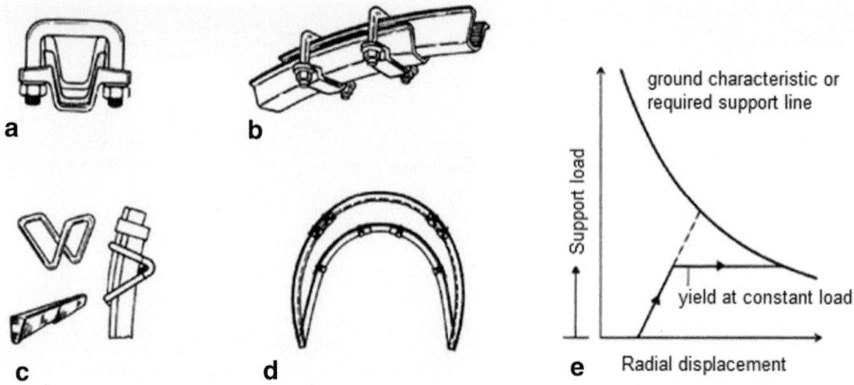

Fig. 6.55 Yielding arch. **a** Cross section. **b** Clamp joint. **c** Alternative joint. **d** Arch configuration before and after yielding. **e** Idealized load-radial displacement response. (Modified from Brady and Brown 2004)

6.9.2 Final Linings

Generally, if the tunnel is built using first-phase linings, they are dimensioned to stabilize the cavity for a certain period, whereas the final lining is aimed at guaranteeing the durability of the work for all its operating life, in the different conditions that it may face. Therefore, it can be stated that the final lining fulfils the following aims:

- Substituting the first-phase lining: considering the aggressive characteristics of the tunnel environment, it can be assumed that either the alteration of the metal components of first-phase stabilization measures, or the alteration of the sprayed concrete or of the stabilization measures would partially or totally jeopardize the strength capabilities of the structural elements.
- Increasing the structural strength: the final lining has to be able to withstand the pressures transmitted by the rock mass during its service life; these can change in time, for example due to creep or squeezing, increase in water pressures or as a result of an earthquake.
- Preserving the work functionality in the long term: the final lining has to limit the cavity convergence so as not to jeopardize the inner spaces of the tunnel that are fundamental for its correct service. The final lining has to remain intact in time. Possible fissures might alter its strength, let the reinforcement corrode, lead to undesired inflows with the resulting alteration of the hydrologic equilibrium of the environment surrounding the work.

6.9.2.1 In Situ Cast Concrete (Unreinforced and Reinforced)

Final linings can be made of in situ cast concrete, either unreinforced or reinforced. The lining shell is created by means of a formwork car. In general, the cast field can

Fig. 6.56 Casting phase of the invert. (By Pizzarotti)

Fig. 6.57 Casting of the vault. (By Pizzarotti)

reach a length equal to the tunnel diameter (but also much longer, as in the case of long and small hydraulic tunnels), Usually, the cast is made in the following steps: at first, the invert is built (where it exists, Fig. 6.56) and then the vault foundations; after that, the vault is cast (piers and crown, Fig. 6.57).

6.9.2.2 Waterproofing and Water Management Systems

The waterproofing of the final lining is carried out on the first phase lining intrados and has to be made before the casting. It is aimed at avoiding (if it is complete on the whole perimeter) or limiting (if it is limited to the vault) inflows (Fig. 6.58) creating a waterproof barrier, usually made by a membrane of PVC or other plastic materials, externally paired to a draining layer and fixed to the vault with suitable anchor systems (Fig. 6.60).

If the waterproofing is partial, the bottom of the waterproof layer is connected to a draining system for intercepted water (Fig. 6.59) to allow its discharge.

Fig. 6.58 Damages caused by inflows in an operating road tunnel

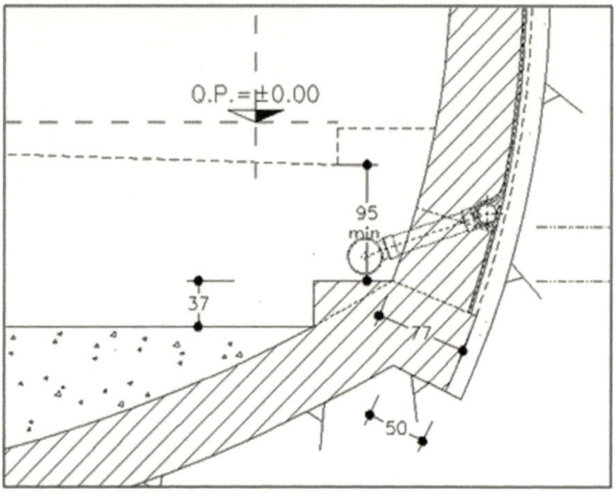

Fig. 6.59 Control of ground mass water in the operation phase: waterproofing with a waterproof membrane linked to its drainage

In some applications, where the laying of the waterproof sheet is difficult due to a particularly complex shape of the walls (cavities intersection or change in the dimensions) sprayed waterproofing membranes can be used.

When the problem is represented by the seal on the inside of a hydraulic tunnel (e.g. hydraulic pressured tunnel), waterproofing can be carried out by means of steel sheets and contact injections (Fig. 6.61).

6.9.2.3 Prefabricated Linings

Prefab Segment Rings for Shielded and Pressurized TBM

Lining in tunnels bored with shielded TMB is realized by precast segments in reinforced concrete. They are placed by means of mechanic devices and constitute a close ring that has the double aim of confining the excavation and hindering the advancing thrust of the machines.

Fig. 6.60 Waterproofing membranes

The precast segments must have adequate characteristics to withstand longitudinal stresses due to TBM thrust and transversal stresses due to the ground and water pressures.

The implementation of the precast segments takes place inside the shield; therefore, their extrados diameter is smaller than the inner diameter of the shield tail. The gap between segment extrados and excavation perimeter is filled as the segments come out of the shield when the TBM advances, in order to avoid soil deformations and material falls on the lining contour that would imply non-homogeneous load and ground reaction. The filling can be performed with the following methods:

- Injections of pea-gravel, later clogged with cement mix through valves placed in the segments, for non-pressurized TBM
- Injections of cement mortar by means of nozzles placed in the tail of the shield or valves placed in the segments, for pressurized TBM

If pressurized TBMs are used, injections are performed with considerable pressures, thanks to the hydraulic seal devices placed on the tail of the shield between the segment and the shield; they allow an effective and immediate confinement action of the cavity (Fig. 6.62).

Usually, a ring is made by 5–7 segments (Fig. 6.63); the way they are divided, the shape and the dimensions of each segment depend on the characteristics of the

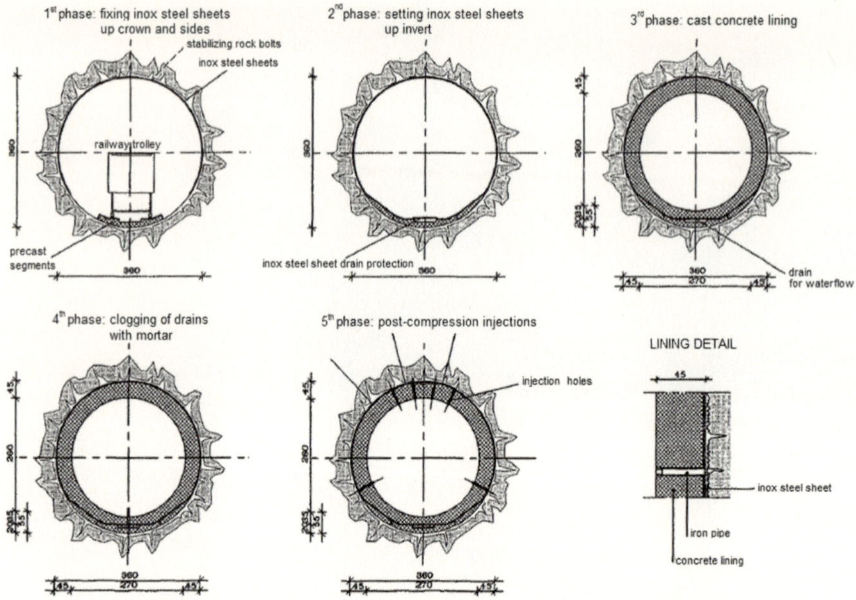

Fig. 6.61 Steel sheets, injection and waterproofing for an hydraulic pressurised tunnel

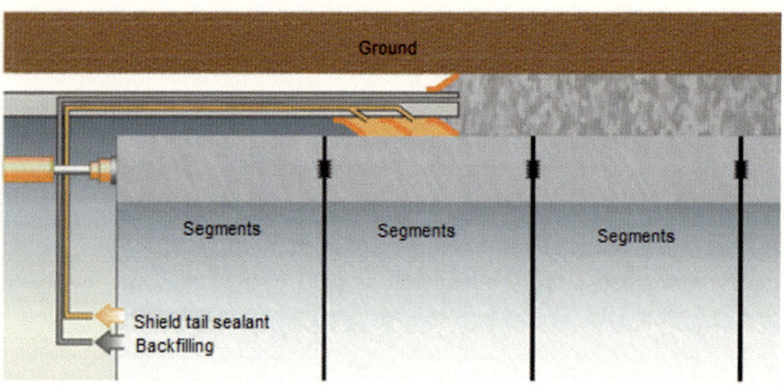

Fig. 6.62 By Hydraulic seal system on the tail of the shield. (By BASF, modified)

tunnel to be constructed. For example, a tapered geometry of the rings allow to follow the bending of the tunnel layout and compensate possible small deviations from the axis of the design layout. For double-shield non-pressurised TBM special hexagonal segments can be used in order to make possible the excavation and the advancement of the machine while erecting the ring, reducing, in that way, the operational time.

To provide a waterproof lining, precast segments are equipped with gaskets along the whole perimeter (Fig. 6.64). Moreover, the segments must be cast with tiny dimension tolerances (a tenth of a millimetre).

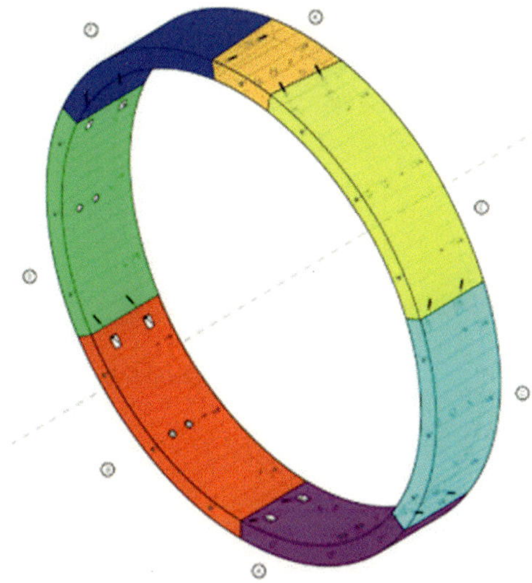

Fig. 6.63 Ring of precast segments for TBM

Fig. 6.64 Detail of the waterproofing gaskets of precast segments

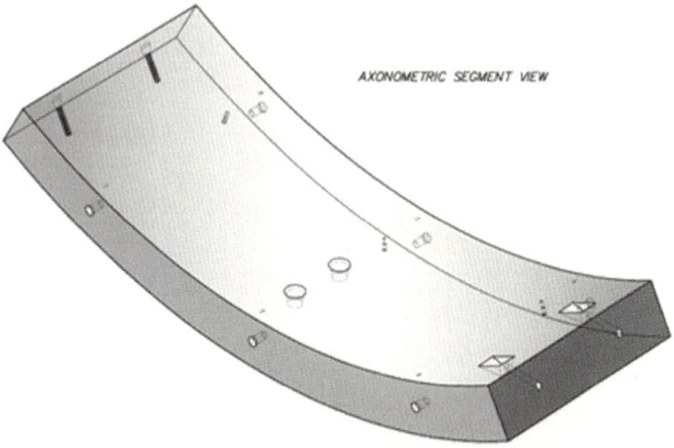

Fig. 6.65 Precast segment; notice the slots for installation of the longitudinal and transversal connections, the erector holes placed in the centre of the segment allow to move and install it

Fig. 6.66 Tunnel lined with precast segments; notice the slots for longitudinal and transversal bolts. (By Pizzarotti)

Within the same ring, the structural continuity of the lining is achieved by connections made by steel bars or bolts, whereas adjacent rings can be connected both with bars or dowels (Figs. 6.65 and 6.66).

Interesting applications of a precast lining system called "articulated vault" have been carried out in large traditional excavations.

Articulated Vaults

In the "articulated vault" system the precast segments are installed at the face, and a dedicated jack system puts them in full immediate action. These elements create a lining that is both first phase and final. The system was successfully applied in the construction of large section tunnels (railway/underground stations or caverns). Usually, it implies the previous realization of the posts of the articulated vault, as in the "traditional" method, with portioning of the excavation if required (Figs. 6.67 and 6.68).

The use of articulated vault allows to make the most of the main advantages of precast elements, among which are quick installation and good quality of the

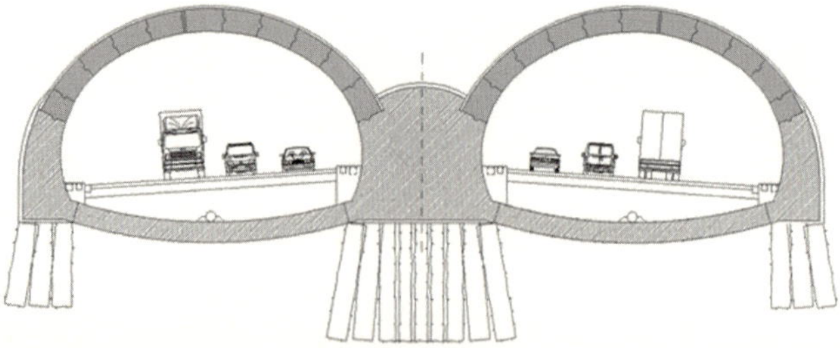

Fig. 6.67 Articulated vault realized with precast segments—cross section

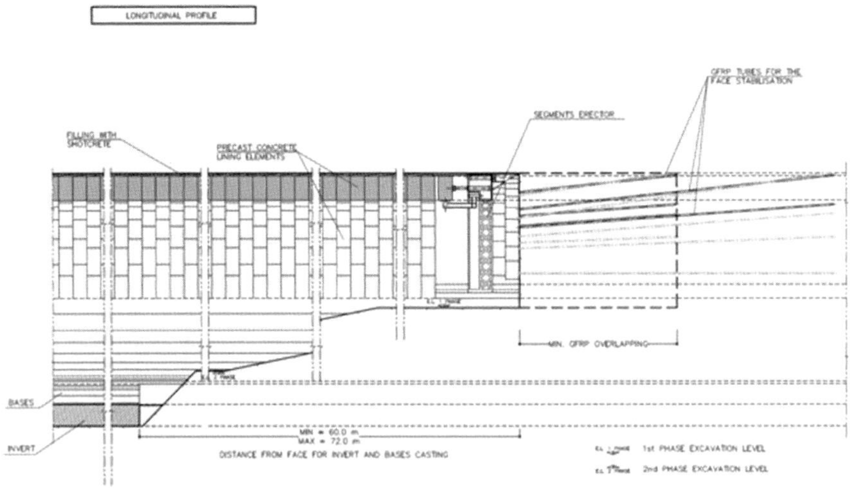

Fig. 6.68 Articulated vault realized with precast segments—longitudinal profile

conglomerate, even when the excavation is performed with traditional techniques. The lining waterproofing is obtained by means of gaskets placed on the segments contour, as described in the previous paragraphs for precast lining in mechanical excavations.

6.9.2.4 Single-Shell (Monocoque) Linings

As previously discussed, usually, the lining with shotcrete is considered a temporary first-phase lining, while it is assumed that the final lining will be implemented in the long term; the two linings are usually divided by the waterproof membrane. On the contrary, the single-shell solution implies the use of shotcrete as final lining that completes or substitutes the structural functions of the first-phase support (Fig. 6.69).

The thickness of the different layers and the suitable solutions (with or without interposed waterproofing) apt to guarantee the lining functionality are chosen according to the presence of water and the loads to be faced.

The main difficulty is to obtain a monolithic behaviour of the different layers and to guarantee at the same time a correct waterproofing of the structure. This technical-operational aspect can be overcome, for example, by realizing sprayed waterproof membranes that offer good shear strength performance. In this way, it is possible to create a composite lining in which the two structural layers (inner and outer) work together because the membrane allows the transmission of normal and shear stresses.

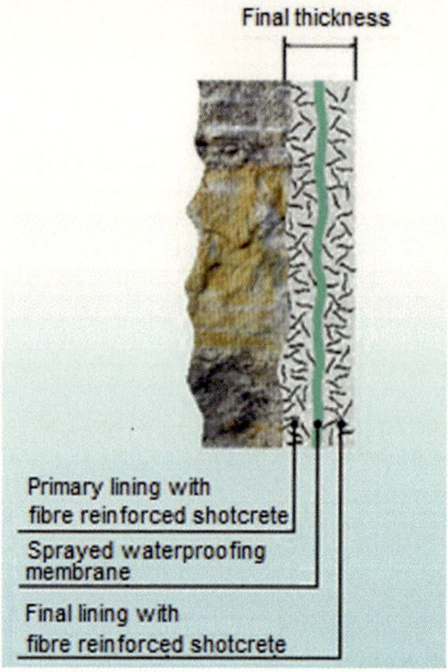

Fig. 6.69 Single-shell lining. (Modified from BASF 2013)

References

Brady BHG, Brown ET (2004) Rock mechanics for underground mining, 3rd edn. Kluwer Academic Publishers, Dordrecht

Kovári K, Fechtig R (1996) Trafori alpini storici in Svizzera. S. Gottardo Sempione Lötschberg, Società per l'arte delle costruzioni. Stäubli AG, Zurigo

Lombardi G, Deere D (1993) "Grouting design and control using the GIN principle". Water Power and Dam Construction, England

http://www.basf.com

www.bauer.de

http://www.belloli.ch

http://www.dywidag-systems.at/en

http://www.herrenknecht.com

Chapter 7
Ground-Structure Interaction

As already described in the previous chapters, the building of a tunnel consists of the progressive removal of the ground and the implementation of stabilization and lining measures ("structure" in following paragraphs) in different phases and with different procedures. The problem of interaction structure-ground is highly hyperstatic and its solution depends on the initial stress field, the characteristics of the ground and of the lining, as well as on the building procedures and time of the work. The 3D aspect of the problem has to be taken into account through emipirical-analitycal asscssments (see Chap. 4, Sect. 3) or numerical methods (see Chap. 7, Sect. 6). In the latter case, two different procedures can be adopted: the creation of a 3D numerical model or the use of a simplified modelling that combines axisymmetric and plane numerical analysis.

Many authors studied the interaction problem between ground and structure, developing more or less simplified methods. The leading ones are presented in the following paragraphs.

7.1 Rabcewicz Theory

Rabcewicz method (1964) allows to take into account the cooperation between the rock ring contouring the excavation and the lining and supporting measures of the cavity.

It can be considered as a limit equilibrium method because it is based on the simple comparison between equilibrium pressures and strength of the complex constituted by rock ring and support measures. Therefore, it has to be necessarily associated to the monitoring of the deformations during the excavation to verify that the support measures can suitably limit them.

It is an old and rarely used method since numerical analysis became more widespread, but it can provide useful preliminary information. It is based on the assumption that, during the redistribution of stresses induced into the rock mass by the opening of a cavity (Fig. 7.1), rock at the sides of the cavity tend to displace along the surfaces of wedges where shear strength is overcome.

The calculation methodology suggested by Rabcewicz and Golser (1973) allows to carry out a check of the stabilization devices of those wedges (Fig. 7.2). Actually, the

P. Gattinoni et al., *Engineering Geology for Underground Works*,
DOI 10.1007/978-94-007-7850-4_7,

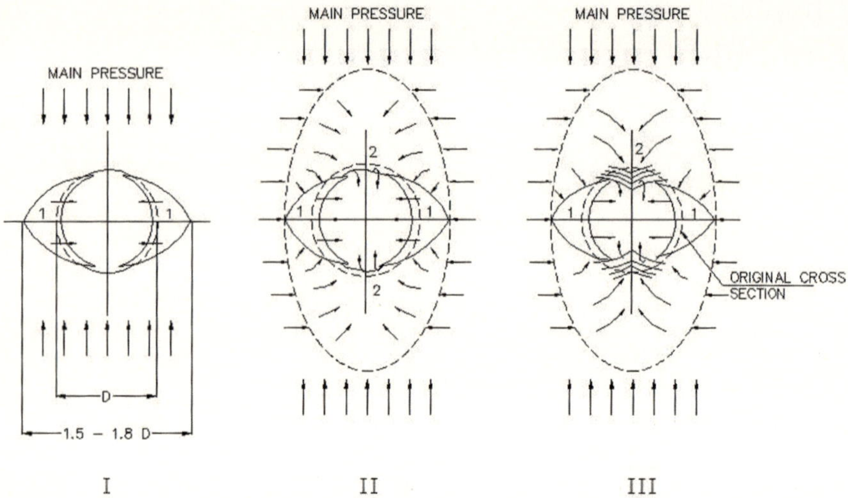

Fig. 7.1 Schematic trend of the failure phases of a cavity contour due to pressure redistribution. **a** Forming shear wedge. **b** Stress increase in crown and invert. **c** Roof Crown failure. (Rabcewicz 1964)

strength offered by the complex "rock- stabilization devices" to oppose to the wedge movement towards the cavity is calculated, assuming for the shear strength of the materials (rock on the contour and lining) Coulomb linear relations ($\tau = c + \sigma \tan \phi$) with adequate values of c and ϕ for each material.

The strength offered by the first-phase lining in each component (shotcrete, mesh and/or steel ribs reinforcement and radial bolting) is calculated. These contributions are summed to the one offered by the cooperating rock ring, thus giving the total strength. Its ratio with the equilibrium radial pressure (that can be calculated in different ways, as described in the following paragraphs) provides the safety coefficient of the structure under construction.

7.2 Method of Hyperstatic Reactions

The method of hyperstatic reactions allows to study the behaviour of lining under the action of external loads, keeping into account the presence of the surrounding ground through a series of elastic or elastoplatic elements reacting only under compressive stress.

In particular, two kinds of loads can act on the lining of a tunnel:

- Active loads: vertical and horizontal loads due to the weight of the rock/soil, the weight of the lining, surface loads, seismic stresses, hydraulic loads, thermal expansions, etc.
- Passive loads: hyperstatic reactions of the ground.

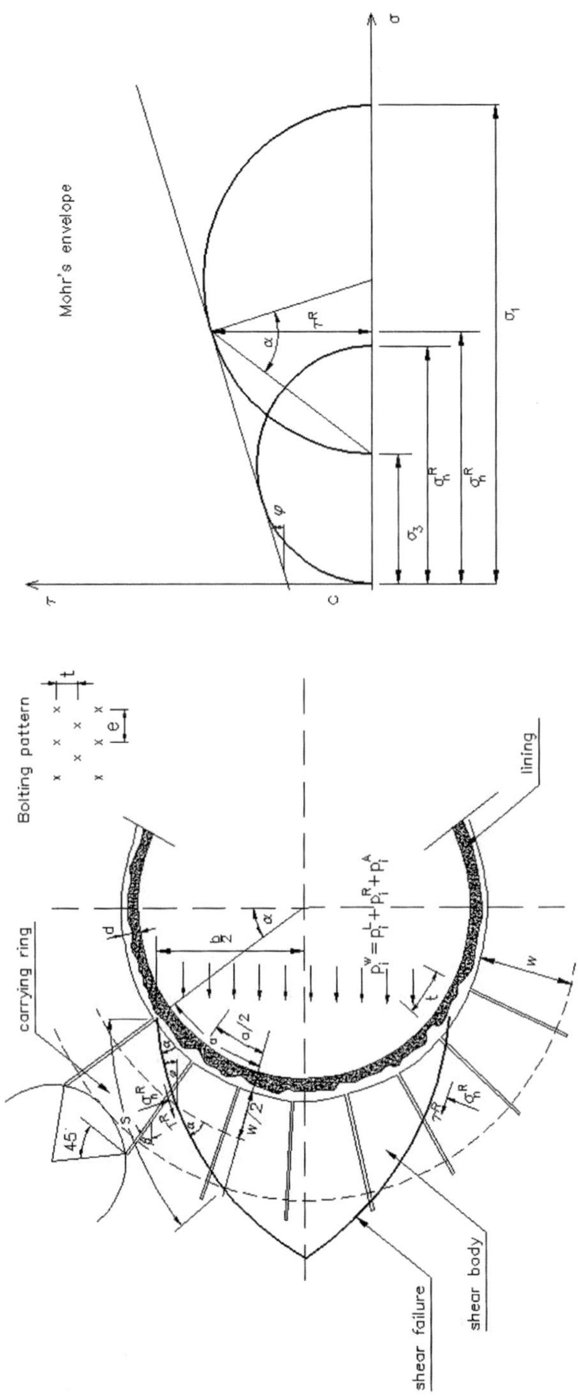

Fig. 7.2 Calculation scheme with the Rabcewicz method. (1964)

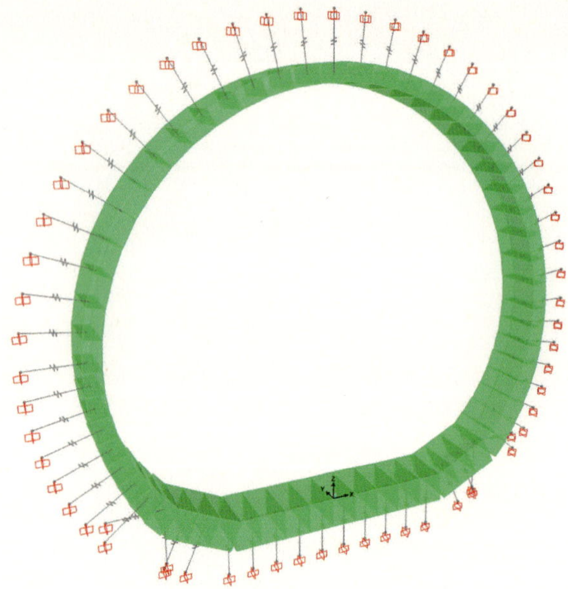

Fig. 7.3 Model of the cross section spring type

The lining deformations are obtained through the structural analysis of the cross section (Fig. 7.3) carried out with the classical methods of the structural mechanics (assimilating the lining to an arch structure, a shell or similar structures) or, nowadays more often, using finite elements numerical models. In the latter case, the structure is discretized through finite elements as beams, of suitable length, in order to obtain a sufficient accuracy of the bent components of the cross section. The thickness of the beams varies according to the structural element (invert, piers, ceiling).

The ground deformations are obtained assuming that the ground has a reaction module, e.g. assuming that the reaction in a point only depends on the deformation of the same point and that it is proportional to that deformation. From the practical point of view, this implies the use of link elements (only acting under compressive stress) placed in correspondence to the nodes of the model, capable of transmitting to the structure a reaction equal to the contact pressure ground-structure. Usually, in the hypothesis of plane deformations, the setting of the stiffness parameters of the links is performed using Boussinesq (posts) and Galerkin formulations (invert and ceiling).

According to those authors, the reaction module of the ground (k) is obtained applying following formulas:

Boussinesq $k_B = \frac{E'}{(1-\nu^2)\cdot B \cdot C_d}$
Being:

E ground Young's module
ν Poisson's coefficient
C_d shape coefficient of the equivalent foundation, function of the geometrical characteristics of the base L and B (respectively length and base of the foundation)

Galerkin $k_G = \frac{E'}{(1+\nu)\cdot R_{eq}}$

R_{eq} being the equivalent radius of curvature

Finally, the reaction that the link can transmit is obtained by multiplying each reaction module of the ground by the influence surface of the node. Therefore, it is possible to assess the stresses acting on the lining.

The hyperstatic reaction method allows to simplify the interaction problem between the structure and the surrounding ground. All ground-dependent factors are transformed in a series of loads on the structure (loads, thrusts, hyperstatic reactions of the springs). Therefore, the validity of the method depends on the possibility to describe adequately the following elements:

- The initial stress state
- The mechanical and deformation characteristics of the ground
- The influence of the different execution phases of the tunnel
- The kind of contact ground-lining

The method undergoes relevant approximations concerning the definition of the reaction module of the ground and the estimate of the behaviour of the interface ground-structure, as well as the definition of the loads. In the design practice, it can be used for the dimensioning of final linings, as an alternative to more complex methods as the numerical ones.

7.3 Evaluation of the Loads Acting on the Linings

There are simplified methods to assess the loads acting on the lining that do not imply the analysis of the excavation phases and a rigorous modelling of the interaction ground-structure (for example the Rabcewicz's method and the method of the hyperstatic reactions). In this case, the pressures to be balanced by the structure must be assessed. Some authors (Ritter 1879; Kommerell 1940) suggest that the vertical load on the linings is generated by an elliptical or parabolic solid (Fig. 7.4). Other authors (Caquot and Kerisel 1956; Terzaghi 1946) adopt different load distributions.

7.3.1 Vertical Loads

7.3.1.1 Soils: Caquot and Kerisel's (1956) and Terzaghi's (1946) Formulations

Considering a round excavation having radius r at an axial depth H (Fig. 7.5), the radial (P_r) and tangential (P_t) pressures on the linings can be calculated using following equations by Caquot and Kerisel (1956), valid for cohesionless soils:

$$P_r = \gamma \cdot r(1 - \cos\vartheta) + \frac{2 \cdot H}{k_p - 2}\left[\frac{r}{H} - \left(\frac{r}{H}\right)^{k_p - 1}\right]$$

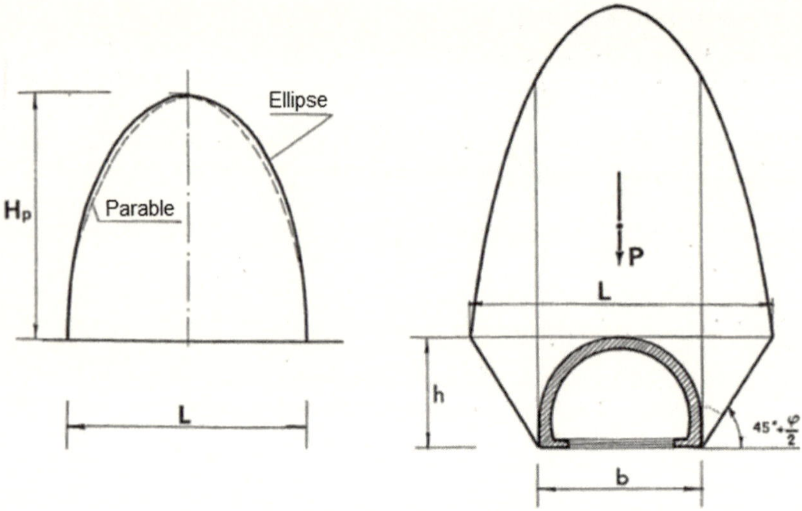

Fig. 7.4 Shape of the detensioned mass after excavation. (Ritter 1879; Kommerel 1940).

Fig. 7.5 Calculation scheme of the vertical load on the ceiling. (Caquot and Kerisel 1956)

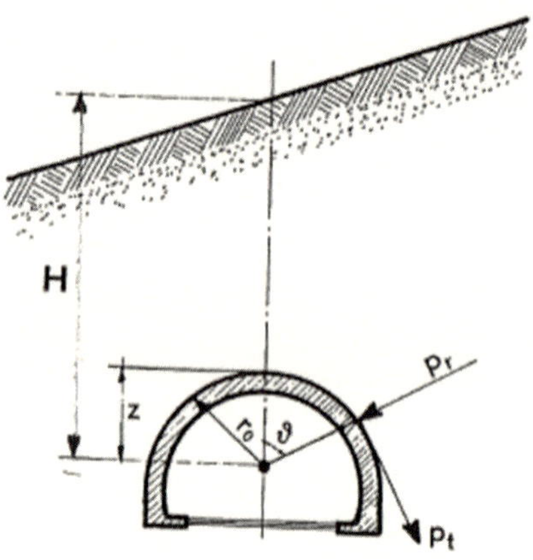

In ceiling $\vartheta = 0$, therefore the vertical load is given by:

$$p_v = \frac{\gamma \cdot H}{k_p - 2}\left[\frac{r}{H} - \left(\frac{r}{H}\right)^{k_p - 1}\right]$$

where the passive thrust coefficient in given by:

$$k_p = tg^2\,(45^\circ + \phi/2)$$

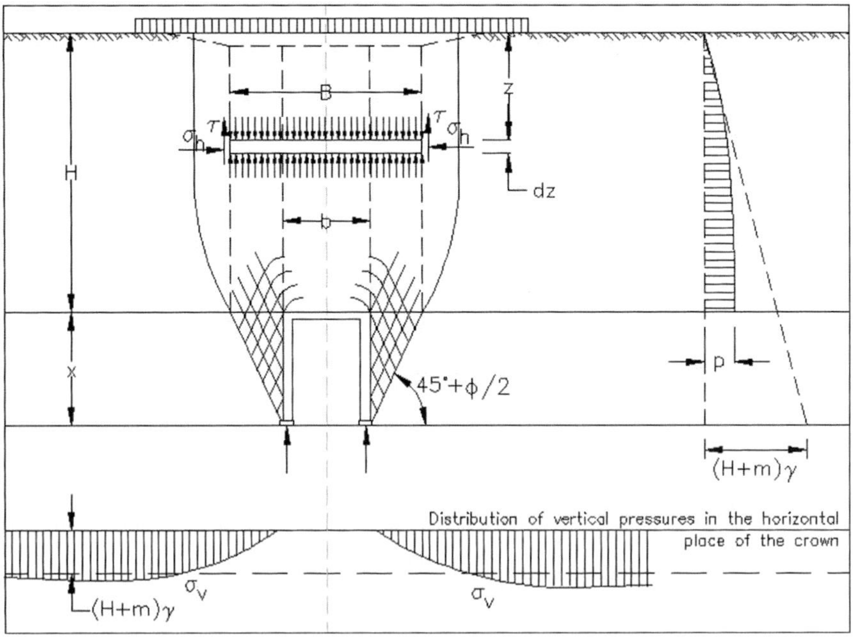

Fig. 7.6 Basic assumptions of Terzaghi's rock pressure theory

The equation for the vertical pressure on the ceiling proposed by Caquot and Kerisel (1956) can be rewritten for cohesive soil obtaining:

$$p_v = \frac{\gamma \cdot H}{k_p - 2}\left[\frac{r}{H} - \left(\frac{r}{H}\right)^{k_p - 1}\right] - \frac{c}{tg\phi}\left[1 - \left(\frac{r}{H}\right)^{k_p - 1}\right]$$

Terzaghi's formulation (1940) for soils takes into account the friction strength developing along the side surfaces of the load prism having height H (Fig. 7.6) and the possible cohesion c.

Considering that the trend of vertical stress is influenced by the presence of the tunnel in a stretch H_2, (Fig. 7.7) whereas the trend is litostatic above the tunnel, the author provides following general equation:

$$p_v = \frac{B\left(\gamma - \frac{2c}{B}\right)}{2 \cdot K \cdot tg\phi}\left(1 - e^{-K \cdot tg\phi \cdot \frac{2 \cdot H_2}{B}}\right) + \gamma \cdot H_1 \cdot e^{-K \cdot tg\phi \cdot \frac{2 \cdot H_2}{B}}$$

where

$$B = b + 2 \cdot H \cdot tg(45° - \phi/2)$$

whereas K is an experimental coefficient almost equal to 1.

Please note that:

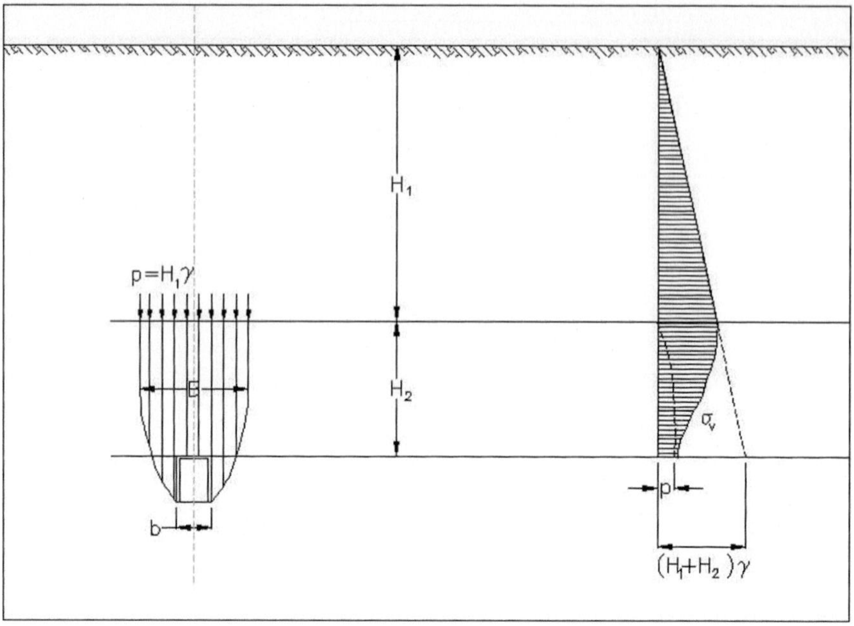

Fig. 7.7 Rock pressures at greater depths (Terzaghi 1946)

- For $H_2 > 2.5B$ there is no longer shear displacement on vertical planes, therefore, the load $H_1\gamma$ completely unloads on H_2.
- Until $H_1 < 4H_2 = 10B$, the exponential terms arenegligible, whereas K can be considered equal to 1.

Comparing Cauqot and Kerisel's (1956) and Terzaghi's (1946) formulations, it can be noted that, at great depth, both provide a solution independent from the depth. In particular, p_v equations are simplified. Considering a case in which cohesion is negligible, it is obtained:

$p_v = \frac{\gamma \cdot r}{k_p - 2}$ Caquot and Kerisel's (1956)

$p_v = \frac{\gamma \cdot B}{2 \cdot K \cdot tg\phi}$ Terzaghi's (1946)

In general, Caquot and Kerisel's (1956) equation leads to determining lighter loads than those obtained with Terzaghi's (1946). It can be noted that the contribution of cohesion is more relevant in Caquot and Kerisel's (1956) formulation. Actually, with quite high c values, the load on the ceiling is cancelled out, whereas it is always present if Terzaghi's (1946) formulation is used. Numerical calculations gave evidences that, between the two theories, Caquot–Kerisel's leads to results closer to reality.

Table 7.1 Terzaghi's rock load H_p in metres of rock on tunnel crown with width B (m) and height H_t (m) at depth of more than $1.5(B + H_t)$

Rock condition		RQD (%)	Rock load H_p (m)	Remarks
1	Hard and intact	95–100	Zero	Light lining, required only if spalling or popping occurs
2	Hard stratified or schistose	90–99	0–0.5B	Light support, mainly for protection against spalls. Load may change erratically from point to point
3	Massive, moderately jointed	85–95	0–0.25B	
4	Moderately blocky and seamy	75–85	0.25B–0.35$(B + H_t)$	No side pressure
5	Very blocky and seamy	30–75	(0.35–1.1)$(B + H_t)$	Little or no side pressure
6	Completely crushed	3–30	1.10$(B + H_t)$	Considerable side pressure. Softening effects of seepage toward bottom of the tunnel require either continuous support for lower ends of ribs or circular ribs
7	Squeezing rock, moderate depth	NA	(1.10–2.10)$(B + H_t)$	Heavy side pressure, invert struts required; circular ribs are recommended
8	Squeezing rock, great depth	NA	(2.10–4.50)$(B + H_t)$	
9	Swelling rock	NA	Up to 75 m irrespective to value of $(B + H_t)$	Circular ribs are required. In extreme cases use yielding supports

7.3.1.2 Rock masses: Terzaghi's (1946) Classification and Approaches Based on Bieniawski's Characterization

Terzaghi (1946) developed a rock classification (Table 7.1) that provides a load height H_p according to the conditions of the rock itself (with references to the geometric magnitudes schematized in Fig. 7.8)

Thanks to this elaboration, the vertical load on the ceiling can be obtained as:

$$p_v = H_p \cdot \gamma$$

More generally, that relation can be expressed as:

$$p_v = \alpha(B + H) \cdot \gamma$$

where

α is a multiplying coefficient that depends on rock conditions (Table 7.1)
H depth of the tunnel
B width of the tunnel

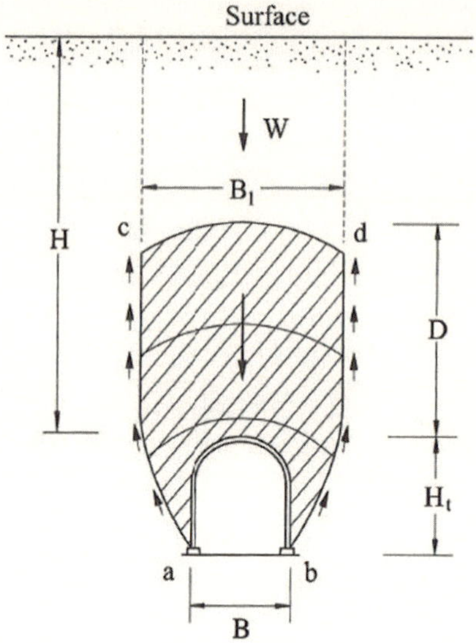

Fig. 7.8 Calculation scheme of the vertical pressure on the ceiling. (Terzaghi 1946)

The crown of the tunnel is assumed to be located below the water table. If it is located permanently above the water table, the values given for types 4, 5 and 6 can be reduced by 50 %.

Approaches based on Bieniawski's characterization The pressure on the crown can be evaluated using RMR. The following equation proposed by Unal (1983) was developed on the basis of coal mines studies for openings with a flat crown:

$$p_v = \gamma \cdot b \cdot \left(\frac{100 - RMR}{100} \right)$$

where b is the tunnel width and it can be assumed that $RMR = GSI$ or $BRMR$ (see Sect. 3.4).

A similar formulation was proposed by Goel and Jethwa (1991) based on pressure values measured in 30 monitored Indian tunnels:

$$p_v = \frac{7.5B^{0.1} \cdot H^{0.5} - RMR}{20RMR} [MPa]$$

where

B opening span in metres
H overburden in metres

According to the authors, the latter equation is valid for arched underground openings in both squeezing and non-squeezing ground conditions (but not in the case of rock bust), in tunneling by conventional blasting methods using steel rib supports for $50\,\text{m} < H < 600\,\text{m}$.

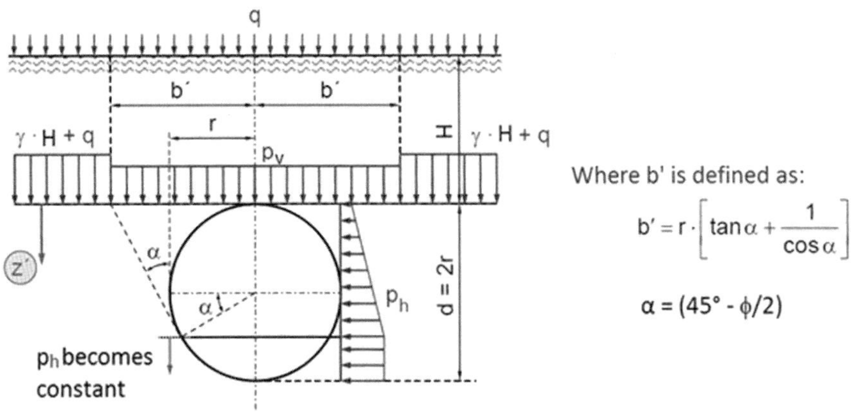

Fig. 7.9 Calculation scheme of the pressure on the lining according to Terzaghi.

7.3.2 *Horizontal Loads*

According to Terzaghi (1951), the deformations developing in deep tunnels are so important that they lead to an active thrust. Therefore, the horizontal load can be expressed by the following equation:

$$p_h = (p_{v,z=H} + \gamma \cdot z') \cdot k_a - 2 \cdot c \cdot tg(45 - \phi/2)$$

k_a being the active thrust coefficient:

$$k_a = \mathrm{tg}^2(45 - \phi'/2)$$

and where

$P_{v,z=H}$ the vertical load at the ceiling
z' vertical axis where zero is at the ceiling and the same direction as the gravity
c cohesion
ϕ friction angle
H overburden

This situation is schematized in Figure 7.9.

Terzaghi's assumption to consider the active thrust conditions was criticized by many authors. Houska (1960), for example, proposed a modification of Terzaghi's method on the basis of experimental results obtained during the construction of the Donnerbühl tunnel (Bern, Switzerland). In particular, Houska (1960) points out that the lateral pressures on the tunnel should be calculated according to the scheme in Figure 7.10.

According to Houska (1960), the lateral pressures can be obtained redistributing part of the load p_v acting on the ceiling on the sides (Fig. 7.10). This implies an increase of the lateral pressures with respect to Terzaghi's solution.

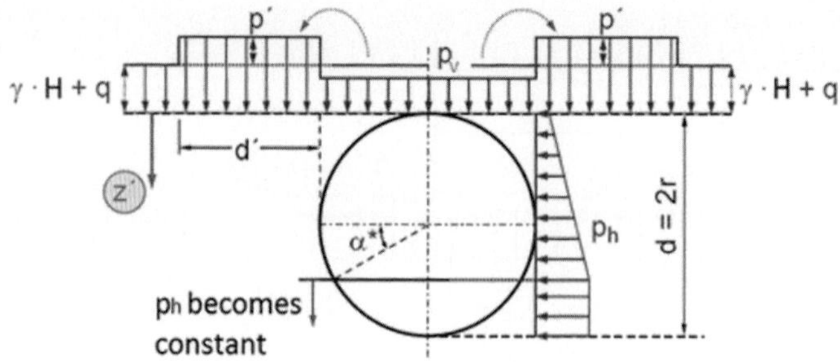

Fig. 7.10 Calculation scheme of the pressure on the lining according to Houska.

From the formal point of view, Houska's (1960) solution is very similar to Terzaghi's and it is based on the assumption to consider only two-thirds of the shear strength:

$$\mathrm{tg}\phi^* = \frac{2}{3}\mathrm{tg}\phi$$

$$c^* = \frac{2}{3}\mathrm{tg}\phi$$

Starting from this hypothesis, Terzaghi's equation of vertical and horizontal loads according to ϕ^* and c^* can be rewritten.

$$p_v = \frac{B\left(\gamma - \frac{2c^*}{B}\right)}{2 \cdot K \cdot tg\phi^*}\left(1 - e^{-K \cdot tg\phi^* \cdot \frac{2 \cdot H_2}{B}}\right) + \gamma \cdot H_1 \cdot e^{-K \cdot tg\phi^* \cdot \frac{2 \cdot H_2}{B}}$$

$$p_h = \left(\gamma \cdot H + q + p' + \gamma \cdot z'\right) \cdot k_a{}^* - 2 \cdot c^* \cdot tg\,(45 - \phi^*/2)$$

Where p' is the "redistributed" load equal to (considering the shape defined in Fig. 7.11):

$$p' = \left(\gamma \cdot H + q - p_{v,z=H}\right) \cdot \frac{r}{d'}$$

7.3.3 Inclined Loads

In case of asymmetrical loads, for example, if homogeneous bedding planes with dip *i*, the load semiellipse or parabola, is inclined in the dip direction of the stratification (Fig. 7.12 a) and the thrusts can be broken up in their vertical and horizontal components. In that case, the height of the load H_p can be obtained through the previously discussed Terzaghi's or Kommerll's methods, whereas the width L is given, according to dip i, by following equations:

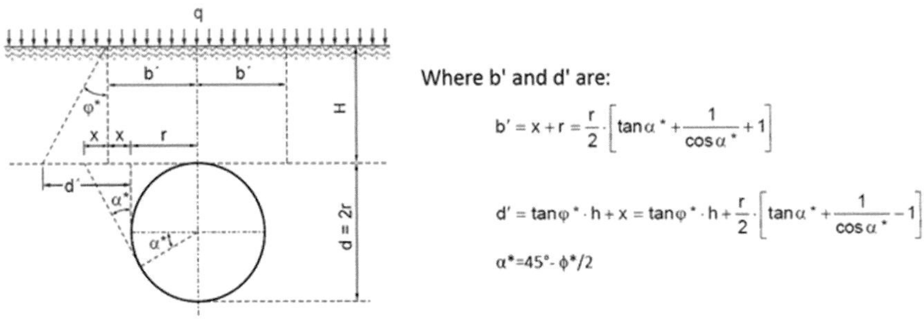

Fig. 7.11 Problem geometry according to Houska (1960). (Angerer and Fillibeck 2012)

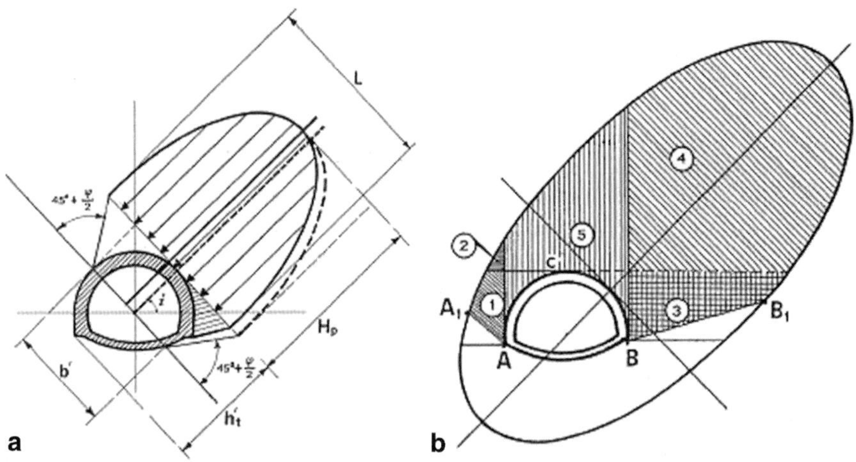

Fig. 7.12 Calculation scheme of the loads on the tunnel in case of inclined loads. **a** Identification of the load semiellipse. **b** Definition of the loads on the tunnel

$$L = b' + h_t' \cdot tg\left(45^\circ - \frac{\phi}{2}\right)$$

Where b', h' are respectively the width and height of the tunnel measured perpendicularly and parallelly to the strike of the thrust (Fig. 7.12a) and both depend on i.

The two displacement surfaces obtained geometrically (stretch A-A_1 and B-B_1) can be used to define the active thrusts (1 and 3 in Fig. 7.12b), their overloads (2 and 4 in Fig. 7.12b) and the vertical load (5 in Fig. 7.12b) on the tunnel.

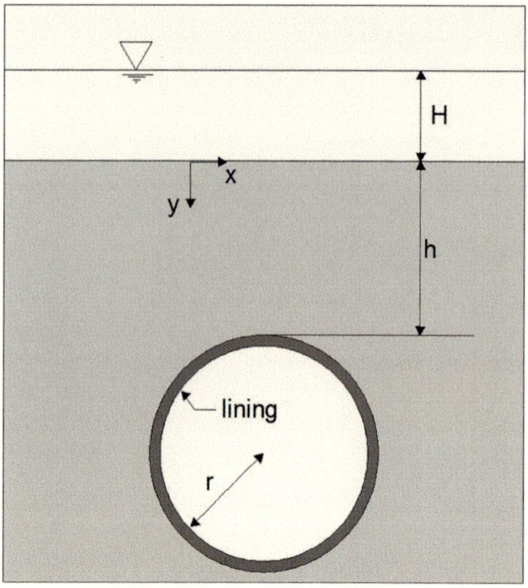

Fig. 7.13 Scheme of vertical load on the ceiling due to the presence of groundwater. (Kolymbas 2007)

7.3.4 *Loads Assessment on the Lining in Case of Tunnel Under Groundwater Table*

Generally, the lining of a tunnel under the water table can be:

- Waterproof (sealed tunnel): this implies the application of the whole hydrostatic load on the lining and the absence of seepage motions;
- Drained (drained tunnel): the lining undergoes a lower load than the hydrostatic one; water pressure might be null.

Several authors provided simplified solutions to calculate the pressure generated on the lining by the presence of groundwater, neglecting the analysis of the interaction ground-structure (see Sect. 4.4).

For example, referring to the shape represented in Figure 7.13, Kolymbas (2007) provides following values of the load on the ceiling:

- For undrained tunnel:

$$p_v = h\frac{\gamma' - \frac{c}{r}\left(\frac{\cos\phi}{1-\sin\phi}\right)}{1+\frac{h}{r}\left(\frac{\sin\phi}{1-\sin\phi}\right)} + \gamma_w\,(H+h)$$

the second term of this equation represents the water pressure.

- In case of a drained tunnel, seepage stress will have to be taken into consideration; assuming a linear distribution of the hydraulic gradient, the pressures becomes:

$$p_v = h\frac{\gamma' - \frac{c}{r}\left(\frac{\cos\phi}{1-\sin\phi}\right)}{1+\frac{h}{r}\left(\frac{\sin\phi}{1-\sin\phi}\right)} + \gamma_w\frac{H+h}{1+\frac{h}{r}\left(\frac{\sin\phi}{1-\sin\phi}\right)}$$

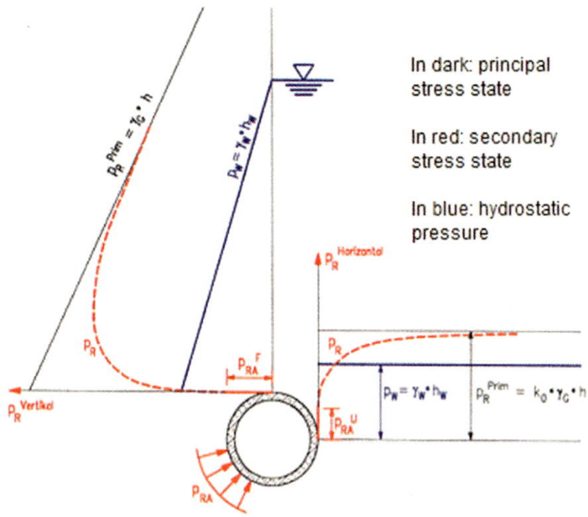

Fig. 7.14 Scheme of horizontal and vertical stresses around the cavity: the primary stress state in *black*, the secondary stress state in *red*, the hydrostatic pressure in *blue*

On the contrary of what is generally assumed, through the last equation the authors also demonstrate that the introduction of a drainage does not allow to annul the pressure of groundwater on the tunnel. In reality, the pressure of groundwater influences the lining through seepage stresses.

In case of shallow tunnels in soft soil, it is widely accepted that the pressure on the lining has to be assessed as the overlapping of pressures due to the effective load of the soil and those due to the water. For deep tunnels in rock, some authors demonstrated that the presence of water can be neglected if its pressure does not overcome the weight of the rock above the tunnel (Schuck 2005). The latter approach is meaningful considering that the increase in the pressure on the ceiling means thicker linings or/and reinforcements, and therefore the work will be more expensive.

The assumptions listed below were considered by Schuck (2005) to assess the influence of the presence of groundwater on the dimensioning of tunnels in rocks:

- The intact rock is impermeable.
- The water only flows along fractures and discontinuities.
- The tunnel is deep (there are no failure mechanisms involving the surface and the stress on the ceiling are almost equal to the stresses on the sides).
- The initial position of the groundwater table is defined with height h_w with respect to the ceiling and it has a hydrostatic trend.

Under these assumptions, the distribution of the stresses around the tunnel can be schematized as in Figure 7.14, where P_{RA} (marked with F or U) is the pressure of the rock acting on the lining after its implementation. The rock surrounding the cavity undergoes compression and the radial pressure P_{R} increases with the distance from the lining up to a value corresponding to the initial $P_{\mathrm{R}}^{\mathrm{PRIM}}$.

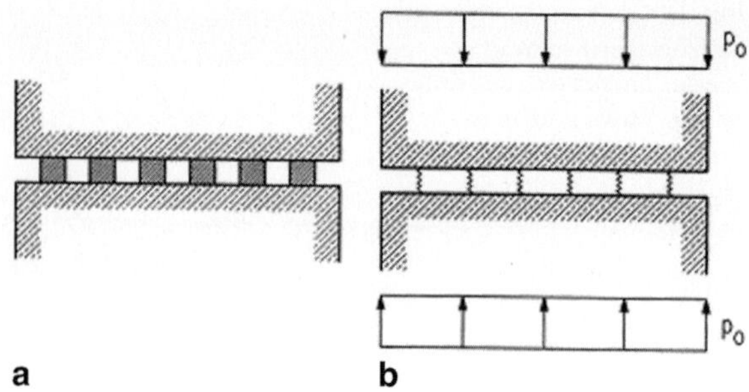

Fig. 7.15 Schematization of the rock mass according to Schuck (2005). **a** Rock bridges. **b** Schematization of rock bridges through springs

The author assumes that, in rocks, pressures are transmitted along the "bridges" of intact rock. This situation can be modelled through two stiff plates separated by a series of springs (Fig. 7.15).

The water pressure among the springs begins to influence the two stiff plates only when it overcomes the value of the rock pressure. At this point, the springs are completely unloaded and only the water pressure is effective (Fig. 7.16).

After the opening of the cavity, the secondary stress state (in red in Fig. 7.17) is generated and the pressures P_{RA} and P_{RB}(con $P_{RA} < P_{RB}$) act respectively on the two surfaces A and B (joint between lining and intact rock and horizontal joint in the rock).

Then, the hydrostatic pressure annulled during the excavation phase is reset, with a value P_{WA} on surface A and P_{WB} on surface B ($P_{WA} > P_{WB}$). P_W is only relevant if it overcomes the value of P_R.

From the effective point of view, the only change in the original stress state that is introduced by the presence of water is $\Delta P_A = P_{WA} - P_{RA}$

To visualize the change of P_W, imagine a disc in an isotropic infinite semispace, stimulated by a stress state with radial symmetry (Fig. 7.17).

Once the primary stress state existing before the opening of the cavity is perturbed, the system develops towards a new equilibrium (without water) characterized by a secondary stress state (P_{RA} e P_{RB}), in red in Figure 7.18. After a certain period, the hydrostatic pressure begins to recreate, if it overcomes P_{RA}, in the joint, only pressure $P_{WA} = g_w h_w$ will act in this joint.

On the contrary, the presence of water will be negligible on surface B with regards to the stress state until P_{WB} is lower than P_{RB}, and the only pressure will be P_{RB1}. It can be demonstrated that this happens in every case.

Let us assume that no hydraulic pressure can develop on surface A (for example, due to previous waterproofing injections around the tunnel contour), but it is present on surface B at a certain distance from the cavity (Fig. 7.19).

In absence of water, an equilibrium of the secondary stress state will also be reached in this case after the opening of the tunnel; this equilibrium is characterized by two stresses on surfaces A and B equal to P_{RA} and P_{RB}, with $P_{RA} < P_{RB}$.

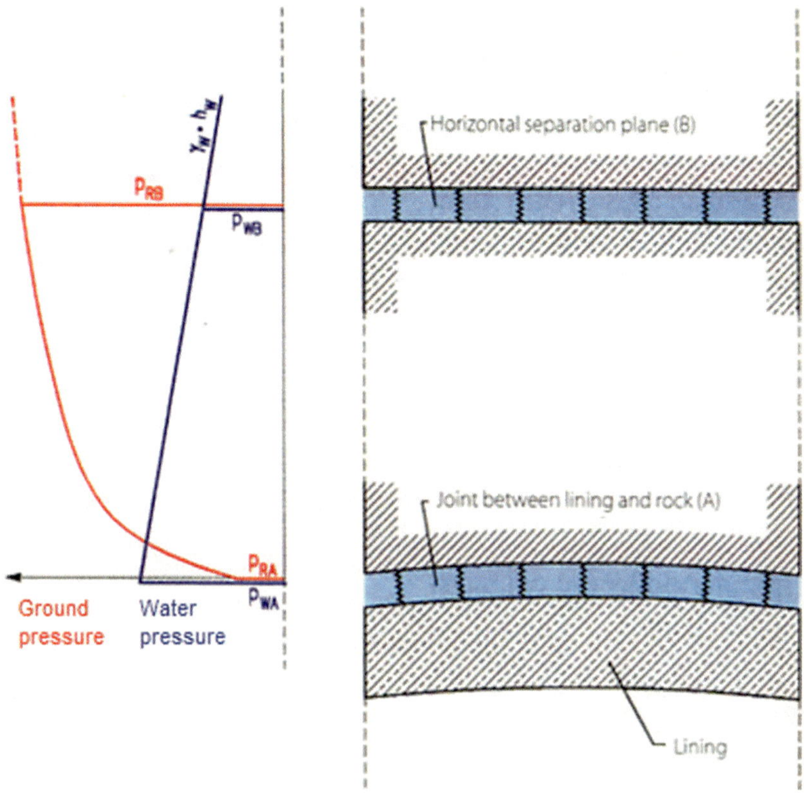

Fig. 7.16 Horizontal separation plane in the roof zone of a tunnel: rock pressure in *red* (PR), water pressure in *blue* (PW)

After a certain period, the hydraulic pressure will reset and it might be higher than P_{RB} in joint B. Then, only pressure P_{WB} will be effective in joint B. In other words, the presence of water will generate an overpressure $\Delta P_B = P_{WB} - P_{RB}$

As a result of the increase of pressure B, an increase in the pressure $D_{PA} \leq D_{PB}$ will be generated on plane A as well as due to equilibrium reasons. The limit case is represented by $D_{PA} = D_{PB} = P_{WB} - P_{RB}$, corresponding to a maximum pressure on the lining equal to $P_{\mathrm{RA}} + D_{\mathrm{PA}}$, always lower than P_{WB}.

The dimensioning of the tunnel lining only requires to take into account the possible portion of hydraulic pressure D_{PB} exceeding the rock pressure P_{RB}.

7.4 Nailing

Passive nailing is connected to the rock mass for its whole length, so the rigorous modelling of its behaviour should be carried out respecting the congruence of deformations between rock and nail. Therefore, the effect induced by nailing can be

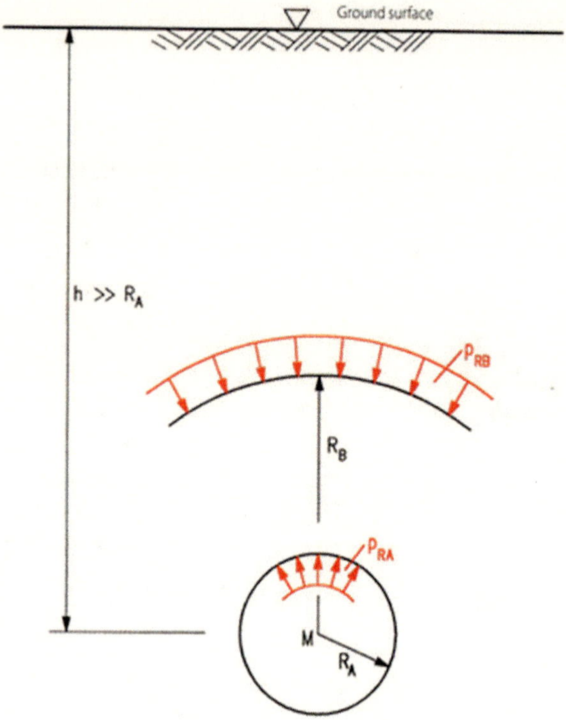

Fig. 7.17 Secondary stress state on the perforated disc model

reduced to a surface action only if an approximation is accepted. Actually, the axial stress is not constant along the nail. On the contrary, it depends on the strain condition of the rock after the ground-nail interaction in each point. Therefore, a complete analysis to model the interaction nail-rock requires the definition of the constitutive law of the interface separating them.

The different methods that can be used for a dimensioning or to check radial nailing measures are characterized by different degrees of precision.

7.4.1 Method of the Confinement Pressure

The confinement effect of nailing is transformed in an equivalent pressure on the cavity contour. The applied pressure is calculated distributing the action of each bar on the relevant surface.

$$P = (A_c \cdot \sigma_s)/(i_l \cdot i_r)$$

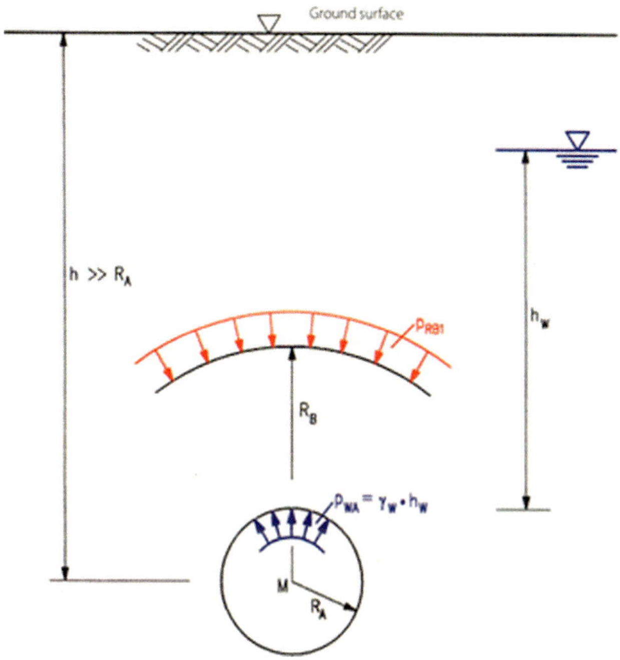

Fig. 7.18 Water pressure on the inner edge of the perforated disc

where:

A_c cross-section area of a nail
σ_S design stress of the nail
i_l longitudinal pitch of the measure
i_l cross pitch of the measure

This simplified method does not allow to take into account the length of the nails and it disregards the contribution to the increase of the ground stiffness. Nevertheless, it is used in the theory of the characteristic lines, thus allowing to detect the equilibrium point of the cavity in terms of pressures and convergences.

7.4.2 *Homogenization Method*

The homogenization method implies the calculation of the nails contribution in terms of equivalent pressure, then it transforms the action of this pressure in a variation of the strength characteristics of the ground. Actually, the effect of pressure P can be transformed in a fictitious cohesion increase (Δc). If the number of nails is optimized (e.g. close to the strictly necessary minimum to guarantee the cavity stability) using

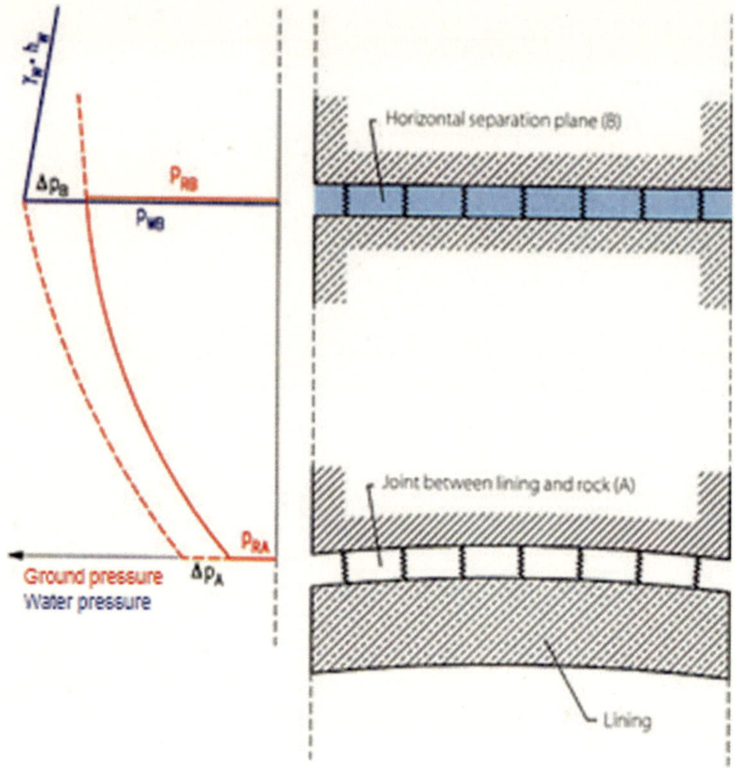

Fig. 7.19 Water permeable horizontal separation plane in the roof zone and water tight joint between lining joint between lining and rock

simple considerations on Mohr's circles, it can be demonstrated that the fictitious cohesion increase caused by nailing is given by:

$$\Delta c = \frac{P}{2} \cdot \sqrt{k_p} = \frac{P}{2} \cdot tg\left(45^\circ + \frac{\phi}{2}\right)$$

Starting from the value of the fictitious cohesion thus calculated, closed solutions are used that provide the equation of the characteristic line of the tunnel. Still, taking correctly into account the nailing efficiency in relationship to the ratio between its length and the size of the yielding zone might be a problem, as well as the fact that there are no anchor points. The solution in closed form of a hole in an elastic-yielding continuum with concentric belts characterized by different mechanical magnitudes is theoretically possible, but it would be quite expensive. It can be tried to adapt the nailing length to the yielding belt thickness, so as to limit its contribution to the improvement of the post-peak parameters of the material. When the solution is reached, the shorter the nail length in relationship to the extension of the yielding stripe, the more approximate is the result.

7.4.3 Modelling of the Cross Section with Continuum Discretization Methods

The methods previously presented are only used for preliminary or approximate analyses. The final design is carried out using finite elements or finite differences methods that allow the rigorous reproduction of the nail behaviour with continuous adherence elements. But only bi-dimensional analyses are possible with short calculation time. The designer can calculate the characteristic line of a section of nailed tunnel, but the designer will have problems in determining the longitudinal effects as the deconfinement rate when increasing the distance from the face. Therefore, the study of the tunnel is carried out integrating some principles of the previously described methods in discretization methods.

The analyses carried out to design a tunnel that can be taken as an example of a method generally used are presented below. The design refers to a tunnel having a 17 m diameter, to be built within shales with a 250 m overburden. The calculation procedure allows to use plane and axisymmetric models to take into account the 3D effect of nailing.

Step 1 Once the characteristics of the soil are known, the analysis of an axisymmetric cross section with uniform nailing on the contour (Fig. 7.20) is performed.

Step 2 The behaviour of the section is assessed finding a correlation (characteristic line) between the convergences and the reduction of inner pressures (Fig. 7.21).

Step 3 A model of identical shape but without nails is analysed, where the mechanical characteristics of the nailed stripe (homogenized section) are increased up to obtaining a characteristic line similar to the one of the nailed section in Step 2 (Fig. 7.22).

Step 4 The mechanical characteristics thus obtained are attributed to the nailed belt in a longitudinal axisymmetric model to infer the function linking the deconfinement rate of the cavity reinforced by the nailing to the distance from the face (Fig. 7.23).

Step 5 This function is used to regulate fictitious excavation stresses in a more rigorous cross-section model with the same shape of the real one and nails modelled as mono-dimensional elements anchored to the soil (Fig. 7.24).

Step 6 According to the analysis results, the suitability of the nailing is evaluated checking the yielding deformation inside the ground and the tunnel convergence, as well as the stress fields of the linings in the different phases. If nailing will not have a permanent function, it can be disabled in the model when the completion phase of the final lining is finished.

The improvement of the mechanical characteristics of the material in Step 3 was obtained according to the principle of equivalent cohesion. As already explained, the equivalent cohesion is determined starting from the pressure obtained by distributing the design load of the single nail on the relevant portion of excavation surface. Still, this value is rounded up because:

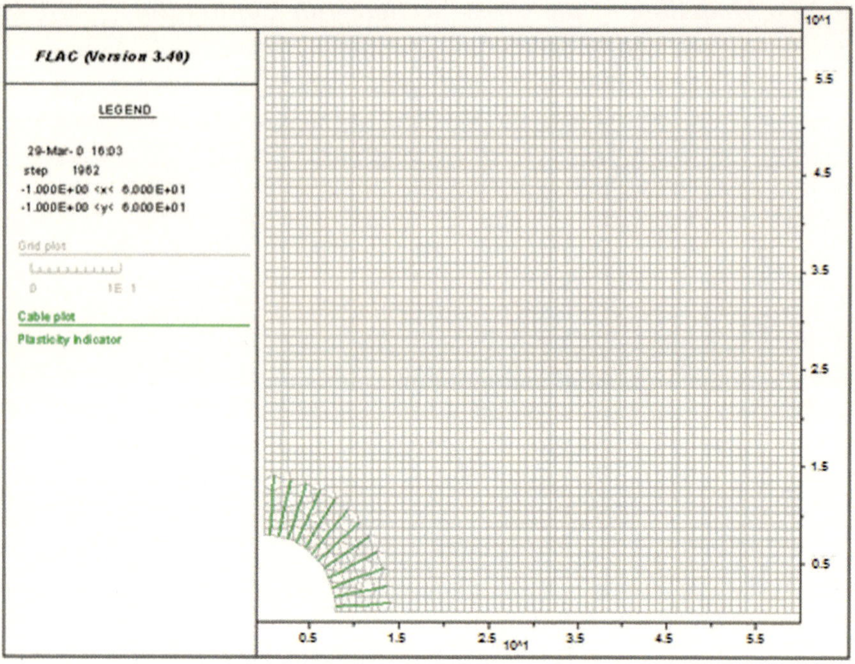

Fig. 7.20 Axisymmetric hole with nailing

- Increasing the distance from the cavity, the nails diverge and the competence area of each one tends to increase.
- It does not take into account the interaction between nail and mass and the chance that the nail does not reach the design load.
- The nail can be shorter than thickness of the plastic (yielding) belt.

In the case study, nails with characteristic yield strength equal to 400 kN were foreseen, installed at a longitudinal distance of 0.8 m, cross distance of 1 m and length equal to 6 m. The design strength is equal to:

$$F_{yd} = 400/1.15 = 347.826 \text{ kN}$$

Table 7.2 shows the values of the tentative equivalent cohesion applied to the nailed belt in the analysis and the obtained values of the confinement pressure p^* that are linked to them by the formula:

$$p^* = \frac{N_y}{i_l \cdot i_r \cdot \left(\frac{R+l}{R}\right)}$$

N_y axial-yielding action of the nail
i_l longitudinal pitch between the nailing rows
i_r distance between the nails measured on the cross excavation perimeter
l radial distance from the excavation face where p^* is calculated

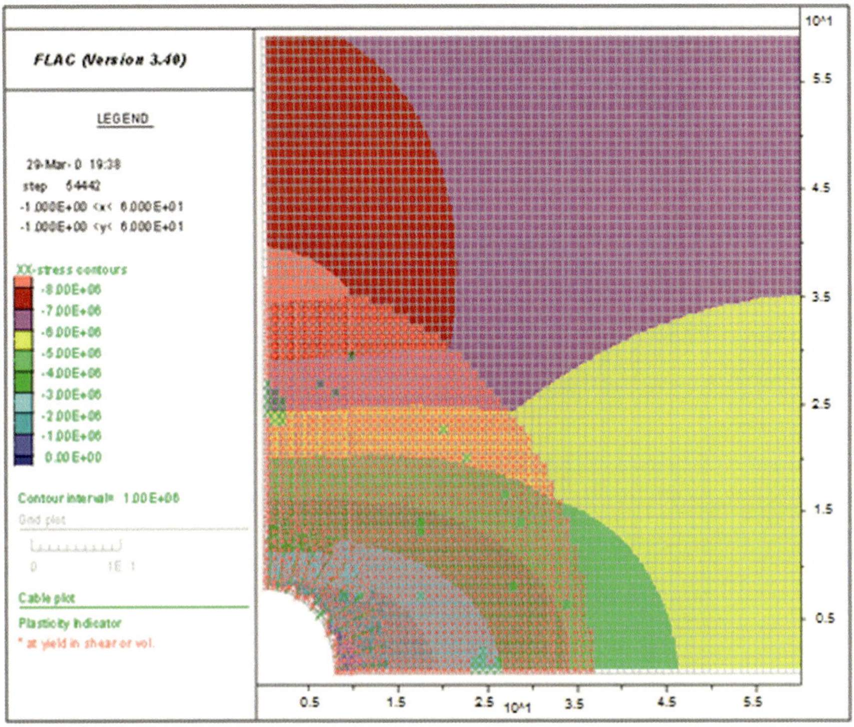

Fig. 7.21 Stress-strain field of the nailed section

The value presented as case 1 is the first tentative value C^*, that is generally adopted in homogenizing analytic methods, referring to the nail effect on the excavation perimeter. The analysis showed that the cohesion value C^* of case 3 provides data very close to the model with nailing. In this case, the confinement pressure can be calculated considering a value of "l" equal to the nail length (it is as if the confinement pressure were distributed on the very end of the nailed belt instead of on the excavation perimeter).

The results of the numerical analysis show that the nailed tunnel benefits from a less abrupt development of convergences close to the face; this allows the optimization of the sequence and of the implementation schedule of the supporting measures. To highlight that aspect, consider that the schedule of the supporting measures referring to the design of the same tunnel foresaw the completion of the first-phase lining and the casting of the invert at a distance that is twice the previous one (8 m instead of 4 m), taking into account or not the effect of radial nailing.

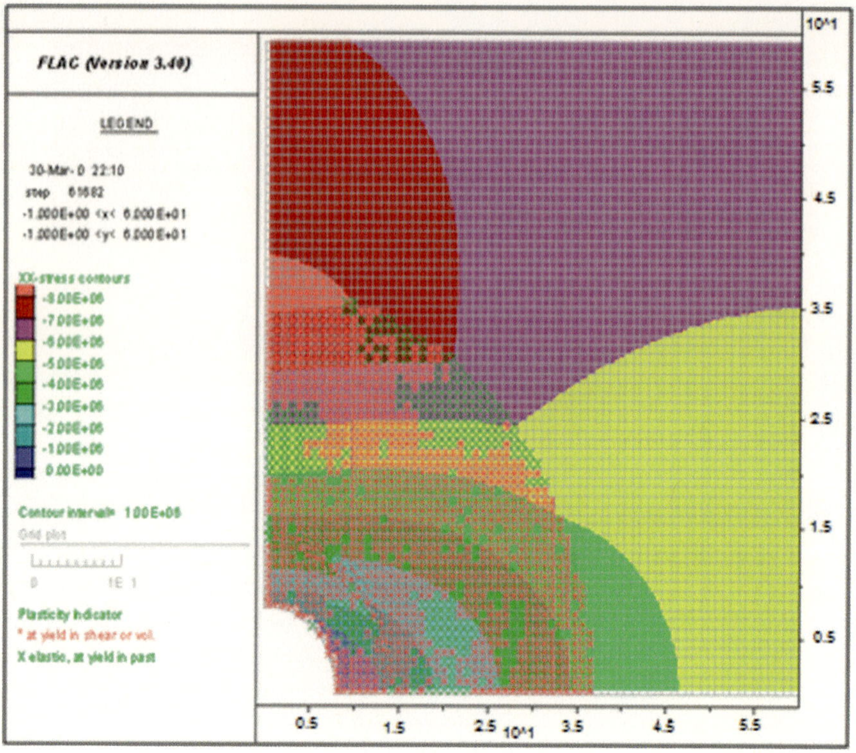

Fig. 7.22 Stress strain field of the homogenized section

Table 7.2 Example of tentative equivalent cohesion calculation

Case	l [m]	C^* [kPa]	p^* [kPa]
1	0	321.738	434.782
2	3	233.991	316.205
3	6	183.850	248.446

7.5 Spiling

Due to its dip, spiling has a double structural effect. The first one is to stabilize isolated mass portions and it is obtained close to the advancing face. The second one is the improvement of the mechanical characteristics of the soil close to the excavation perimeter and is always present, but it develops its effects mainly when convergences increase, that is to say when the distance from the face increases.

Therefore, the nails have to be dimensioned in order to accomplish both functions. Their effect on the overall behaviour of the tunnel is discussed in the paragraphs about radial nailing. Here it is just reminded that the element of the axial action of the normal nail contributes to the confinement effect of the cavity; the effectiveness of the nail in this sense decreases with its dip.

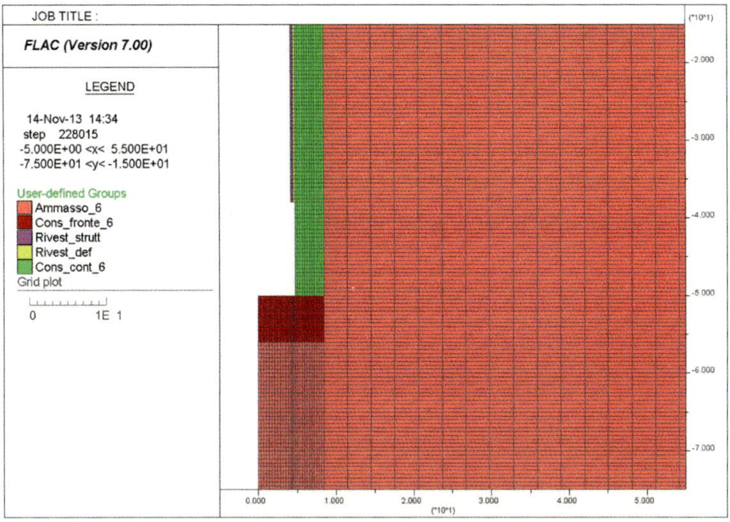

Fig. 7.23 Longitudinal axisymmetric model

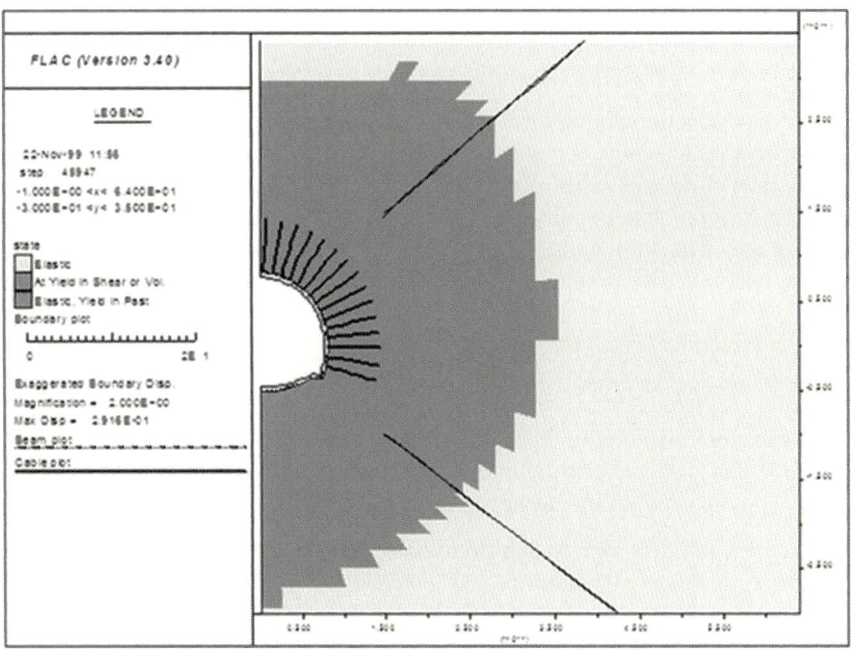

Fig. 7.24 Nailed section

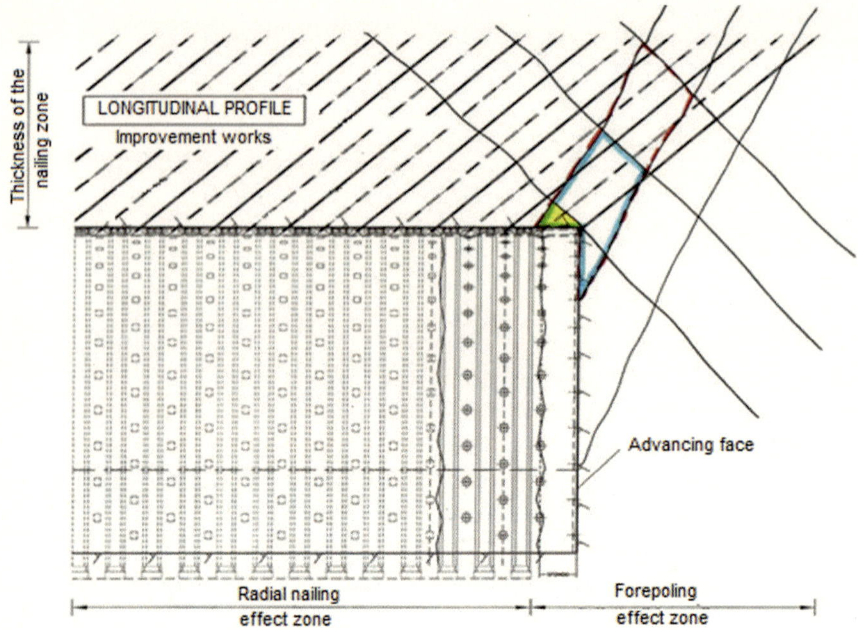

Fig. 7.25 Potential kinematism in the ceiling while advancing

The nail dimensioning in relationship with the stabilizing function of the ceiling while advancing is more complex. When the joint orientation changes, the strength mechanism of the nail varies as well as the stress actions on it. As already discussed in Chap. 6, Sect. 7, the stressing actions on a nail can be axial, shearing or composite.

As it is evident, the identification of the stressing actions on the single nail is complicated also when the geometry of the problem is known. It is even more complicated in the design phase, when the fracturing state of the rock is not thoroughly known. Particularly critical cases that might occur are studied, considering what is known about the fracturing state of the rock; the dimensioning is performed according to this information.

For example, in Figure 7.25, a system of potentially unstable blocks can be observed that is generated by a limited number of joints and stabilized by spiling.

Two possible instability mechanisms can be identified:

- The rock fall of a rock wedge from the ceiling, in the area not supported with first-phase lining with steel ribs and shotcrete.
- The sliding of rock portions that tends to displace along a joint having the same dip direction of the face with respect to the face.

Figures 7.26 and 7.27 represent the equilibrium states of unstable portions, highlighting the shear and axial action components for each nail.

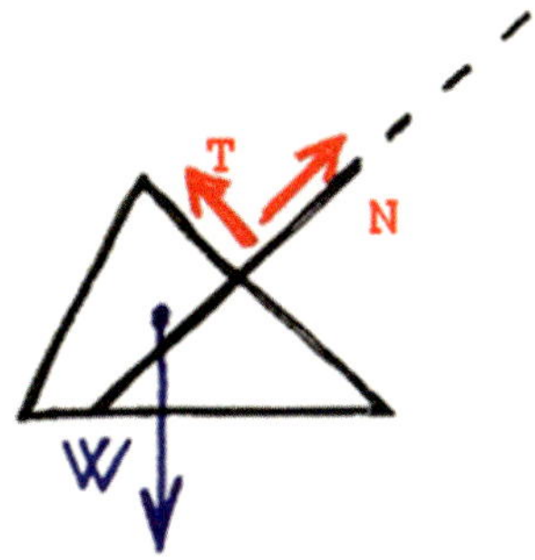

Fig. 7.26 Safety measures preventing the rock fall of a block in the ceiling area

The dimensioning can be carried out using suitable software that analyses the detachment chance of the blocks according to the orientation and the mechanical characteristics of the joints that characterize the mass.

7.6 Forepoling

Usually, elements with good shear and bending strength are used for the implementation of forepoling, because the beam in subhorizontal position has to resist mainly to loads transversal to its axis. Reinforcement elements (pipes, self-boring bars) for forepoling have to be chosen according to the tunnel dimension, the length of the forepole, the distance from the face where the first-phase lining is realized and the geomechanical characteristics of the mass.

Generally, the behaviour of an element that is inserted subhorizontally can be compared to that of a beam because of mainly the bending and shearing nature of the strength stressing it. The structural calculation of this beam is complicated by the difficulty to identify with precision its static scheme and the loads stressing it.

The main function of the element is to protect the advancing area from the fall of material above it. Nevertheless, it is much longer than this area, having a back restrain, on the first-phase lining already assembled, and a front one, above the face core. In first analysis, the elements could be schematized as a beam on multiple supports, even deformable, distributed along two stretches with different elastic characteristics (Fig. 7.28). Simplified models can be derived from this first approach that allows a quicker dimensioning because they provide promptly a value for the maximum stress in the elements according to the stiffness and the loads (Fig. 7.29).

If the element is designed as a beam on two supports, part of the lining support in the advancing area will be charged on the ground area just beyond the face. Generally, when the support of the excavation while advancing is needed, the stability conditions of the face are not good enough to sustain load increases. Therefore, it is better to ignore the contribution of the beam portion beyond the face and design the forepole as a fixed in the tunnel lining (cantiliver static scheme); its length is equal to the whole advancing area. As a consequence, the checks have to consider the acceptability

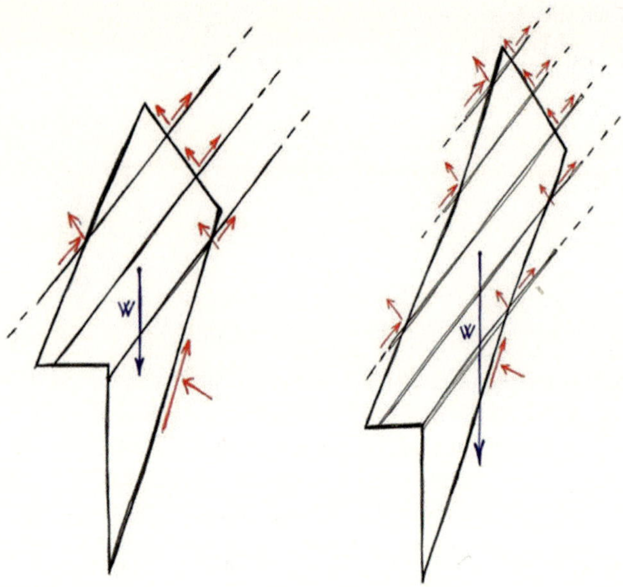

Fig. 7.27 Nailing of rock blocks tending to slide toward the cavity

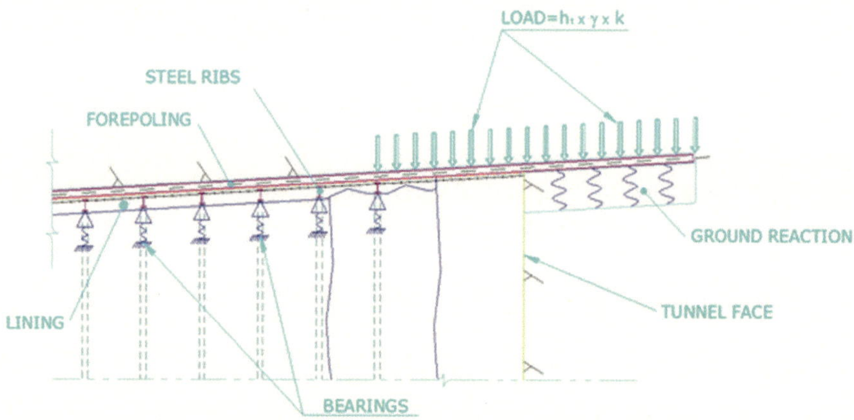

Fig. 7.28 Model of idealized bond of the forepoles

of the momentum and of the shear in correspondence of the restrain, taking into account only the contribution of the elements that are sufficiently fixed in the first-phase lining. In normal designing conditions, the cooperation of the restrain to the beam stiffness is neglected, trusting only the bending strength of the steel section.

The loads definition on the structure is another very uncertain factor that has to be considered with great care. Those loads can be determined applying the limit

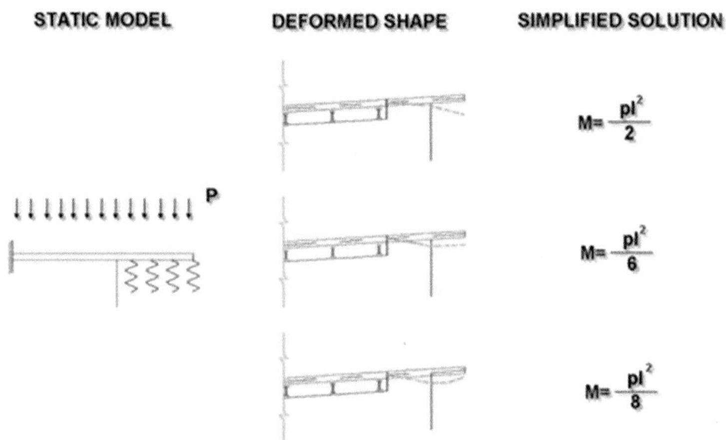

Fig. 7.29 Simplified static schemes

equilibrium methods, similarly to what is generally done for the lining dimensioning. Figure 7.30 shows the hypothesis of a mechanism to determine the load acting on the distance between the face and the first-phase lining.

Using the cantiliver scheme, the moment at the fixed end would be:

$$M = \frac{p \cdot l^2}{2}$$

Generally, Terzaghi's approach (see Sect. 7.3 Terzaghi's vertical loads) is used to determine the load *p*.

The load value thus determined is referred to a model in plane deformations and it is very conservative as it neglects the 3D effects due to the closeness to the face. Therefore, in the design practice, a share of it is considered in the calculation.

The check is carried out in the hypothesis that the poles are fixed at the level of the rib closest to the face. It must be verified that the length of the beam portion overlapping the first-phase lining is long enough not to tip over under the effect of this maximum moment. In practice, it can be said that the forepole is adequately bond if it overlaps two steel ribs. Indicatively, it can be considered that the minimum overlapping length between the different forepole fields has to be at least 3 m.

7.7 Stabilization of the Excavation Face: Number and Length of the Forepoles

The methods to assess the face stability discussed in Sect. 4.8 lead to the identification of pressure σ_3 that, when applied to the face, brings it back to stability conditions. Starting from this, the minimum number of elements N_{min} can be calculated:

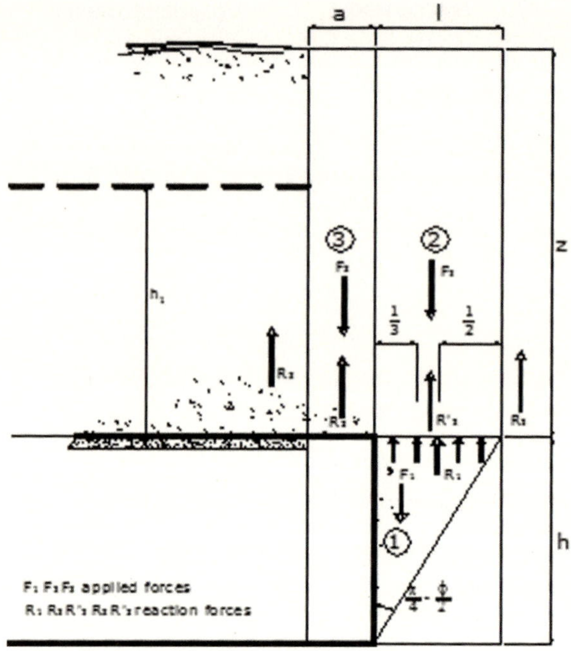

Fig. 7.30 Limit equilibrium of the solids involved in the failure kinematism. (Tamez 1985)

$$N^{\circ}_{\min} = \frac{\sigma_3 \cdot A}{T_{Rd}}$$

where

T_{Rd} design strength of each element
A section area

The minimum length of the elements is assessed starting from the minimum length of the anchorage given by:

$$L_{anc\,min} = \frac{T_{Rd}}{\pi \cdot D \cdot \tau_{Rd}}$$

D being the drilling diameter which will contain the element.

The length of the excavation field has to be added (the elements at the face are implemented while advancing and they involve a stretch beyond the face that is equal to the length of the element; the length of the excavation field is equal to the length of the stretch that can be excavated safely before repeating the supporting measure while advancing, Fig. 7.31); moreover, the elements have to be anchored by friction as they do not have a plate transmitting the traction to the face. The length thus obtained has to be incremented by a further unit to allow the development of the friction. As a consequence, the minimum length of the element is given by:

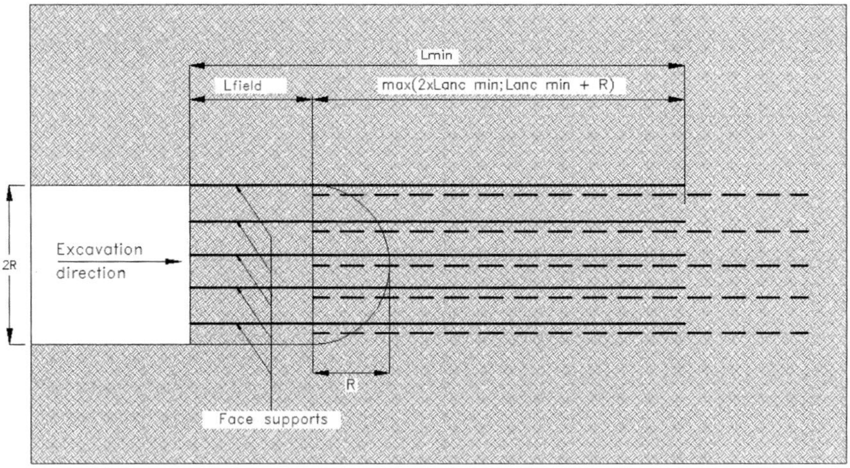

Fig. 7.31 Minimum length of the element—scheme

$$L_{\min} = \max(2 \cdot L_{\text{anc min}}; L_{\text{anc min}} + R) + L_{\text{field}}$$

7.8 Characteristic Lines: Analysis of the Linings

As already discussed, the dimensioning of a tunnel lining is a very complex problem of ground-structure interaction that depends on the shape, the initial stress state, the construction methods and the stiffness of the lining.

The method of the characteristic lines allows to evaluate the suitability of a lining by means of an analysis that takes into account the ground-structure interaction, even if only in the field of simplifying hypotheses that are parts of the method. The first evaluations can be made by plotting the behaviour law of the lining on the characteristic line graph of the cavity pressure–convergence. Figure 4.10 shows the characteristic curves of the linings:

- 0-P_1: stiff elastic lining: the lining placed at a certain distance from the face for which part of the convergence has already occurred ($u_r(P=0)=u_{r0}>0$) has a linear elastic behaviour until the equilibrium is reached in the coordinate point (P_1, u_{r1}), the slope of the stretch 0-P_1 represents the stiffness of the lining.
- 0-P_2: elastic lining reinforced in a second phase: similarly to the preceding case, the lining is installed at a certain distance from the face so the convergence $u_r(P=0)=u_{r0}>0$, has an elastic linear behaviour characterized by a lower initial stiffness; this stiffness increases when the reinforcement of the lining is implemented.

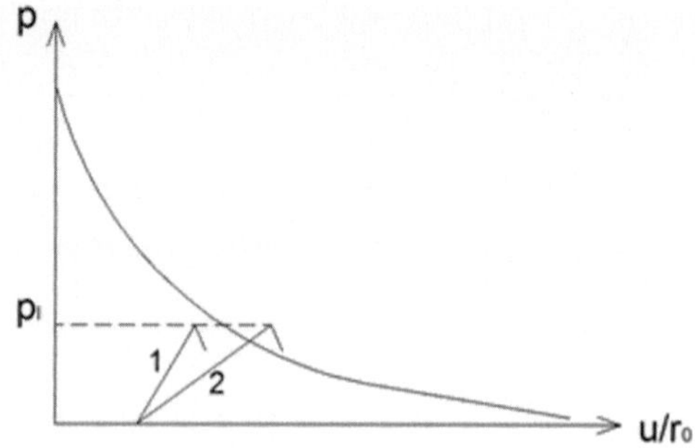

Fig. 7.32 Characteristic curves of the linings: case of two linings having the same failure load p_l but different stiffness. (Lane 1957)

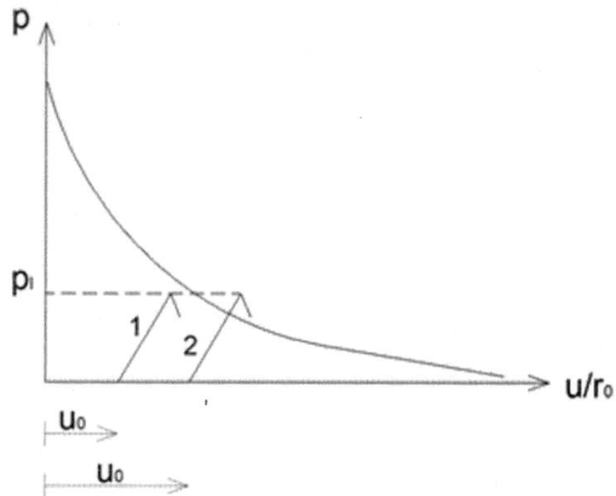

Fig. 7.33 Characteristic curves of the linings: The influence of the distance from the face where the lining is installed. (Lane 1957)

- 0-P_3: elastic-plastic lining: the lining installed at a certain distance from the face so $u_r(P=0)=u_{r0}>0$, yields after an initial elastic behaviour, the equilibrium (defined by the coordinates point (P_3, u_{r3})) is reached in the yielding field.

Let us consider the case of two linings having the same yield load, but different stiffness and brittle behaviour (Fig. 7.32).

The influence of the lining stiffness is evident: the stiff support (case 1) cannot bear the ground-induced pressure and collapses; whereas the less stiff support (case 2) can bear the applied load at the expense of the convergence increase.

The previously discussed cases show the importance of the support stiffness; another important aspect for the assessment of the behaviour of the system support—cavity is the distance from the face where the support is installed (Fig. 7.33). The entity of the initial convergence u_0 depends on it; the lining can be sufficient or insufficient according to u_0: if the value of u_0 is low (case 1), the support cannot bear the convergences induced by the cavity and it leads to the collapse; if the value of u_0 is higher, the same support provides a sufficient strength. Figure 7.34 shows the characteristic lines of some real linings.

If the supports are dimensioned using the characteristic lines method, if k stands for the elastic stiffness of the lining, the elastic part of the characteristic curve of the lining can be identified by following equation:

$$P = k \cdot u$$

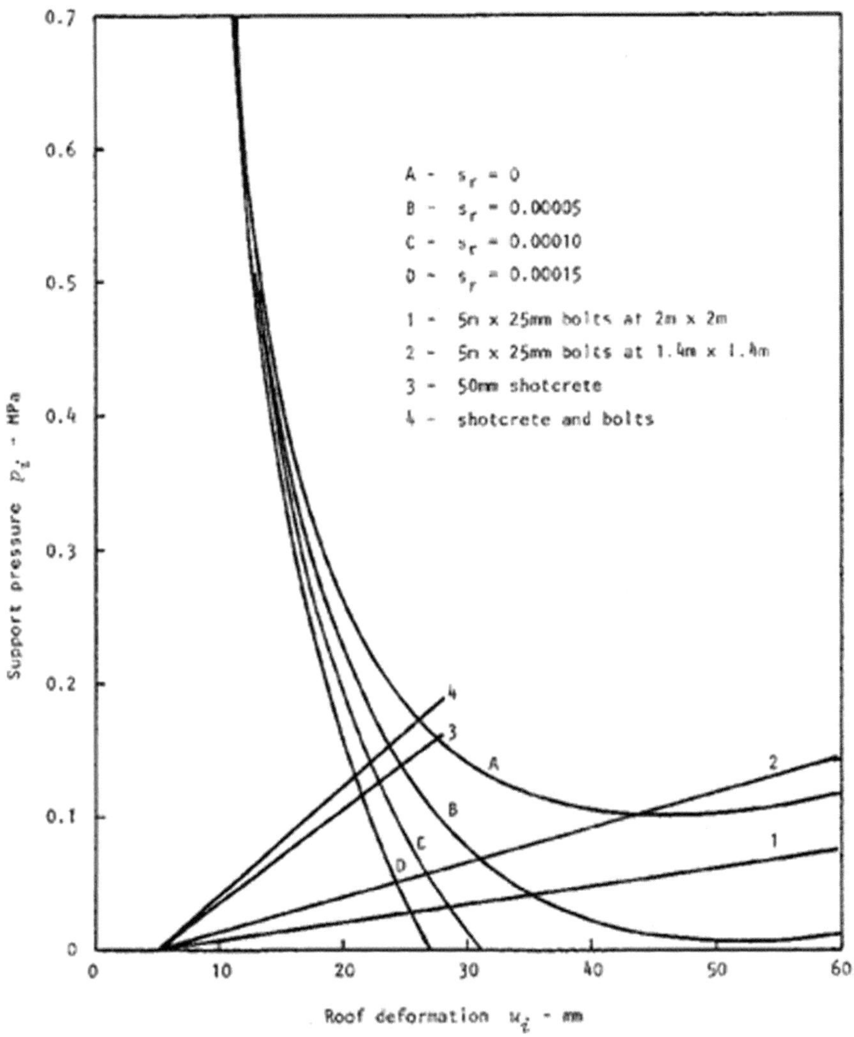

Fig. 7.34 Characteristic curves of some real linings. (Hoek and Browm 1982)

where:

P external pressure on the lining
k lining stiffness
u deformation or the radial displacement on the cavity contour

Note that the unit of measure for stiffness k is pressure divided by length (for e.g. MPa/m).

When the lining is made by two support systems having different stiffness (e.g. projected concrete and steel ribs), the literature suggests to calculate the stiffness of the whole system as the sum of the single stiffness of each component.

The stiffness of the components of the lining are calculated by means of following formulas (assuming the lining has circular shape). Following relation is used for shorcrete:

$$k_c = \frac{E_c \cdot \left(R_i^2 - (R_i - s_c)^2\right)}{(1 + \nu_c) \cdot R_i \cdot \left((1 - 2\nu_c) \cdot R_i^2 + (R_i - s_c)^2\right)}$$

where:

E_c Young's modulus of concrete
ν_c Poisson's modulus of concrete
R_i excavation radius
s_c width of the projected concrete ring

whereas the following expression is used for ribs:

$$\frac{1}{k_s} = \frac{S \cdot R_s^2}{E_s \cdot A_s}$$

where:

k_s rib stiffness
R_s equivalent radius of the rib
S rib spacing
A_s area of the resisting section of the rib
E_s Young's modulus of the rib

Assuming a circular geometry for the lining (Hoek and Brown 1980), the maximum strengths of the components, steel ribs and shotcrete can be evaluated separately as follows.

The maximum shotcrete support pressure can be evaluated with the following equation:

$$P_{\max,c} = \frac{1}{2} \cdot f_{cd} \cdot \left(1 - \frac{(R_i - s_c)^2}{R_i^2}\right)$$

where:

$P_{max,c}$ maximum support pressure of shotcrete
f_{cd} design concrete strength
$R_i = R_{eq}$ equivalent radius, it is the radius of a cavity having the same area of the excavation section being studied
S_c thickness of the shotcrete lining

The maximum steel rib support pressure can be evaluated with following equation:

$$P_{\max,s} = \frac{A_s \cdot f_{sd}}{S \cdot R_i}$$

A_s steel rib cross area
f_{sd} design steel strength
S steel rib longitudinal step
$R_i = R_{eq}$ equivalent radius, it is the radius of a cavity having the same area of the excavation section being studied

After identifying the stiffness of the two components, the two maximum deformations allowed are calculated separately, as if each measure would act by itself, using:

$$U_{\max} = \frac{P_{\max}}{k}$$

where:

$U_{\max}$ max convergence
$P_{\max}$ max pressure value on the lining
k stiffness of each element of the lining

Between the two maximum convergences, the lower is chosen because the support with lower u_r value identifies the max pressure the linings can bear when acting together. In that way, the maximum pressure that can act on the whole lining is obtained (that is, considering the two stiffnesses).

$$P_{\max} = U_{\max} \cdot (k_c + k_s)$$

Note that the stiffness of the structures obtained by applying the formulas above (valid for closed circular geometries) are extremely high and can be used only in relative terms for the comparison between the behaviour of the different elements of the structure, but they can be applied to real cases in absolute terms only if "behaviour coefficients" are taken into account that consider the effects of:

1. Coupling between ground and structure
2. Deferred development of the strength of the shotcrete with respect to the installation and subsequent progressive increase of the deformation modulus in time
3. Not perfectly circular geometry
4. Lack of immediate closure of the ring with invert

On the other hand, they neglect the contribution of radial nailing and/or other measures while advancing. The presence of those measures in real situations provides a contribution to the harmonization of the mass behaviour on the excavation contour, stabilizes potentially unstable blocks and cooperates with the first-phase lining (during the implementation phases where the lining has not reached its final configuration or strength), and contains instability phenomena of the structure in case of peak loads.

7.9 Numerical Methods

The calculation methods previously discussed introduce strong assumptions which are: simple geometry, homogeneous grounds and isotropic stress states. Moreover, the study of the structure-ground interaction is carried out separately, taking only partially into account the influence they have on each other. Those methods allow a qualitative forecast of convergences and loads on the structures and represent an important, easy-to-use decision-making tool to identify possible difficulties developing as a consequence of the cavity excavation and/or during the work's life (high convergences, high stress on the linings, etc.).

Nowadays, the numerical methods commonly used in design allow to study underground works thoroughly evaluating the behaviour of the system structure-ground using a single calculation model. As already pointed out, numerical models can be either 2D or 3D. Three-dimensional models take into account at the same time the 3D factor, the time-dependence and the static indeterminacy of the problem. All the advancing phases can be schematized: excavation, face reinforcement, installation of the lining, decaying of the material characteristics in time, etc. Considering the great modelling and calculation difficulties characterizing 3D models, they are only used in very peculiar situations (e.g. tunnels intersections, large cavities etc.) requiring very detailed evaluations. Usually bi-dimensional numerical models are used, always taking into account the 3D aspect of the problem through a combination of axisymmetric and plane strain analysis. Following paragraphs are focused on the description of the procedure to adopt for a two-dimension analysis.

Generally speaking, the numerical modelling is carried out as follows:

1. Determination of tunnel cross-section equivalent radius

   $R_{eq} = \sqrt{A_{cs}/\pi}$
   with A_{cs}: cross-section area

2. Determination of the characteristic "pressure–convergence" curve (Fig. 7.35): generally by means of an analytic solution (e.g., by Ribacchi and Riccioni (1977) formulation); axialsymmetric model (disk geometry) can be used in particular if radial supports are envisaged (see previous example).
3. Plot of "convergence–distance from face" curve (Figs. 7.36–7.38) derived from an axialsymmetric numerical model with and/or without stabilization structures (if no stabilization structures are envisaged, the curve can be compared with analytical solution, e.g. Panet and Guenot (1982)) .
4. Determination of virtual support pressure by plot of "fictitious forces of excavation (EFF)–distance from face" curve (Fig. 7.39).
5. Plain strain modelling:
   a. Geostatic state of stress modelling (Fig. 7.40)
   b. Excavation-phase modelling (Fig. 7.41): removing of ground elements inside the excavation shape that are substituted by EFF, initially equal to the geostatic ones

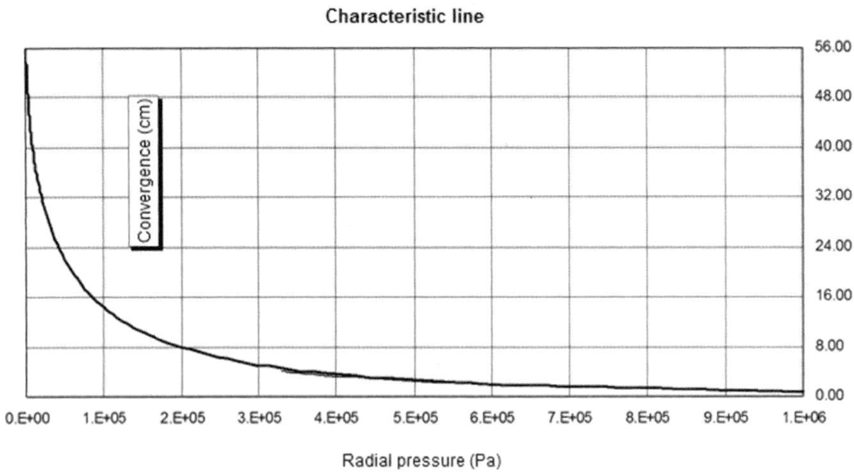

Fig. 7.35 Characteristic "pressure–convergence" line obtained with Ribacchi and Riccioni's (1977) formulation for two different dilatancy values (ref. point 2)

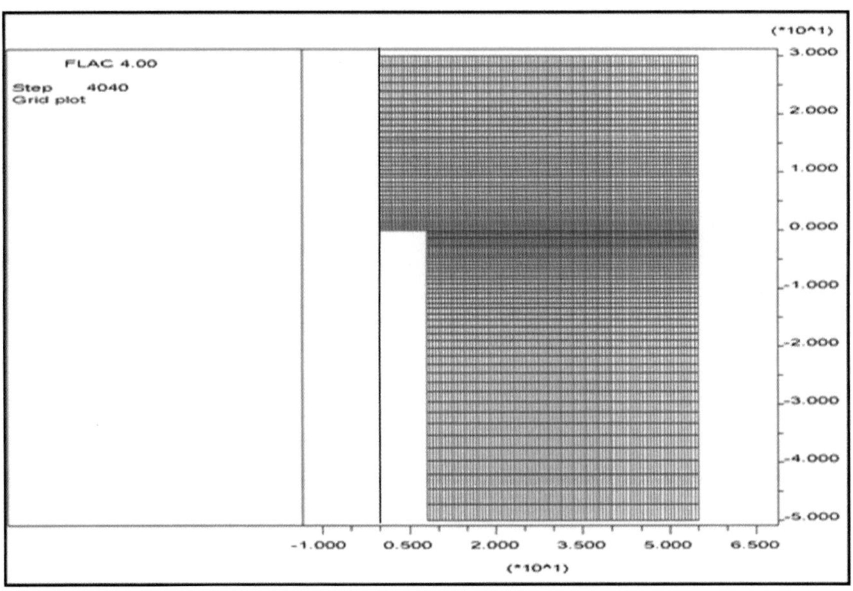

Fig. 7.36 Determination of "convergence–distance from face" line: axisymmetric longitudinal model of the tunnel (ref. point 3)

c. Modelling of the increasing distance from the face through gradual decreasing of virtual support pressure (EFF) and activation of lining elements (Figs. 7.42–7.51)
d. Long-term condition modelling (e.g. decreasing of the strength characteristics of the ground, changes of the groundwater table position, decrease in the lining strength, etc.)

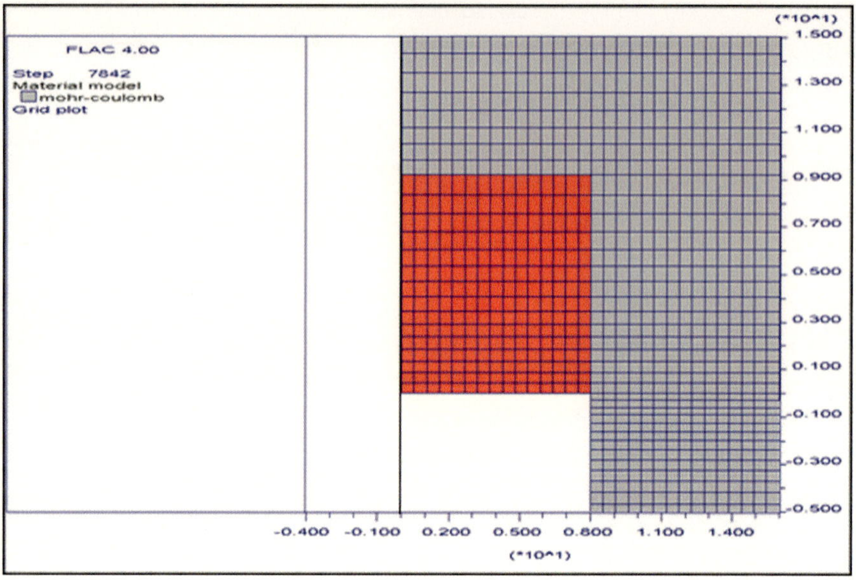

Fig. 7.37 Excavation with face consolidation by VTR, modelled as an equivalent increase of cohesion beyond the face (ref. point 3)

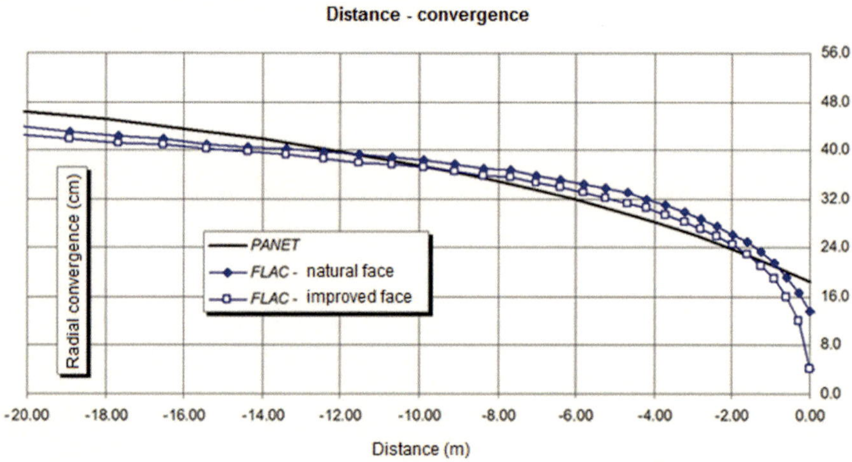

Fig. 7.38 Characteristic "convergence–distance from face" curve obtained with Panet and Guenot's (1982) formulation and with numerical models with and without face stabilization (ref. point 3)

A good design praxis includes the comparison, whenever possible, of the results obtained through numerical modelling with the analytic solutions, as to "adjust" the numerical modelling and reduce the possibilities of error.

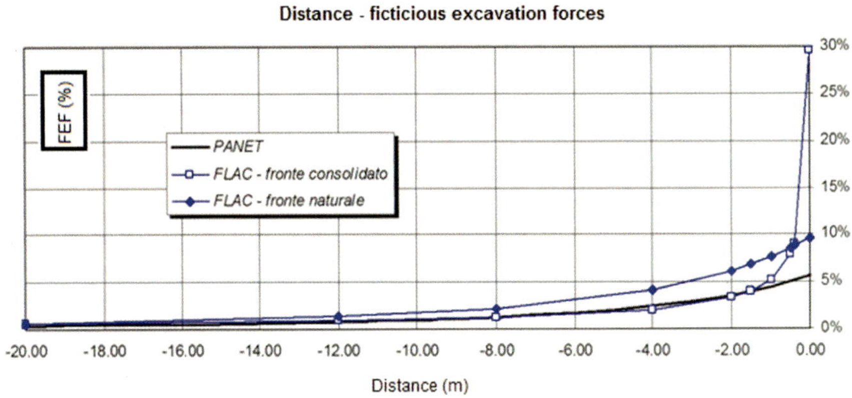

Fig. 7.39 Characteristic "fictitious forces of excavation (EFF)–Distance from face" curve obtained with Panet and Guenot's (1982) formulation and with numerical models with and without face stabilization (ref. point 4)

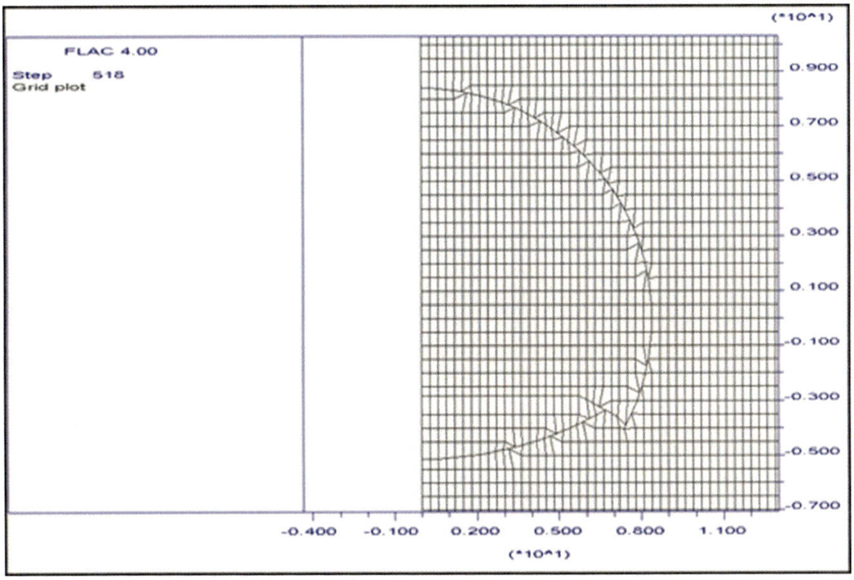

Fig. 7.40 Geostatic equilibrium (plot of the mesh near excavation perimeter). The model is undeformed due to the initial conditions imposed (ref. point 5)

7.10 Seismic Aspects

As already mentioned in Chap. 1, Sect. 10, a seismic event that is often fatal for other structures, is usually less relevant for underground works. Exceptions are represented by particular cases of lithologic and structural discontinuities in the mass

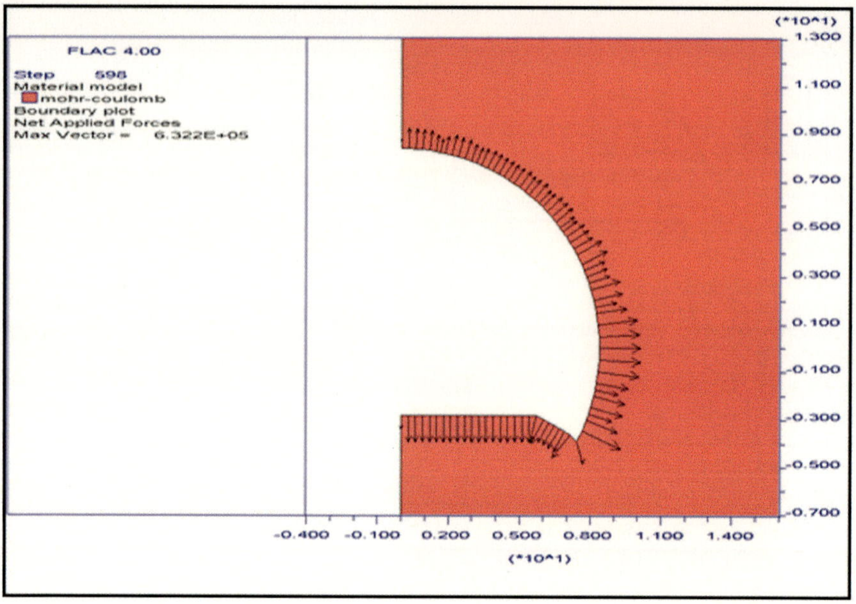

Fig. 7.41 Excavation boundary is restrained and reaction forces are recorded. They will be used as FFS (ref. point 5.b)

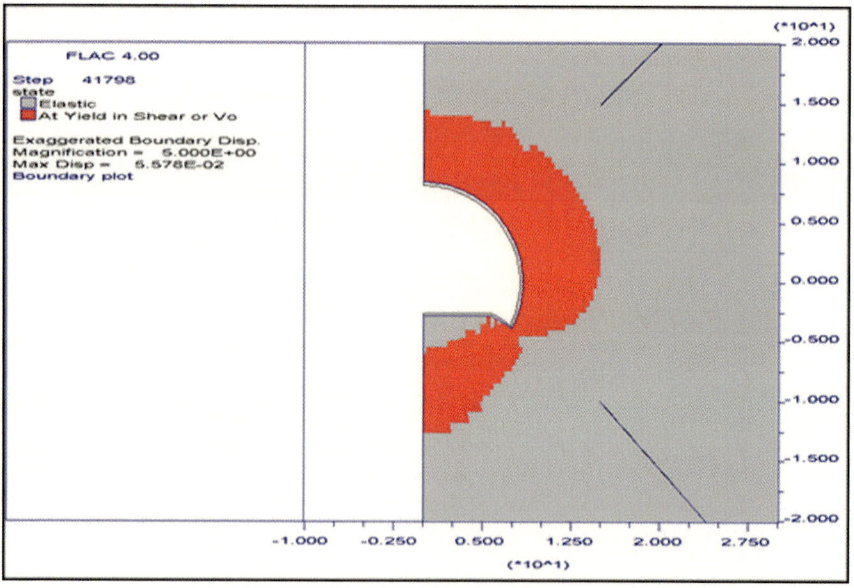

Fig. 7.42 Plastic elements and displacements at excavation face ($x = 0$, EFF $= 30\,\%$) (ref. point 5.c)

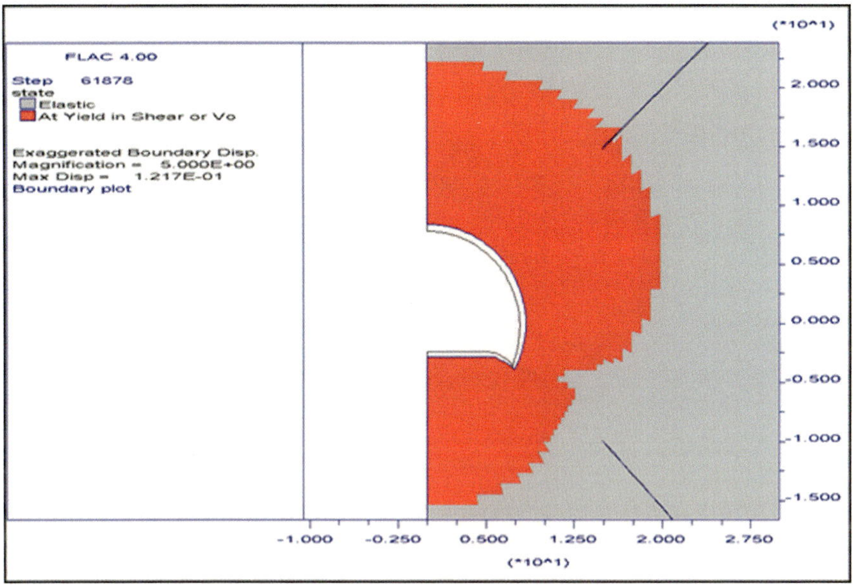

Fig. 7.43 Plastic elements and displacements at first-phase lining application ($x = 0.8$ m, EFF = 11 %) (ref. point 5.c)

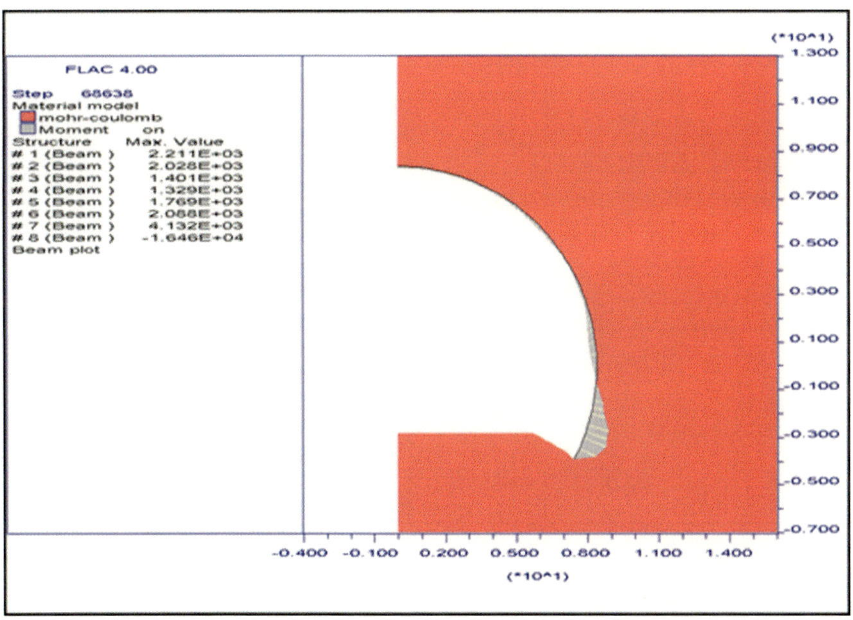

Fig. 7.44 Bending moment on first-phase lining ($x = 2$ m, EFF = 6.1 %) (ref. point 5.c)

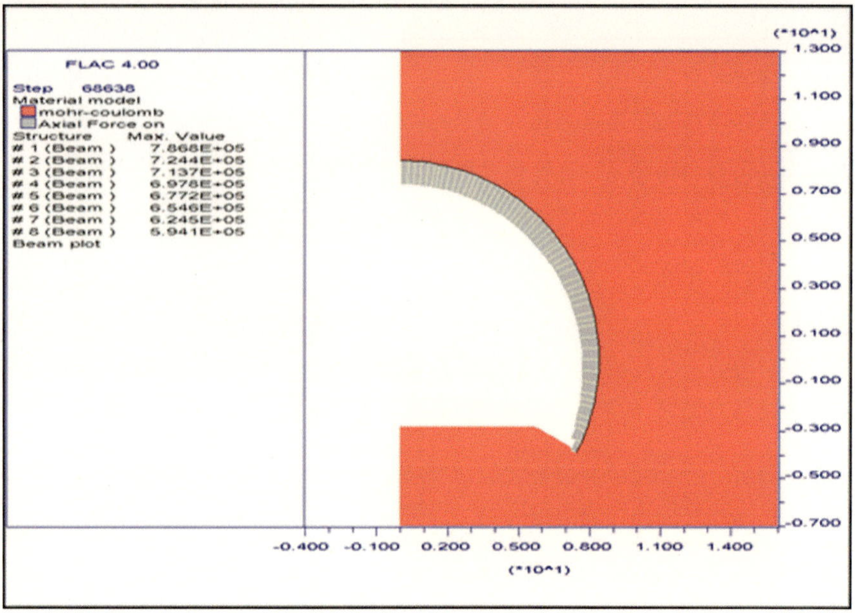

Fig. 7.45 Axial force on first-phase lining ($x = 2$ m, EFF = 6.1 %) (ref. point 5.c)

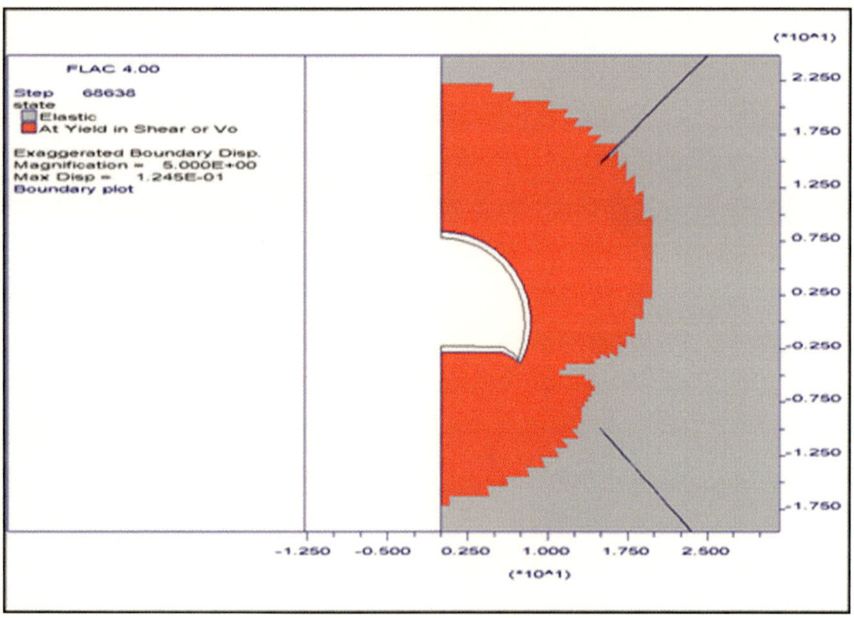

Fig. 7.46 Plastic elements before invert excavation and casting ($x = 2$ m, EFF= 6.1 %) (ref. point 5.c)

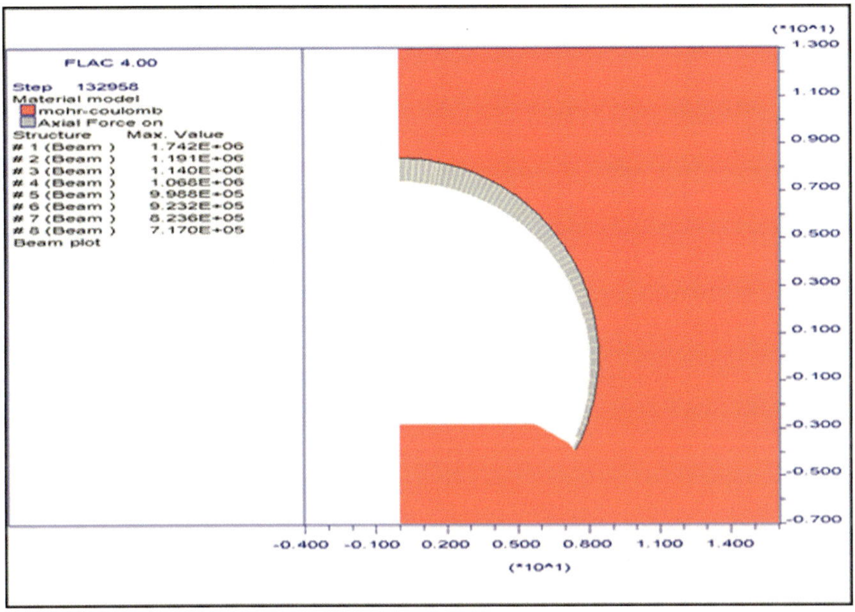

Fig. 7.47 Axial force on shot concrete before invert excavation and casting (EFF = 3.9 %) (ref. point 5.c)

being excavated or very shallow works. Usually, damages caused by the earthquake are limited and lead to local functionality losses (fissuring, dislocations) but not to collapse.

The study of the underground work behaviour affected by an earthquake is carried out on the basis of imposed deformation, caused by the effect of compression seismic waves (waves *P*, for which the particles movement is parallel to the propagation direction) and shear waves (waves *S*, for which the particles movement is perpendicular to the propagation direction).

In particular situations of lithologic and structural discontinuities of the ground or very shallow works, numerical analysis or simulations have to be carried out, whereas for homogeneous situations, quicker calculation methods can be used that allow an immediate evaluation of the increase in stress and strain of the lining due to the earthquake.

A possible simplification of the problem consists in assuming that there is complete connection between tunnel and ground and the displacement of the structure is equal to that of the ground (Panet 1986; Abatelli 1989), thus neglecting the ground-structure interaction. This assumption is close to reality for structures whose stiffness is negligible with respect to the ground (tunnel in good-quality mass). On the other hand, in the case of structures with no negligible stiffness, the assumption of ground-structure connection is conservative, as it implies that the calculation considers greater deformations (those of the ground free to deform) with respect to those undergone by

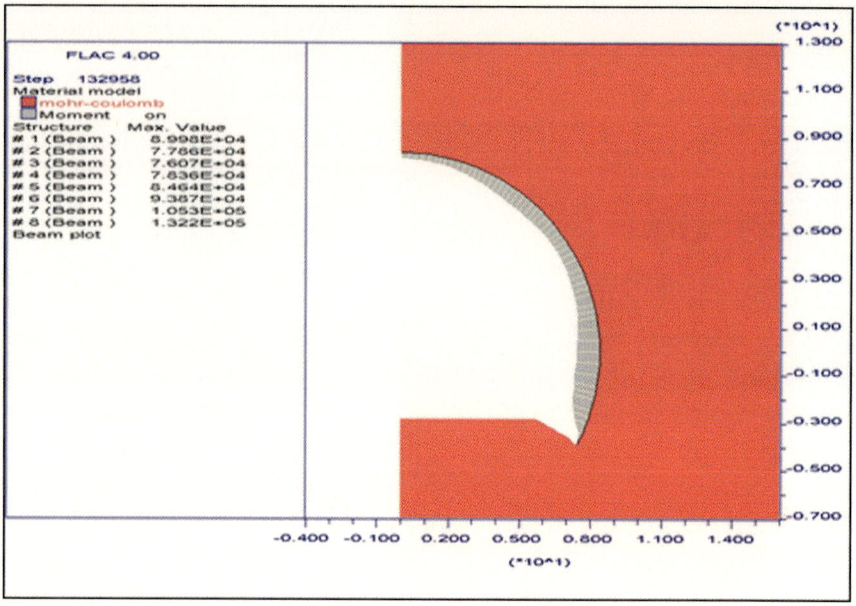

Fig. 7.48 Bending moment on shotcrete before invert excavation and casting (EFF = 3.9 %) (ref. point 5.c)

the lining. In reality, the structure hinders the ground displacement and, as a consequence, the deformation is smaller. Moreover, the assumption of total connection between structure and ground is in favour of safety, as a relative displacement between them would decrease the ground capability to transmit deformations to the structure.

The modalities to determine the actions increasing the stress on the lining due to the earthquake are presented below:

- Contribution of the earthquake to the longitudinal axial action: the seismic wave transit produced alternate compressions and tensile stresses equal to:

$$\sigma_{\max} = \pm\, \varepsilon_{\max} \cdot E_{\text{cls}}$$

being:

$$\varepsilon_{\max} = V_{\max}/C_p$$

$V_{\max}$ max speed of the ground particles
C_p propagation speed of waves P

- Contribution of the earthquake to the bending axial action: the seismic wave transit produce alternate curving with stresses corresponding to:

$$\sigma_{\max} = \pm\, \varepsilon_{\max} \cdot E_{\text{cls}}$$

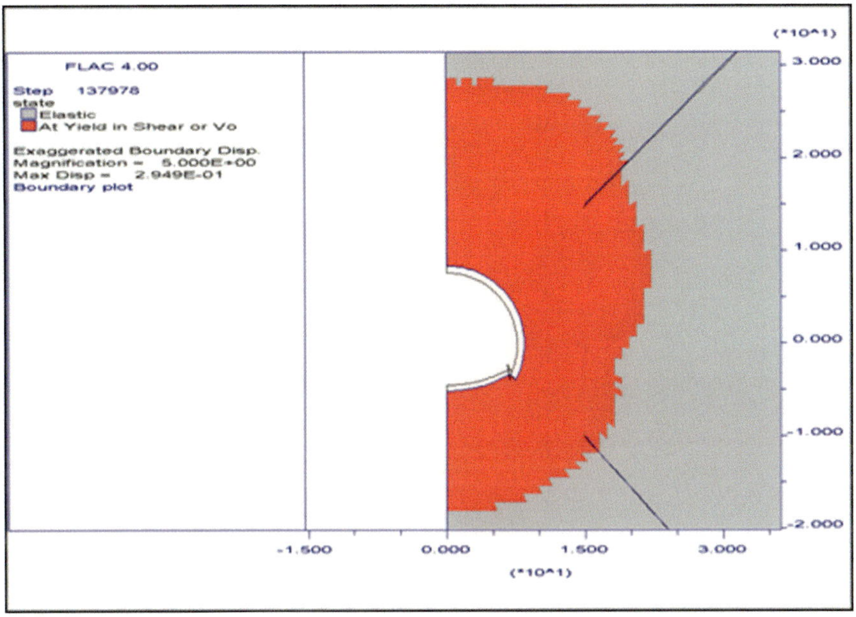

Fig. 7.49 Plastic elements and displacements just before final lining vault casting ($x = 32$ m, EFF = 0.22 %) (ref. point 5.c)

being:

$$\varepsilon_{\max} = (R \cdot A_{\max})/C_s^2$$

C_s propagation speed of waves *S*
$A_{\max}$ acceleration on the ground variable according to the seismic area
R tunnel radius

- Contribution of the earthquake to the cross action: due to the seismic motion, the cross section becomes elliptical and it can be represented by relative displacements between the different points of the structure according to the expression:

$$S(\mathrm{dx}) = V_{\max}/\,(\omega \cdot C) \cdot \sin(\omega \cdot \mathrm{dx})$$

being:

$\omega = 2 \cdot \pi / \lambda$ with λ wave length
C_p propagation speed of waves *P*

The distortion *S*(dx) of the calculation will be compared with the limit values, in an order of magnitude equal to 1/150, that can cause structural problems and loss of equilibrium.

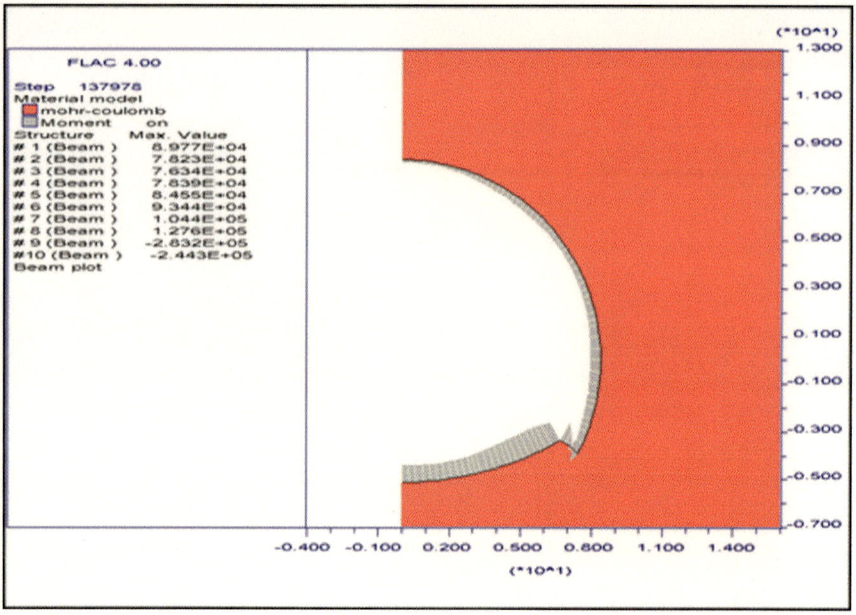

Fig. 7.50 Bending moment on supports just before final lining vault casting ($x = 32$ m, EFF $=$ 0.22 %) (ref. point 5.c)

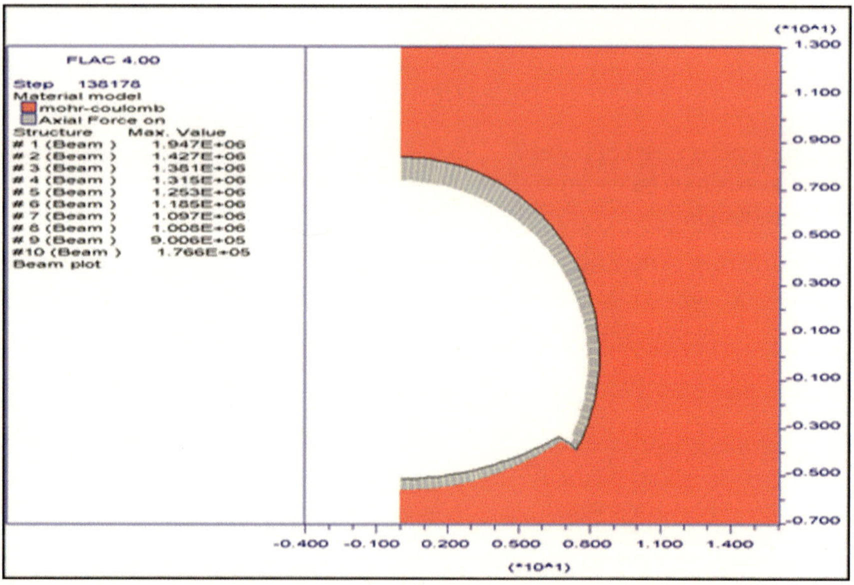

Fig. 7.51 Axial force on final lining ($x = \infty$, EFF $=$ 0 %) (ref. point 5.c)

7.11 Final Considerations

Some final considerations can be drawn considering following points: the ground reaction curve, the analyses conducted to evaluate the face stability, the comparisons between loads and strengths of the linings used to define the structures and actions to be provided in order to ensure the stability of the cavity. In general, in order to successfully face the advancing phase in difficult grounds, designing strategies can be evaluated that are based on two alternative principles:

- "Strength"
- "Escape"

From the theoretical point of view, in case of the "strength" principle, the aim is the containment of the ground deformations, in the limits allowed by the strength of the confinement structures. This approach is based on the idea that excessive deformations imply an uncontrollable increase in the thrusts on the linings through a progressive decrease of the strength characteristics of the ground. The tools used in this design and operational approach are the full section excavation, the reinforcements of the excavation face, the application of "stiff" first linings, even closing the invert, the casting of the invert at a very small distance from the face, the closing of the lining at a very small distance from the face as well.

This approach implies relevant design and operational consequences. Loads on the first-phase lining increase, in consequence of the high share of hindered convergence. Therefore the casting of the final lining has to be more and more anticipated. The final lining gets more and more loaded and has to be made more and more resistant introducing reinforcements in very short times.

The limits of this approach are represented by the fact that, in extreme applications, the series of underground implementation steps set in shorter length, thus implying smaller and smaller working areas. The development of the strength of concrete castings is not quick enough to face the thrusts transmitted to the lining. The quantity of reinforcement is excessive. The strength required to the first-phase lining implies the installation of heavier and heavier ribs and thicker and thicker layers of shotcrete. The intensity of the face reinforcement increases exponentially.

The alternative approach based on the "escape" principle intends to limit the stresses on the confinement structures as much as possible and in the limits allowed by the ground deformability. It is assumed that the decay of the strength characteristics of the ground can be controlled, thus limiting deformations within acceptable thresholds, without an uncontrollable increase of the thrusts on the linings.

This design and operational approach implies the implementation of deformable first-phase lining, the casting of the invert and the lining closure at a not very short distance from the face, the excavation of a wider tunnel in order to absorb convergence. In particular, it implies most of all the use of radial nailing or of face supports to rule and control the evolution of deformations and to give ductility to the ground and first-phase lining.

Design and operational consequences are represented by the fact that deformable structural elements have to be used and the length and intensity of nailing have to be increased with the increase of accepted deformations.

The limits of this method in extreme applications are the length of nailing, its effectiveness in low-quality grounds and the real possibility to create deformable linings that guarantee the stability of the cavity and the uniformity of convergences. Monitoring is always required due to the difficulty to assess beforehand, the acceptable deformation level. Often the acceptable deformation level is small; this requires compromises with the approach based on the "strength" principle.

Even the latter requires continuous monitoring to avoid the excessive stress of the structures and, in the praxis, it can be convenient to combine both the methods (the strength approach and the escape one) to get the best result.

Therefore, a rigid application of both methods is often counterproductive.

References

Assadi A, Sloan SW (1991) Undrained stability of shallow square tunnel. J Geotech Eng 117(8):1152–1173

Attewell PB, Boden JB (1971) Development of the stability rations for tunnels driven in clay. Tunnnels Tunnelling Int 3:195–198

Broms BB, Bennemark H (1967) Stability of clay at vertical openings (ASCE). J Soil Mech Found Eng Div, SMI 93:71–94

Bhasin R (1994) Forecasting Stability Problems in Tunnels constructed through clay, soft rocks and hard rocks using an inexpensive quick approach Gallerie e Grandi Opere Sotterranee, 42 (1994), pp 14–17

Caquot A, Kerisel J. (1956) Traile de mécanique des sols. Gauthier Villars, Paris

Carranza-Torres C (2004) Computation of Factor of safety for shallow tunnels using Caquot's lower bound solution—Summary Report (May 2004)

Cornejo L. (1988) El fenomeno de la inestablidad del frente de excavation y su repercussion en la construccion de tuneles. Tunnels and Water, Serrano, vol. I. Balkema, Rotterdam, pp 79–87

Davis E, Gunn M, Mair R, Seneviratne H (1980) The stability of shallow tunnels and underground opening in cohesive material. Geotechnique 30(4):397–416

Ellstein AR (1986) Heading failure of lined tunnels in soft soils. Tunnels & Tunneling 18:51–54

Fuoco S, Lucarelli A, Pasqualini E (1997) Contribution to the definition of tunnel face stability of deep tunnel in continuous media. ITA Conference, San Paolo

Goel RK, Jethwa JL (1991) Prediction of support pressure using RMR classification. In: Proceedings of the Indian Geotechnical Conference. Surat, pp 203–205

Grasso P, Russo G, Xu S, Pelizza S (1993) Un criterio per la valutazione speditiva del comportamento di gallerie allo scavo mediante classificazione geomeccanica Gallerie e Grandi Opere Sotterranee, n.39

Hoek E, Brown ET (1980) Underground Excavations in Rock. Institution of Mining and Metallurgy, London

Hoek E, Marinos P (2000) Predicting tunnel squeezing problems in weak heterogeneous rock masses. Tunnels and Tunnelling International, pp 45–51: part one; pp 33–36: part two

Hoek E, Marinos P, Benissi M (1998) Applicability of the geological strength index (GSI) classification for very weak and sheared rock masses. The case of the Athens Schist Formation. Bull Engg Geol En 57(2):151–160

Houska J (1960) Beitrag zur Theorie der Erddrücke auf das Tunnelmauerwerk. Schweizerische Bauzeitung 78(1960):607–609

Kolymbas D, Wagner P (2007) Groundwater ingress to tunnels—The exact analytical solution. Tunnelling Underground Space Technol 22(1):23–27

Kommerell O (1940) Statiche Berechnung Von Tunnelmauerwerk. W. Ernst & Sohn, Berlin

Lane KS (1957) Effect of limiting stiffness on tunnel loading. Proceedings of the Fourth Internationa Conference on Soil Mechanics and Foundation Engineering, London, 12–24 August 1957

Leca E, Dormieux L (1990) Upper and lower bound solutions for the face stability of shallow circular tunnels in frictional material. Géotechnique 40(4):581–606

Leca E, Dormieux L (1992) Contribution à l'étude du front de taille d'un tunnel en milieu cohent. Revue Française de Géotechniquen 61:5–16

Leca E, Panet M (1988) Application du calcul à la rupture à la stabilité du front de taille d'un tunnel. Revue Française de Géotechnique 43:5–20

Panet M (1995) Calcul des tunnels par la méthode convergence-confinement. Presse de l'Ecole des Ponts et chausses. Paris, pp 13–14

Panet M, Guenot A (1982) Analysis of Convergence behind the face of a tunnel. International Symposium "Tunneling 82". Brighton, pp 197–204

Rabcewicz Lv (1964) The New Austrian Tunnelling Method, Water Power. Parte one November 1964, pp 453–457. Part Two, December 1964, pp 511–515

Rabcewicz Lv, Golser J (1973) Principles of dimensioning the supporting system for the "New Austrian Tunneling Method", Water Power, March 1973, pp 88–93

Ritter (1879) Statik der Tunnel gewlbe, Berlin. Springer

Sakurai S (1997) Lessons learned from field measurements in tunnelling. Tunnellin and underground space technology. Pergamon, Oct–Dec, Volume 12 Issue 4, pp 453–460

Schuck W (2005) Rock and Water Pressure when dimensioning tunnels in rock. Tunnel 3:43–45

Tamez E (1985) Estabilidad de tuneles excavados en suelos. Curso Victor Hardy 85, Mexico

Terzaghi K (1951) Mécanique théorique des sols. Dunod Ed., Paris

Terzaghi K, Proctor RV, White TL, (1946) Rock tunneling with steel supports with an introduction to Tunnel Geology, Youngstown

Unal E (1983) Development of design guidelines and roof-control standards for coal-mine roofs. Ph. D. Thesis, Pennsylvania State University

Chapter 8
Monitoring

8.1 Introduction

The present chapter describes the characteristics and the equipment of the monitoring activities, which should be performed during an underground excavation. Generally, monitoring has two main aims:

- Performing an aid to the design, by verifying if planned previsions match with the actual behaviour and conditions of the ground, registered during construction.
- Ensuring that the structures will be able to carry out the function for which they were designed, both during construction (first-phase linings) and during its service life (final linings).

For these purposes the following main surveys, measurements and controls can be performed:

- Geomechanical surveys of the excavation face or walls
- Convergences
- Deformation of the ground
- Strain and loads on linings
- Water-pressure measurements
- Acoustic emissions
- Working parameters recording and specific geophysical surveys for TBM excavation
- Surface settlements and surrounding infrastructures monitoring

During the work progress, the monitoring system should collect as much information as possible and as relevant for the design in order to evaluate the effectiveness of ground improvements and confinement systems, and the suitability of the construction methods adopted.

As a rule, for tunnels, the position of measuring stations should not be strictly predetermined but they should be placed where either geomechanical changes or the cross-section variations dictate. In a similar way, the monitoring outside the tunnel has to be foreseen close to the tunnel layout, within the subsidence area identified

P. Gattinoni et al., *Engineering Geology for Underground Works*,
DOI 10.1007/978-94-007-7850-4_8,

by the tunnel design, paying particular attention to existing buildings, works, slopes with potential landslides etc.

Usually, the monitoring frequency varies in time. The first reading (reading zero) takes place immediately after the instrument installation; then, the closer is the excavation face, the more frequent will be the readings. Moreover, in case of relevant changes in the measured magnitude, it would be better to increase the reading frequency or even to install further monitoring instruments in order to obtain a more complete picture of the evolution of the phenomenon in time. As a rule, the measurements should be performed until complete stabilization is reached (e.g. convergence measured with optical targets < 0.1 mm/d) and, in case of instruments placed on first-stage linings, a final reading is to be carried out before placing the final lining. Note that the zero reading is only aimed at providing information about the initial state of the parameter measured and being a reference for following readings. Therefore, that measure has to be performed very carefully. For example, a more detailed procedure is used for the first reading of the inclinometers with respect to the following paragraphs.

In summary, a minimum measurements frequency and indicative positions can be provided in design, but they must always be adjusted to the actual response of the rock mass to the excavation.

Often instruments characterized by a data acquisition system with sensors are used, which perform an automatic reading, thanks to the placing of a peripheral data collector. Data can be conveyed in a multichannel data collector placed outside the tunnel. The use of this kind of station avoids the direct reading at the relevant section. The memory of the data acquisition system has to be suitably calibrated on the frequency of reading.

8.2 Geomechanical Surveys

The geomechanical surveys of excavation face provide information concerning the characteristics of lithological, structural and geomechanical properties of rock mass (understood as a complex consisting of the rock matrix and of the discontinuities); they also provide information concerning geotechnical and stratigraphic characteristics of soils.

Two types of surveys are possible:

- *Detailed face surveys:* to be carried out in case of any significant change in structural-geological, geomechanical conditions or significant changes in the geometry of the excavation section.
- *Quick face surveys:* to be carried out systematically.

During the excavation, the typical section (in terms of rock mass reinforcements and linings) to be adopted can be confirmed, adjusted or redefined, in accordance to the results obtained from face surveys.

In general, the face mapping requires following equipment:

- Geological compass to measure the planes orientation in space
- Disc (diameter 30 cm) for the support of the compass

- Tape measure and roller tape measures
- Schmidt's Hammer
- Point Load Strength Test
- Profilometers (Barton's shape tracer)
- Pocket penetrometer
- Digital camera

Laboratory tests such as uniaxial compression of rock may also be necessary.

The face survey activities must be carried out by geologists and geomechanical engineers having specific training and experience. The operators complete a separate report sheet for each survey type: geostructural and geomechanical.

The report sheets (Figs. 8.1 and 8.2) usually contain following information:

- For the geostructural survey:
  - Excavation advancement
  - Typical section, first- and second-phase lining description
  - Lithology, rock characteristics, tectonic structures and the fracturing conditions
- For the geomechanical survey:
  - Rocks classification (rock mass rating system, e.g. after Bieniawski)
  - Situations of risk and anomalies found in the first-phase lining
  - Behaviour of the rock mass due to the excavation
  - Improvement/deterioration trend during advancement

For the detailed face survey, both report schedules are to be filled in, whereas the quick face surveys must include photos and a brief description of the observed face conditions.

8.3 Measurements of Convergence

Automatic recording theodolites and electronic equipment can be used to measure the convergences. They must allow the execution of the distance measurements of the instrument from the geodetic spades with an error < 1 mm for distances up to 80 m, in normal visibility conditions in tunnel, and < 3° for directions. The execution of the convergence measurements requires the ongoing employment of an experienced surveyor and an assistant.

The geodetic spade consists in optical targets (reflectors) mounted on a standard convergence bolt which is cemented in the rock and has a length of at least 0.5 m. The optical targets should be installed at the minimum possible distance from the face.

The data obtained allow:

- To calculate the position of the optical targets in a local coordinate system (x, y, z)
- To calculate the differences between the coordinates of the optical targets in the successive readings

Geological survey report				
Document n°: 56				
Distance: 11164.70 m	Survey n°: 56	Date: 15/06/2012	Time: 13:00	Operator:
	Excavation method: Explosive	Pull length: 2.20 [m]	Section type:	Overburden: 628.2 [m]
Lithology description: Tonalite with oriented minerals, visible grains of minerals (up to 1cm): mica altered to chlorite and quartz, the latter has a flattened shape having the same orientation of micas. Presence of isometric crystals of red-orange garnet with some mm diameters				
Rock mass description: Rock mass is very blocky. Two major joint families are present. Small discontinuities and irregular fissures (length of some dm and thickness <1mm) filled with quartz				
Hydraulic conditions: Dry				
Rock mass behavior – stability: Two perpendicular systems of joints and a zone of mylonite on the right side, possible overbreak of some cubic meters				
Potential risks:				
Discontinuity description: On the left side of the face a fault with core in mylonite is noted, this is linked to damaged zone of strongly fractured tonalite whose planes are parallel to the alignment of the fault core (200°/80°)				
Type of rock mass: 100% Pusteria Fault	Type of rock mass behavior: TCR 4	Overbreak [m³]:	Trend: improvement	
Reinforcement structures:		Anomalies in lining:		

Geological survey report				
Document n°: 56				
Distance: 11164.70 m	Survey n°: 56	Date: 15/06/2012	Time: 13:00	Operator:
	Excavation method: Explosive	Pull length: 2.20 [m]	Section type:	Overburden: 628.2 [m]

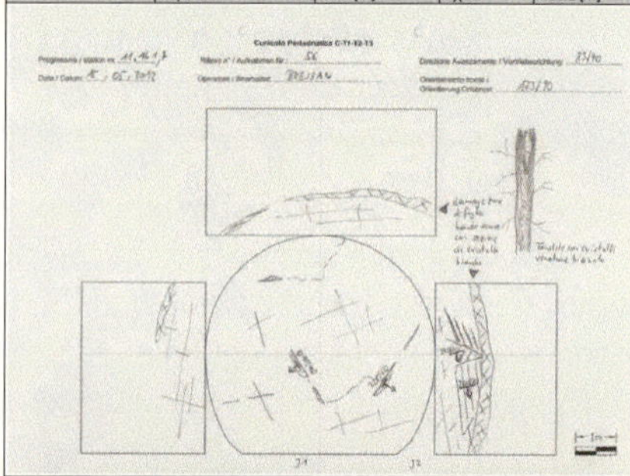

Geological survey report				
Document n°: 56				
Distance: 11164.70 m	Survey n°: 56	Date: 15/06/2012	Time: 13:00	Operator:
	Excavation method: Explosive	Pull length: 2.20 [m]	Section type:	Overburden: 628.2 [m]
Photo				

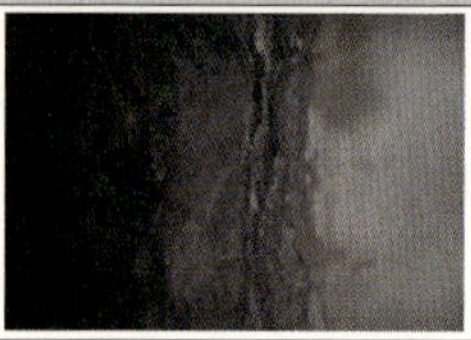

Notes:

Dark-light mylonite bands on the right side of the tunnel

Geological survey report									
Document n°: 56									
Distance: 11164.70 m			Survey n°: 56		Date: 15/06/2012	Time: 13:00		Operator:	
			Excavation method: Explosive		Pull length: 2.20 [m]	Section type:		Overburden: 628.2 [m]	
SURVEY DATA									
Lithology	%	Col.	Grains dimension [mm]	Weathering	Water pressure variation	Schmidt hammer [MPa]	Fract. degree	Sampling	
Tonalite	100	...		Slightly weathered	none				
Structural type	Strike(°)/Dip(°)		Spacing [mm]	Persistence [m]	Roughness (JCR)	Fill	Weath.	Opening [mm]	
1) Joints	010/40		600-2000	<1	0-2	none	slight	<0.1	
2) Joints	200/80		600-2000	1-3	0-2	none	slight	<0.1	
Water	Type/Loc.		Flow trend	Flow [l/s]	Temp. [°C]	pH	El. cond. [µS]	Sample	
-	-		-	-	-	-	-	-	
DATA PROCESSING									
Discontinuities – graphical representation				Block stability					

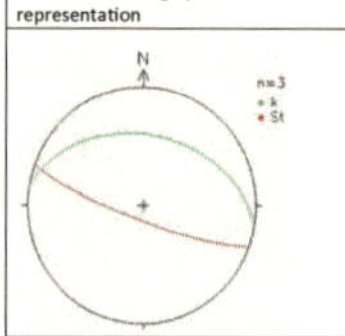

Block volume [m³]: 18.2

Block weight [t]: 49.0

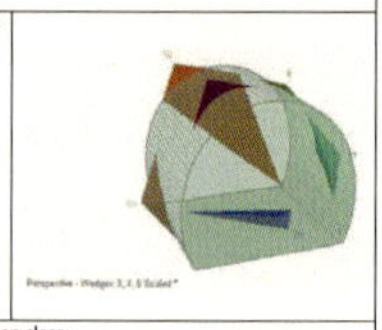

Geological strenght Index (GSI): 50
GSI Structure: Very blocky
GSI Surface condition: Fair

Water hendicap class: -

Geological survey report						
Document n°: 56						
Distance: 11164.70	Survey n°: 56		Date: 15/06/2012	Time: 13:00	Operator:	
	Excavation method: Explosive		Pull length: 2.20 [m]	Section type:	Overburden: 628.2 [m]	
RMR classification						
Classification Parameter		**Range of values**				
Rock strength	Point load test	> 10 MPa	4 - 10 MPa	2 - 4 MPa	1 - 2 MPa	
	UCS	> 250 MPa	100-250 MPa	50-100 MPa	25-50 MPa	5-25 1-5 <1
		15	12	7	4	2 1 0
RQD		90-100%	75-90%	50-75%	25-50%	<25%
		20	17	13	8	3
Discontinuity spacing		> 2 m	0.6 -2 m	200-600 mm	60-200 mm	< 60 mm
		20	15	10	8	5
Joint conditions	Persistence	< 1 m	1-3 m	3-10 m	10-20 m	> 20 m
		6	4	2	1	0
	Opening	none	< 0.1 mm	0.1-1.0 mm	1-5 mm	>5 mm
		6	5	4	1	0
	Roughness	Very rough	rough	Slightly rough	Plane	Smooth
		6	5	3	1	0
	Fill	none	hard<5mm	hard>5mm	plast.< 5	plast.>5
		6	4	2	2	0
	Weathering	none	weak	moderate	high	decomp.
		6	5	3	1	0
Water	General condition	none	humid	wet	dripping	flow
	Q 10m tunnel	0 l/min	< 10 l/min	10-25 l/min	25-125 l/min	>125 l/min
		15	10	7	4	0
Orient. Joints-Tunnel		very favorable	favorable	fair	unfavorable	very unfavorable
		0	-2	-5	-10	-12
					RMR - Index:	48
					Rock mass class:	III

Fig. 8.1 Survey of the excavation face: descriptive form with hand drawing and picture with relevant elaboration

- To draw a graphical plot of the results as specified below

For each measurement station, the obtained results must be reported by plotting at least on following graphs:

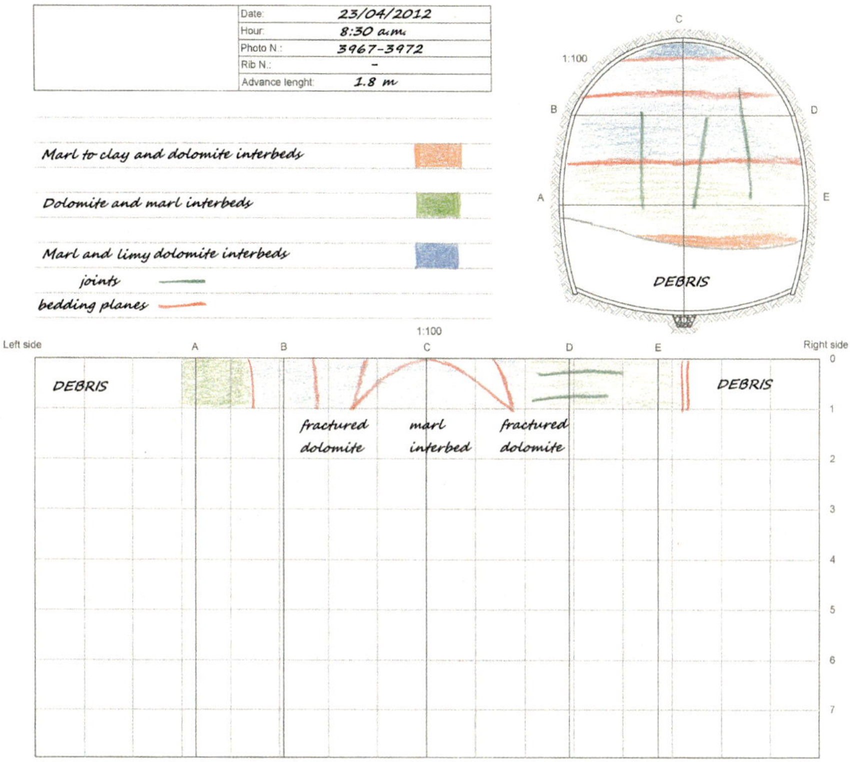

Fig. 8.2 Survey of the excavation face: form filled during the survey and relevant elaboration

- Displacement vector of each optical target versus time, with an indication of the activities developed in the tunnel
- Displacement vector of each optical target versus distance from face, with an indication of activities developed in tunnel
- Displacement at the measuring section with an indication of theoretical profile
- The x, y and z components of displacement vector of each optical target versus time, with an indication of activities developed in tunnel
- The x, y and z components of displacement vector of each optical target versus distance from face, with an indication of activities developed in tunnel

Figures 8.3, 8.4, 8.5 and 8.6 represent typical measuring stations with five optical targets (three along the crown, and two at the sides) and the relevant results.

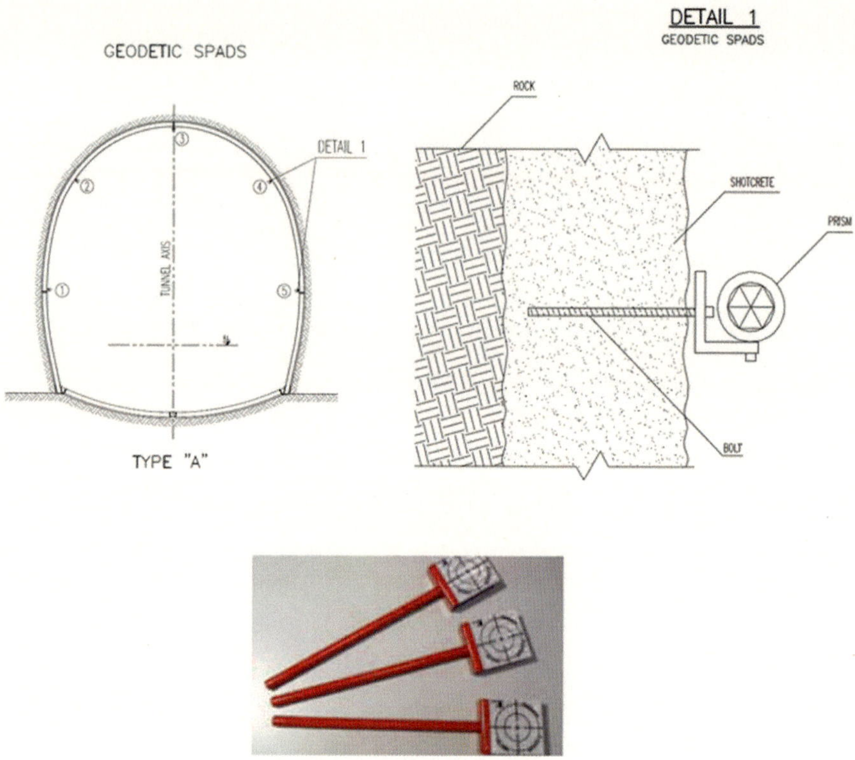

Fig. 8.3 Optical targets: location on the monitored section, installation detail

Fig. 8.4 Optical targets. (by Pizzarotti)

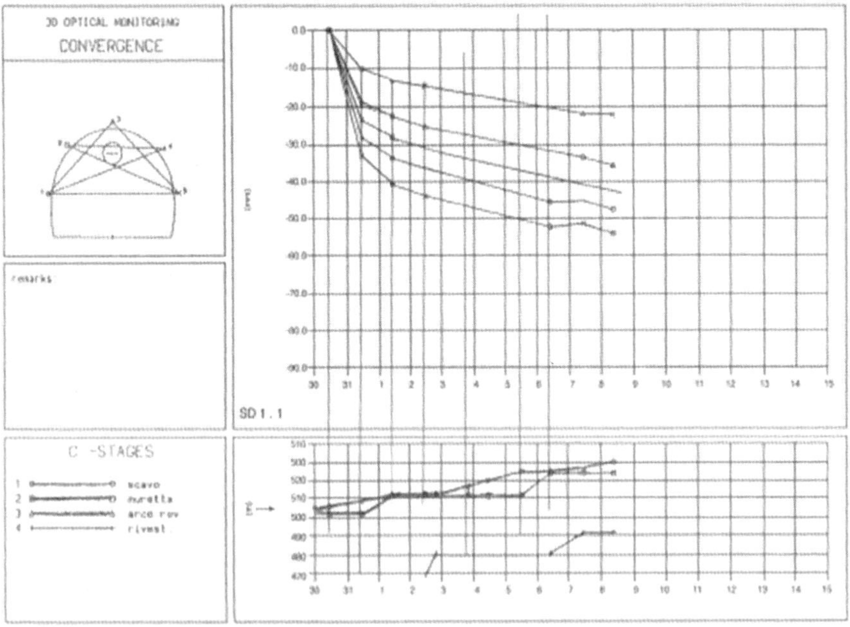

Fig. 8.5 Typical graphical plots of convergence readings by mean of optical targets

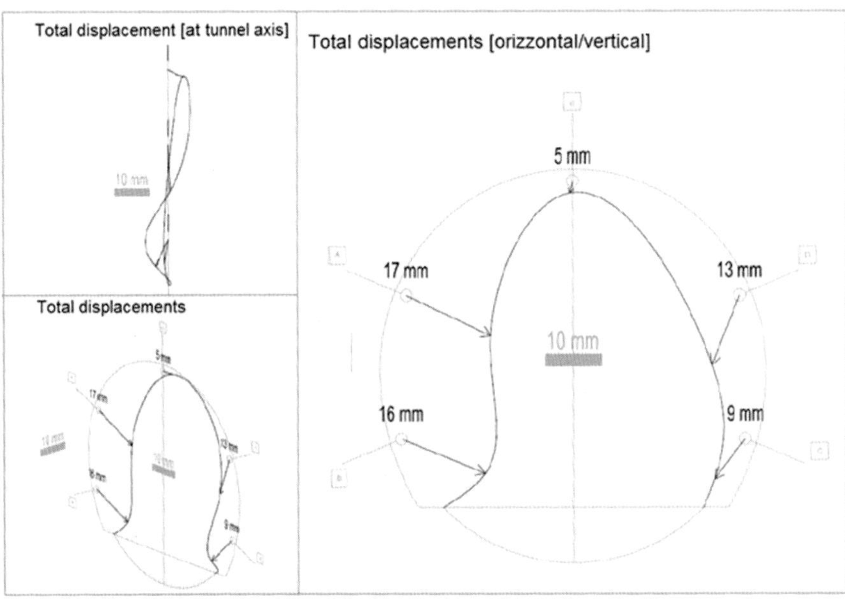

Fig. 8.6 Elaboration of the results of convergence readings

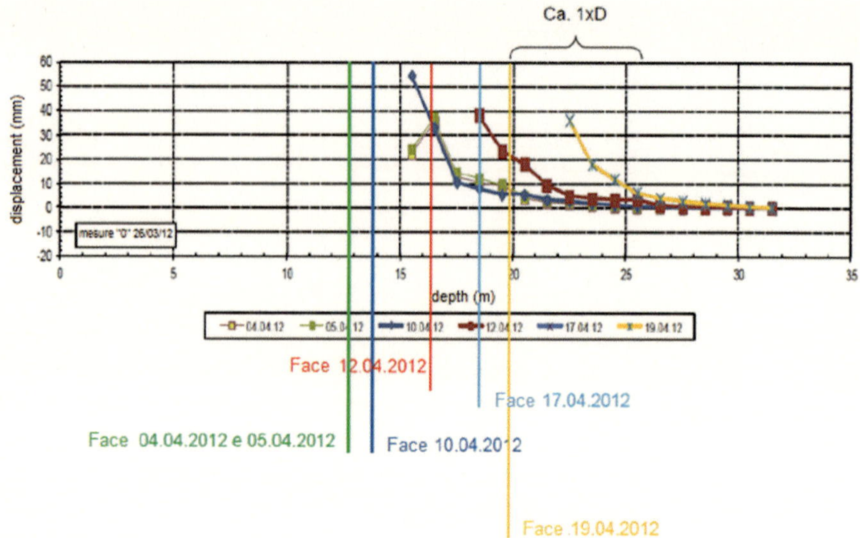

Fig. 8.7 Results of the measurements of the face extrusion through an extensometer. The position of the excavation face on different dates of measurement are also shown

8.4 Measures of Rock Deformations

8.4.1 Face Extrusion

The longitudinal displacements that may occur at the excavation face are usually monitored through incremental extensometers (extensometers) or optical targets installed on the tunnel nucleus.

The incremental extensometers are placed inside the sub-horizontal boreholes and allow the continuous monitoring of the axial displacements concerning a series of measurement bases. Measurements are carried out in a PVC pipe, whose length is usually equal to at least 4–5 times the tunnel diameter, fitted with suitable metal-ring anchors placed at a distance of 1 m from each other. Anchors are connected to the surrounding ground through a concrete injection, but they are free to move along the pipe. The measure of the changes in distance between pairs of nearby anchors are carried out using a probe that detects the exact position of the anchors while passing through the hole.

The probe is fitted with a high precision potentiometric sensor to detect the position of the anchors. This records the positions of the metal rings placed along the pipe.

These instruments allow the measurement also if the PCV pipe has been partially demolished while advancing; therefore, they are particularly suitable for the monitoring of the excavation face (Fig. 8.7).

Fig. 8.8 Head of a multibase extensometer

Similar instruments are available, as well, that do not require the use of the measuring probe. In this case, the reading of the anchors' position is carried out through cables connected to a data logger placed at the bottom of the instrument.

Like the convergence measurements, the face extrusion can also be measured using optical targets. This discontinuous monitoring method can be applied, for example, to monitor the face extrusion while the works are suspended for the weekend, for the implementation of reinforcement while advancing or in case of prolonged interruptions due to other factors.

8.4.2 Radial Deformations

The incremental extensometers previously described for the face monitoring can also be used to measure radial deformations; moreover, as the instruments are not damaged during their function, the measures of deformations on the cavity contour can be made also using multibase extensometers. They allow to evaluate the deformations along the measurement beams connected to deep anchors, with respect to the tool head placed on the first-phase lining (or on the surface, or in a cavity close to the one to monitor) (Fig. 8.8), whose displacement has to be detected with precise topographic measurements. The comparison between the displacement of the intermediate beams with respect to the longer one and/or with respect to the head allows to determine the differential displacements.

Usually, a measurement station is made by at least three multibase or incremental extensometers, one placed at the ceiling (vertical) and two on the sides (horizontal). Figure 8.9 shows a section monitored with five extensometers having six measure bases placed at a depth of 15 m, 12 m, 9 m, 6 m, 3 m and 0.5 m, one at the ceiling and two on each side.

The use of multibase or incremental extensometers can be foreseen in the sections monitored with vibrating string extensometers (see following paragraph); in this way, the deformations of the monitored lining can be linked to those of the rock mass.

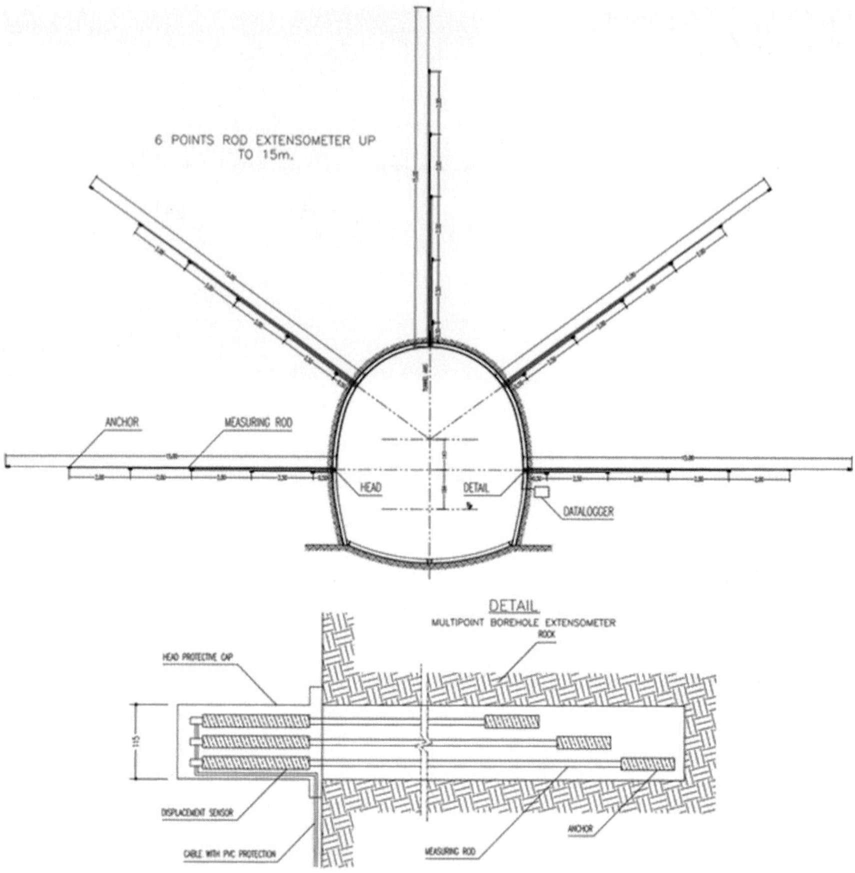

Fig. 8.9 Section monitored with extensometers with 6-measure bases placed at a depth of 15 m, 12 m, 9 m, 6 m, 3 m and 0.5 m

If they are installed from the surface, multibase or incremental extensometers can provide useful information also in case of monitoring of slopes or vertical displacements occurring around an underground work.

8.5 Measures on Linings

8.5.1 Assessment of the Strain with 'Strain Gauges'

Strain gauge bars with vibrating strings are made by a harmonious iron string stretched between two blocks, fixed at the rib wings or at the reinforcement bars through bolting or resin. The vibrating frequence of the iron string is a function of the deformation of the element in the section considered.

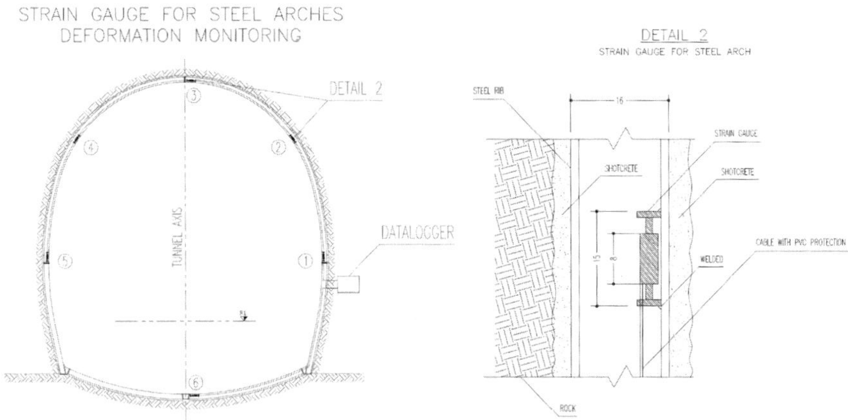

Fig. 8.10 Rib monitoring through strain gauges

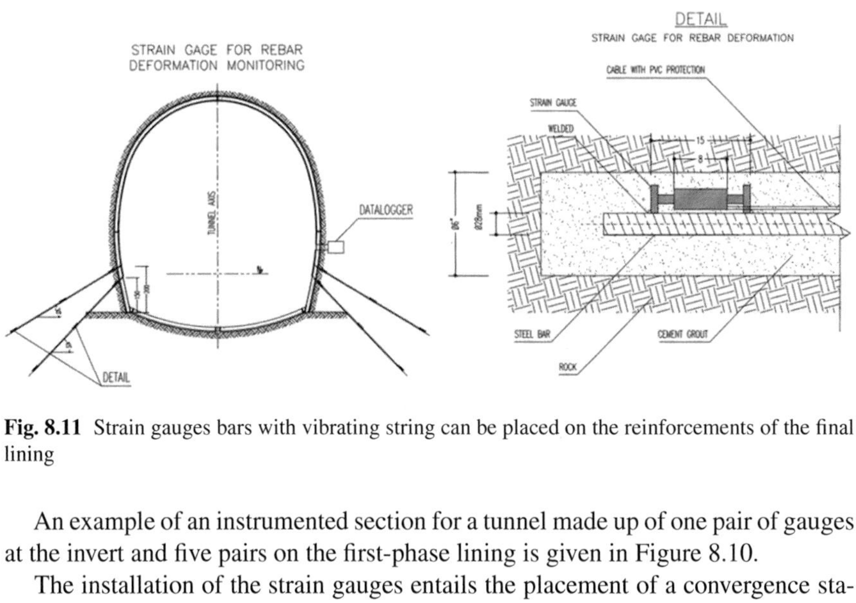

Fig. 8.11 Strain gauges bars with vibrating string can be placed on the reinforcements of the final lining

An example of an instrumented section for a tunnel made up of one pair of gauges at the invert and five pairs on the first-phase lining is given in Figure 8.10.

The installation of the strain gauges entails the placement of a convergence station. Sections monitored through strain gauges are systematically distributed along the tunnel; moreover, they have to be located at most loaded stretches (highest overburdens, crossing cavities, enlargements etc.).

When both types of instruments (strain gauges and optical targets) are installed, the measurements of strain gauges must be carried out with the same frequency of the convergence one.

Strain gauge bars with vibrating strings can be installed on the ribs, as shown in Figure 8.10, on steel reinforcement protected by shotcrete or in the final lining (Fig. 8.11), on nails used to stabilize the ground (Fig. 8.12), or on precast segments of the lining rings of an excavation by TBM (Fig. 8.13).

Fig. 8.12 Nail monitoring through strain gauges

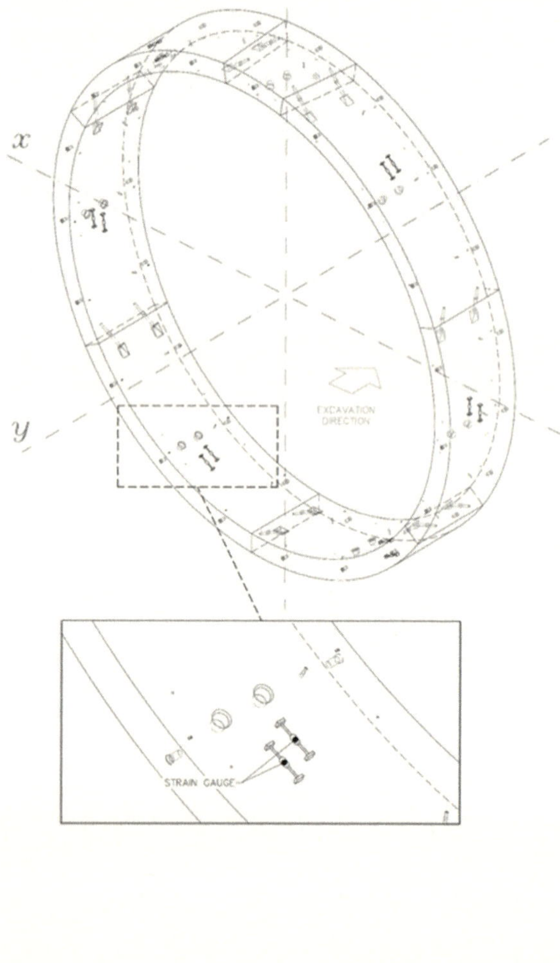

Fig. 8.13 Placement of the strain gauge on the lining with precast segments for TBM

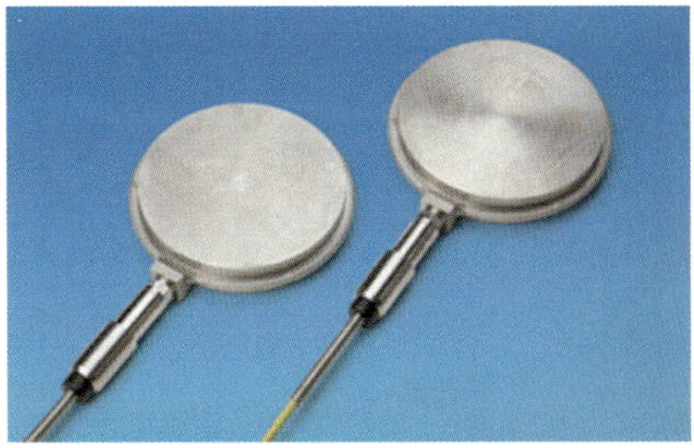

Fig. 8.14 Load cell

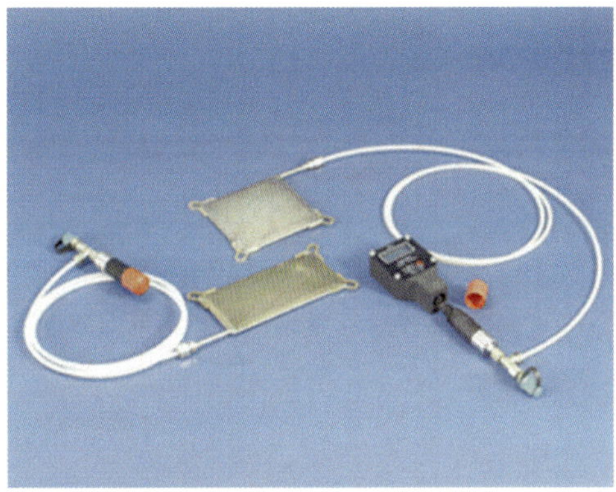

Fig. 8.15 Pressure cell

Knowing the deformation, the Hooke's law ($\sigma = \varepsilon E$) allows the identification of the stress field at the instrument, taking into account the Young's modulus of the element whose stress has to be calculated (shotcrete, concrete, steel). Moreover, the effects of the temperature on the instruments are always to be considered.

8.5.2 Assessment of the Stress

The measurements that allow the direct survey of the stress field in the linings can be made with load or pressure cells placed under the bases or between the connecting plates of the steel ribs (radial measures), between the first-phase lining and the excavation wall (tangential measures) or under the anchor plates for the stress measure.

Load cells (Fig. 8.14) are constituted by a central element in steel and a number of strain gauges grids applied to the inner surface of the element and isolated. A steel plate allows the homogeneous distribution of the load on the whole cell. The deformation due to the cell load is detected by the strain gauges, transformed in an electric signal proportional to the load, and transferred to the reading control unit.

The pressure cells (Figs. 8.15 and 8.16) are made by two steel sheets welded along the whole perimeter and separated in the inner part by a thin cavity that is saturated

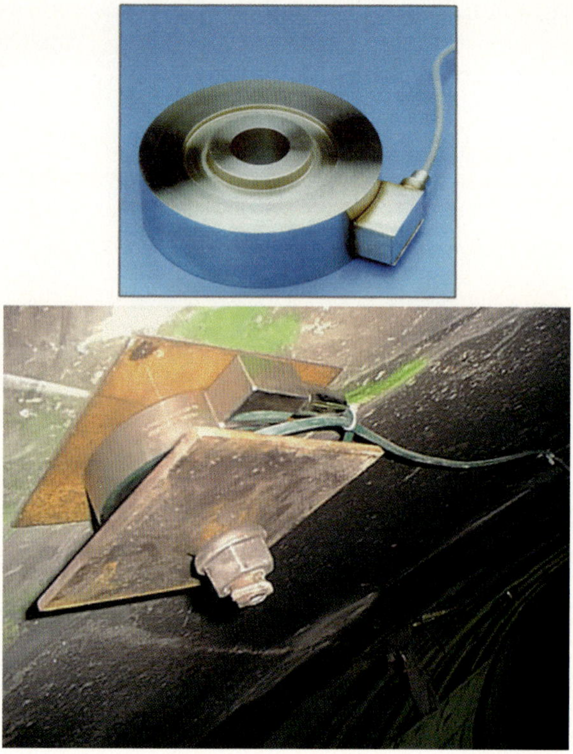

Fig. 8.16 Toroidal load cell. The particular shape with central hole allows to place the cell under the anchor plate to monitor the strength

with vacuum disareated oil that guarantees the maximum stiffness; the change in the pressure due to the load can be measured with a manometer linked to the cell through a pipe.

8.6 Measurements of Pressure and Flow Rate

The measure of the flow rate, conductivity, temperature and pH of tunnel water inflow, with a cyclical repetition in time, is finalized at understanding the hydrodynamic water regime and the correlated flow systems and draining paths. The sampling of some of the main flows can be foreseen as well as the chemical and isotropic analysis aimed at a better understanding of the aquifer.

Piezometers and thermistors can be placed in correspondence to the main water flows; the latter are used for the continuous measurement of the water temperature.

Flow rate measurements have to be carried out behind and beyond the main aquifer areas, for example with a spillway, aimed at understanding the connectivity among the different aquifer systems intercepted by the tunnel, in order to reconstruct the mass hydrodynamics.

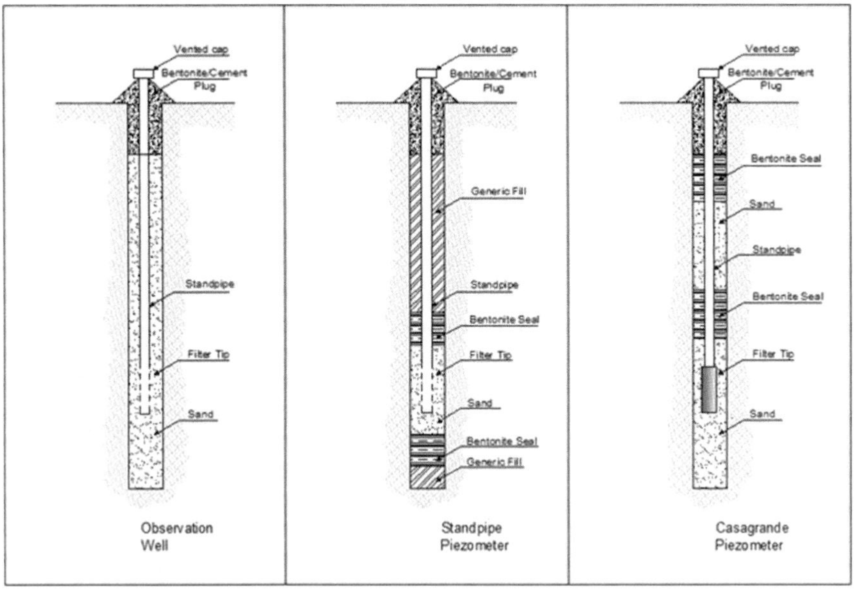

Fig. 8.17 Installation modes of fix-pipe piezometers

In general, if water flows from the prospection holes while advancing and/or draining holes, the water pressure has to be measured with a manometer, and the flow rate with graduated containers. Measurements will have to be repeated until the flow rate and pressure values are completely stabilized.

8.6.1 *Piezometers*

The measures for the monitoring of the water-table level along the tunnel layout and of the interstitial pressures around the tunnel can be performed using piezometers. A piezometer is an instrument made by a pipe placed in a borehole through which the water interstitial pressure is measured in a well-defined ground layer. Generally speaking, a difference can be made between piezometric cells fitted with a pressure transducer (pneumatic, with vibrating string or electric resistance) characterized by a membrane separating a pipe stretch containing the transducer and the piezometers where the membrane is not present (with single or multiple open pipe). The choice of the instrument depends on the hydrogeological characteristics of the ground where it is placed and on the speed of the correct detection of the changes in pressure (the so-called instrument time-lag). In general, instruments with longer time-lags are suitable for monitoring in high-permeability grounds ($K > 10^{-6}$ m/s), whereas those characterized by a prompt response are also suitable in little permeable grounds.

In the open-pipe piezometers schematized in Figure 8.17, the measure of the piezometric level occurs lowering a galvanometric probe (also called freatimeter), or using pressure transducers: pneumatic, with electric resistance or vibrating string,

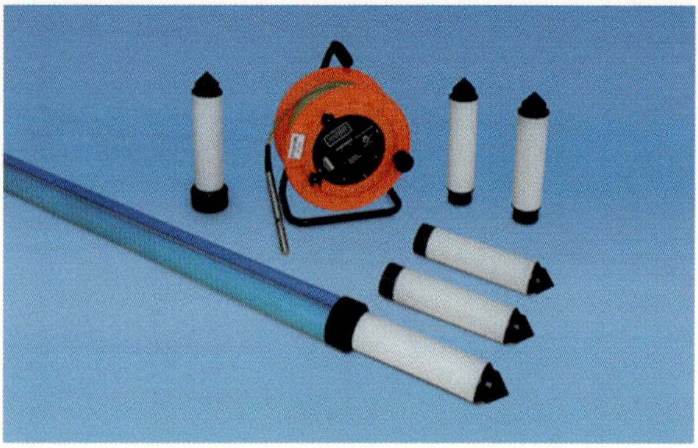

Fig. 8.18 Casagrande piezometers and galvanometric probe

hung below the lowest possible piezometric level. The transducers can be left in situ and collected during the periodical calibrations; their use allows the reading from far away locations also. Open-pipe piezometers are made by a stiff PVC or metal pipes, fissured in its final part and, if necessary, with a tissue–non-tissue sleeve covering the part in the groundwater (this in order to limit the presence of fine material inside the pipe). The upper part of the pipe is sealed with concrete and bentonite and closed with a cap to prevent water inflows from the surface. The cap has a hole, connecting with the outside so that the pressure inside the pipe is always equal to the atmospheric one.

According to the installation mode, a distinction can be made among:

- Monitoring well: The cavity between pipe and hole is filled with sand and gravel, thus creating a connection between the different ground layers. For this reason, observation wells are normally limited to permeable homogeneous ground ($K > 10^{-6}$m/s) where water pressure increases equally with depth.
- Open piezometer with fix pipe: The cavity is sealed with concrete and bentonite above the layer to monitor so that the porous-seeping element measures the interstitial pressure of the precise soil layer. It is installed using a temporary pipe (sleeve pipe) that allows, when removed, to create waterproof and filtering gauge according to a scheme shown in Figure 8.18.
- Casagrande piezometer: It is a fix-pipe open piezometer, but the filtering element is represented by a cylinder in porous stone or in high-density porous polyethylene. The upper end of the porous stone is connected to a PVC small-section pipe (or by two pipes) for the connection on the surface. Also, two Casagrande piezometers can be placed in a borehole at different heights.

Piezometric cells (Fig. 8.19) may use different pressure transducers:

- Pneumatic tranducer: It is made by a compensation chamber connected to a steel membrane that compresses under the water thrust and stretches according to the gas (nitrogen) introduced through a pneumatic control unit. The equilibrium point between the two pressures can be read on the unit display and it corresponds exactly to the interstitial pressure. The piezometric cell and the reading unit are connected by means of a special pneumatic cable.

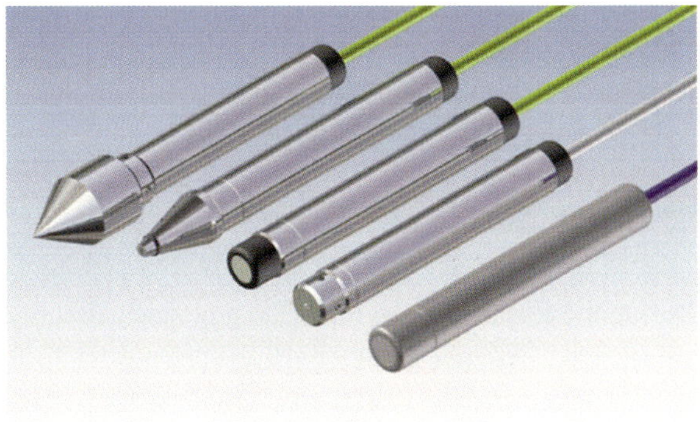

Fig. 8.19 Electric piezometers

- Vibrating string transducer: It is fitted with a metal membrane separating interstitial water from the measuring system. A strained metal wire is attached to a middle point of the membrane; the membrane deflection causes changes in the wire tension. The wire is stressed in the middle point with an electromagnet and the oscillation frequency varying with the wire tension is measured. Starting from the frequency measured, the wire tension is obtained and, therefore, the interstitial pressure.
- Electric resistance transducer: It works in a similar way as the vibrating string transducers, but the parameter measured is the electric resistance of the wire. In particular, the instrument considers that the electric resistance varies with the conductor length, which varies according to the stresses acting on it.

8.7 Measures of Acoustic Emissions

The monitoring of acoustic emissions arising from the deformation induced in the rock masses by the opening of the cavity, allows to foresee the rockburst phenomenon thanks to the acoustic anomalies that precede it.

The change in the stress field due to excavation may trigger microfissures in the rock mass that originates the noises. The measuring system works in a well-defined frequency range. The acoustic emissions detected by a pressure transducer are transformed in amplified and filtered electric impulses. All frequencies outside the working range are considered as sounds not produced by the rock. The definition of the reference range is difficult and it has to be evaluated case by case according to the characteristics of the rock involved in the excavation.

Acoustic emissions are counted, recorded and printed electronically. The trend of the broken line graph represents the number of emissions per time unit. According to that parameter, the ground stability is assessed.

The equipment for the acoustic monitoring is usually installed in continuum in the tunnel stretches with higher rockburst risk. In case of shallow overburden, the installation takes place in correspondence of the connections between massive rock and the damaged zone of the faults, or where deformation behaviours are observed that usually precede the rock burst (spalling, splitting, slabbing).

8.8 Monitoring in Excavation by TBM

During the advancing of a TBM, the previously described monitoring can be carried out as well as systematic monitoring represented by following activities:

- Survey of the machine parameters
- Geophysics surveys
- Geoelectrical surveys

The following paragraphs provide a short description of the procedures for the acquisition and elaboration of the obtainable data during the excavation by means of a cutter head.

8.8.1 Measure of the Machine Parameters

The most interesting data refer to:

- Daily working processes
- Advancing and functioning of the machine

The first ones concern the thrust, displacement and stopping times, the detail of the substitution of cutting tools, notes on the operations carried out at the working site.

The 'performance parameters' are recorded continuously from the data acquisition system situated on the TBMs that systematically provides following parameters:

- Rotating speed of the head (round/minute)
- Instant advancing speed (m/h)
- Head penetration (mm/round)
- Head thrust (t)
- Absorbed power (kW)
- Excavated volume (m^3)
- Specific excavation energy (kWh/m^3)

The last parameter is obtained from the changes in the absorbed power and advancing speed and it corresponds to the energy (or work) necessary to excavate rock volumes. Therefore, it represents an index of the strength opposed by the rock mass to the excavation.

A filter adjusted on the advancing intervals has to be used on the data gathered, managed and organized in the database in order to determine the average values on each thrust effected.

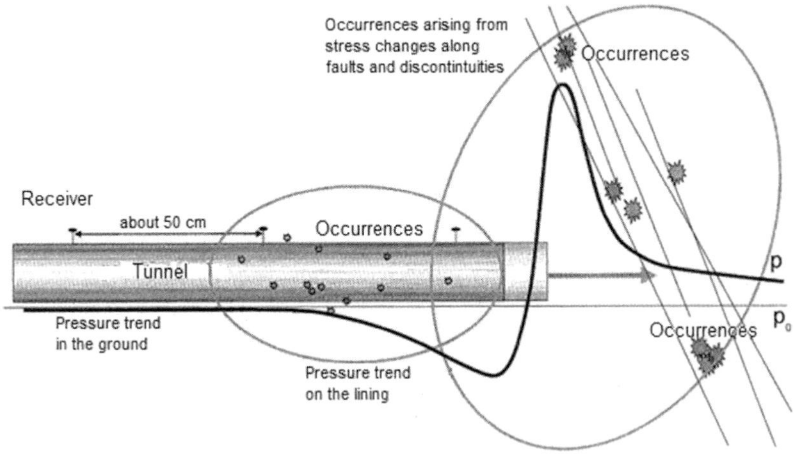

Fig. 8.20 Detection of events to forecast rock burst. (by DWT GmbH&Co, modified)

The data thus determined constitute the base for a series of elaborations aimed at comparing the rock masses with similar excavation behaviour, determining the geomechanical quality according to the interaction between the cutting head and the rock.

Usually, the elaboration of the cutting head parameters is carried out on three levels:

- Graph representation
- Statistic elaboration
- Determination of the geomechanical parameters

In the first level, the trend of the parameters recorded along the tunnel layout is represented; the second level consists of a statistical search for a correlation between the cutting head parameters and the geomechanical characteristics of the rock mass. The third level of elaboration occurs after the identification of the most meaningful correlations and it consists of the determination in continuum of those geomechanical parameters useful for the control of the design geomechanical model. This is possible, thanks to the data of the cutting head (Figs. 8.20 and 8.21).

If the functioning parameters of the TBM detect the crossing of rock areas with poor quality characteristics or with characteristics that are considerably different from those in the area where the TBM previously was, a face survey should be carried out, whenever possible.

8.8.2 Geophysical Seismic Surveys

If a TBM is used, it is particularly important to know the rock mass characteristics beyond the excavation face in advance to avoid crossing critical areas without

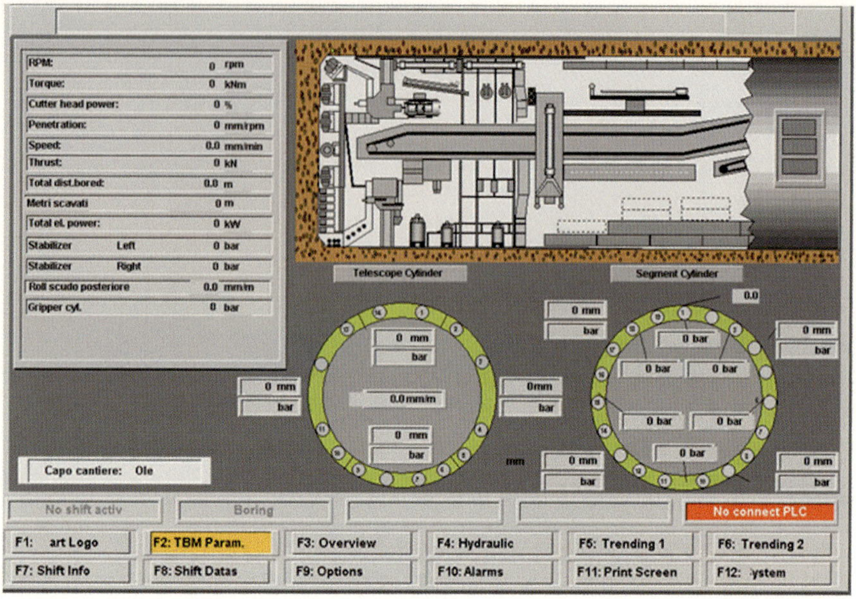

Fig. 8.21 Main screen of the programme managing the monitoring of the TBM parameters. (Monitor Pro software, by Schneider-electric)

a suitable preparation. But the usual speed of a mechanized excavation makes it difficult to acquire promptly the information using the more common techniques of direct survey (drilling tests). Therefore, indirect prospection methods by means of geophysics are often used.

In particular, seismic reflection allows to perform an indirect survey of the portion of the rock mass near the advancing face from the geological and geomechanical point of view. The results of the survey can contribute to the verification of the ground model where the tunnel is being excavated and, if necessary, to the update, thus helping in meeting prompt decisions about the need to adjust the advancing strategy to the real geomechanical conditions.

The survey provides a penetration capacity of some decametres; the overlapping of the surveyed areas has to be guaranteed and it is assured, thanks to perimetral measures. In particular, holes are created in the wall (at a distance of about 50 m, with length 2 m, diameter about 40 mm) in which the induced seismic signals near the face and reflected by the geological structure are measured through geophones or acceleration detectors. A measure section including at least three borings on the perimeter is equipped with three monocomponent sensors. At least three active sections are needed for a correct measure, and in general, six are used.

The energisation occurs every 5–6 m through explosive inserted in other holes having the same diameter and length. The amount of the charge needed depends on the kind of explosive and on the local conditions and it is decided experimentally at the beginning of the measuring campaign. As an alternative, and after adequate

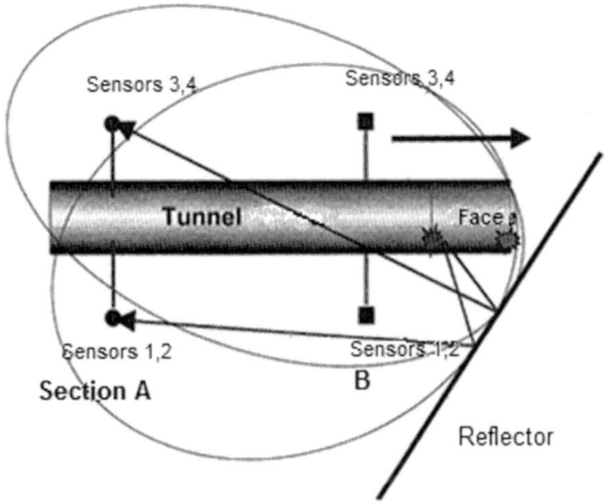

Fig. 8.22 Geometry of the seismic ahead survey—longitudinal profile. (By DWT GmbH&Co, modified)

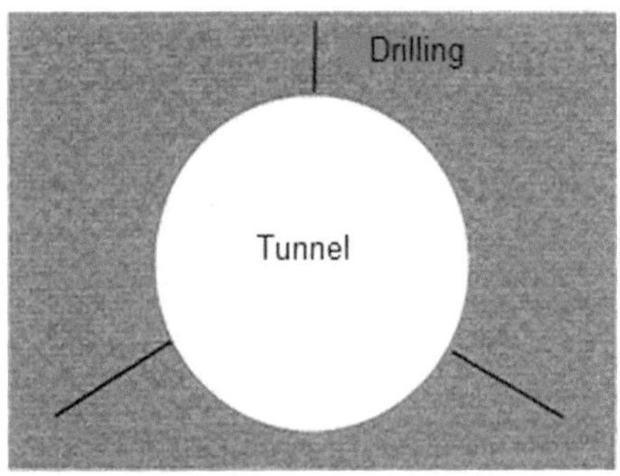

Fig. 8.23 Geometry of the seismic ahead survey—cross section. (By DWT GmbH&Co, modified)

tests, high frequency sources can also be used in order to reduce the work for the preparation and carrying out of the measures.

Each measurement of a new stretch of the tunnel can be carried out daily during the break for the maintenance operations. The time required just for the measurement (without drilling) is about 60–90 min. The data elaboration and the representation of the results take place every day.

As a standard result, the analysis provides the distance of the reflectors detected (fault zones, fractures, discontinuities etc.) from the advancing face with respect to the tunnel axis and the information about the azimut and the dip of the reflectors. Reflectors parallel to the tunnel axis cannot be detected (Figs. 8.22, 8.23, 8.24 and 8.25).

8.8.3 Geoelectric Surveys of the Cutting Head (Shielded TBM)

A continuous monitoring system while advancing can be implemented in order to protect the workers and the cutter head from sudden worsening of the stability conditions of the ground, due to unexpected change in the fracturing and/or alteration or unfavourable hydrogeological conditions from the head of a shielded TBM.

Fig. 8.24 Spatial positioning of the fault zones. (By DWT GmbH&Co)

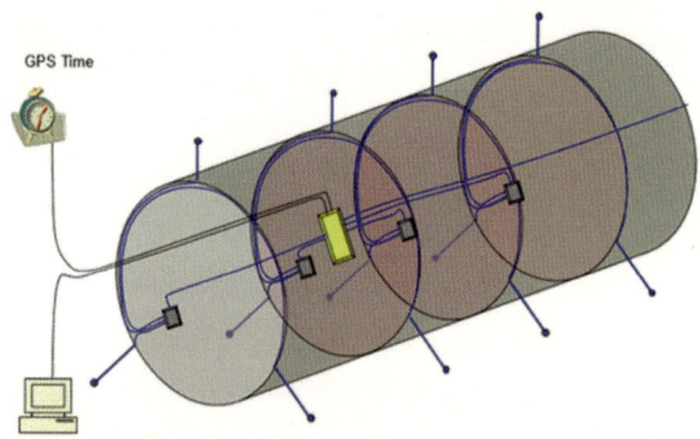

Fig. 8.25 Scheme of the coupled monitoring system rock burst + geoseismic (By DWT GmbH&Co). Equipment of the system detecting rockburst. The sensors and cables for the transmission of the signals from the sensors are in *blue*. The measure detector is in *yellow*. The data server and the timing server are located outside the tunnel and connected to the measure detector through cable and optic fibres (in *black*)

The survey methodology BEAM (Bore Tunnelling Electrical Ahead Monitoring) is a system constituted by the cutter head that emits a weak alternate power that enables to survey a rock prism for a depth of about 20 m beyond the cutter head. The cutter shield constitutes an electrode generating the power required to force the power toward the prism to be studied. The polarization induced in the rock is measured (expressed as percentage frequency effect: PFE) that represents the capability of the medium to store electricity and it is respectively linked to the medium porosity.

- PFE 0: reference value for water and air
- PFE ↓: highly fractured or karstic rocks or coarse soil
- PFE ↑: massive or not fractured rocks

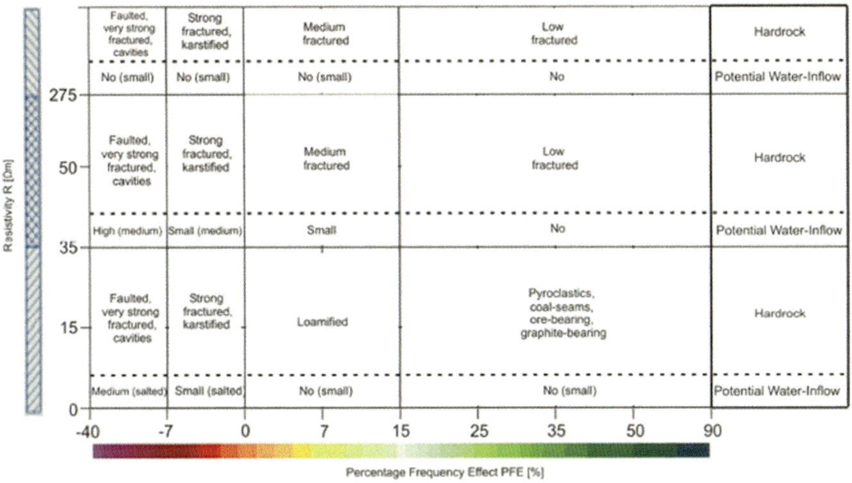

Fig. 8.26 Comparison between the PFE and Resistivity and characteristics of the rock mass. (By BEAM: Bore-tunnelling Electrical Ahead Monitoring)

Moreover, the resistivity of the crossed medium is measured, that parameter provides further information on the presence of fractures and cavities and on the possible presence of water (Fig. 8.26).

The BEAM system is composed by:

1. A BEAM unit placed in the TBM cabin; the measured values are represented graphically and made visible on the monitors (Fig. 8.27)
2. An electrode of measure placed on the cutter head, which detects the values during the excavation and pause phases
3. A surveying electrode placed on the shield
4. An electrode return signal to set a point with known resistivity and to gain control on the inner data

8.9 Surface Settlements and Surrounding Infrastructures Monitoring

The rock mass deformation due to the tunnel excavation, in particular in the case of shallow tunnels and portal areas, can reach the surface too. Moreover, the work processes outside the tunnel, as open-air excavations, can trigger settlements or undesired displacements of the surrounding ground. Excessive deformations of the ground can cause damages:

– To existing structures (very relevant aspect in case of tunnel excavation in urban areas)

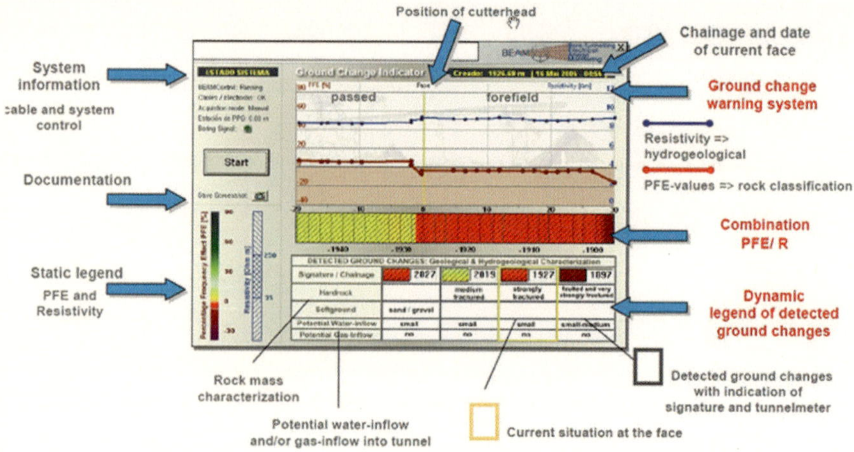

Fig. 8.27 The graph appearing on the display in the operator cabin synthesizes the recorded data. (By BEAM: Bore-tunnelling Electrical Ahead Monitoring)

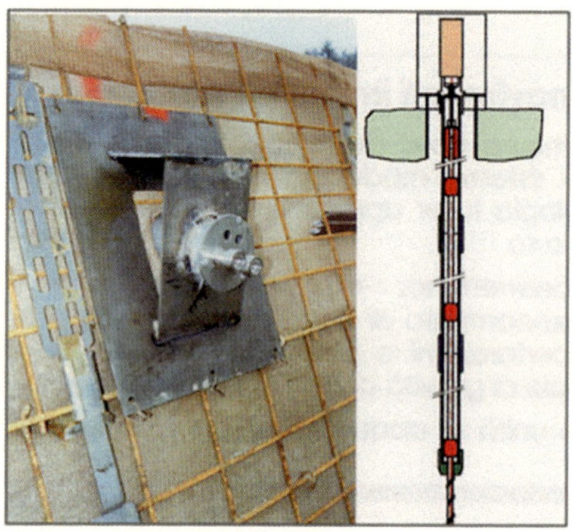

Fig. 8.28 Multibase extensometer to monitor the horizontal displacements of an entrance wall. (By Agisco S.r.l. Italy)

- To the structures in the working phase (for example, to the bulkheads in the portal areas)
- To the surrounding area (leading, for example, to the triggering of landslides)

Therefore, it is clear that a monitoring plan of the areas outside the tunnel represents an essential part of the design. Actually, the entity of the ground deformation, estimated in the design phase, has to be checked during the construction phase through the monitoring in order to confirm and, if necessary, adjust the measures included in the design.

The deformations of the ground can be monitored, for example, through the installation of optical targets, clinometers, inclinometers, incremental extensometers, multibase extensometers (Figs. 8.28, 8.29 and 8.30).

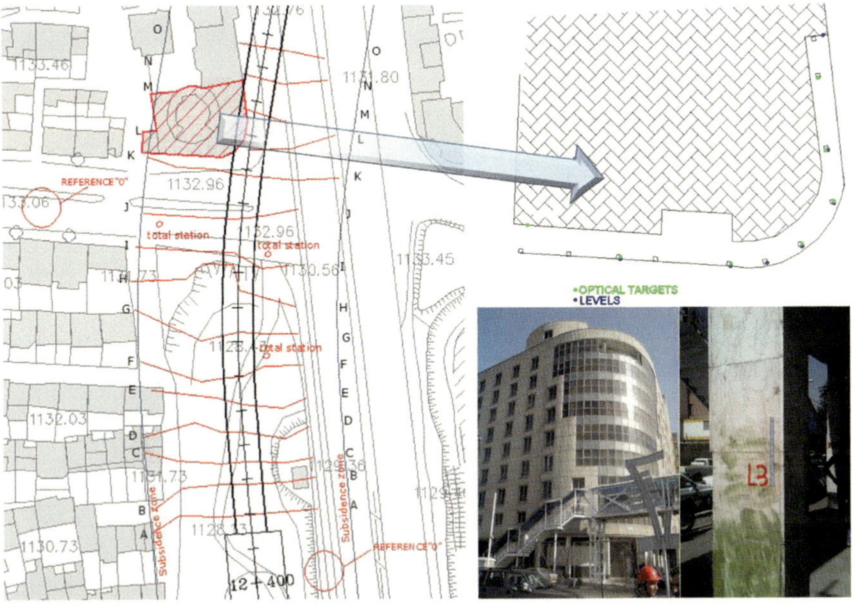

Fig. 8.29 Topographic measures of settlements in urban areas, induced by the excavation of a shallow tunnel with a TBM. In order to monitor the surface settlements and check their compatibility with the design, a number of sections have been identified, transversally to the TBM route, where topographic measuring instruments were placed. Figure shows the planimetric position of monitored sections. The tunnel layout passed under the foundation of an important building that was equipped with the monitoring instruments on the structure pillars

8.9.1 Settlement Gauges and Multibase Extensometers

Settlement gauges are measuring systems devoted to the monitoring of the absolute and/or differential displacements of the grounds. A settlement gauge works like an extrusometer or an incremental extensometer. The leading pipe is equipped with ring anchors at a distance of 1 m from each other; these are the measuring points. The second detector records the position of the different anchor points that move according to the settlements of the surface uplifts. The altitudes of the different anchors are confronted with those of the head that must be detected topographically.

The same can be said for the previously described multibase extensometers.

8.9.2 Inclinometers

Inclinometers measure the horizontal displacements of the ground. Generally, an inclinometric pipe is installed at the level of the portal area of natural tunnels (in an area that is not interested by the tunnel section) and in correspondence of peculiar points (buildings, existing infrastructures etc.).

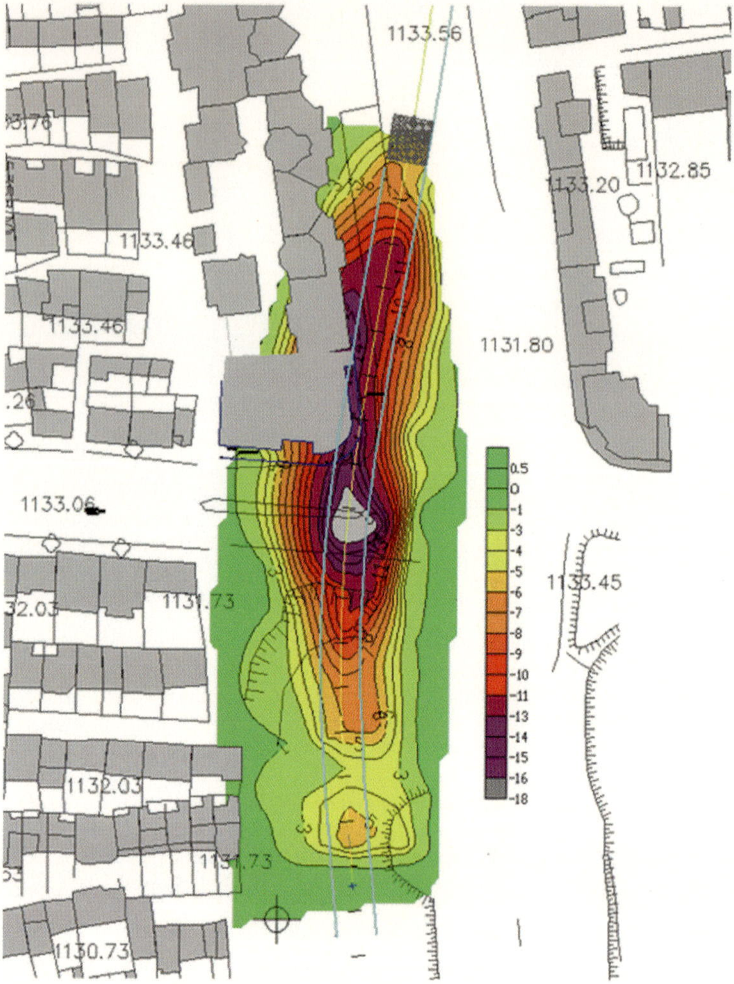

Fig. 8.30 Subsidence measured in the case described in Fig. 8.29

The pipe is equipped with reference and sliding guides disposed on two diameters, orthogonal to each other. The inner diameter of the pipe is about 8 cm. The inclinometric probe is lowered inside the pipe (Fig. 8.31).

The pipes have to be installed vertically in the ground through a borehole. The minimum diameter of the boring is about 14 cm. The dimensional regularity of the hole has to be guaranteed during the boring, avoiding material fall from the walls.

The gap with the verticality of the boring axis or with the positioning cannot be over 2 % and the space for a second 20-metre long measuring probe has to be guaranteed.

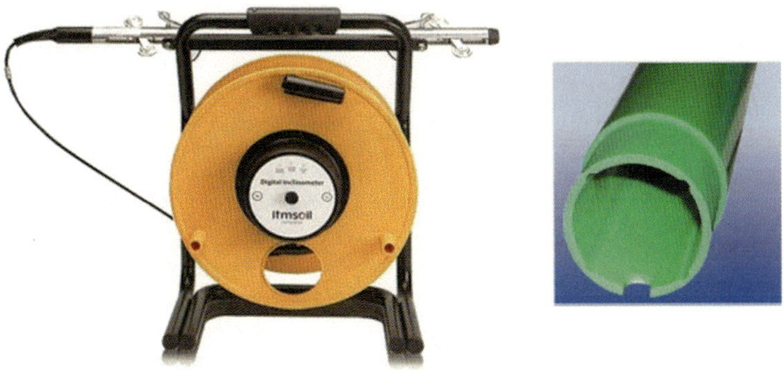

Fig. 8.31 Inclinometric probe and inclinometric pipe

Fig. 8.32 Inclinometer: working scheme

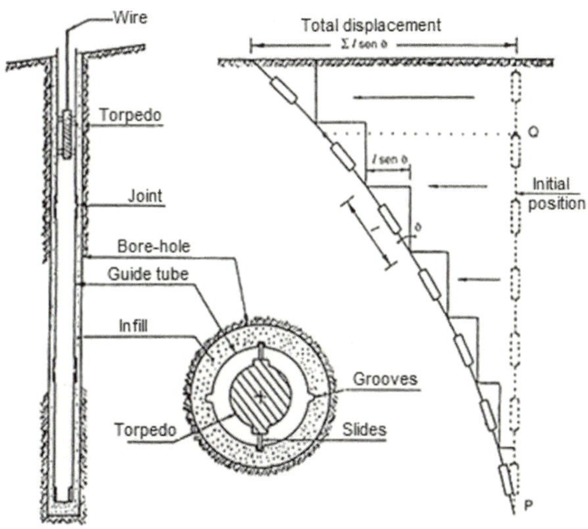

The installation of the pipe occurs assembling it while it is downed in the hole. All the pipe junctions have to be riveted (in the middle point between the measuring guides) and carefully sealed.

The measuring pipe has to be installed, keeping one of the pairs of measuring guides perpendicular to the work axis to avoid twisting. The cementing of the inclinometric pipe occurs through the injection of grout suitable to the characteristics of the ground (mixture of water, concrete, bentonite) through at least two injection pipes, one placed at the bottom and one at the middle of the hole. During the cementing phase, the injection pipe can be retrieved at regular intervals. Simultaneously, the inclinometric pipe is filled with clean water to reduce the external hydrostatic thrust produced by the grout. If the hole walls are not self-supporting, the lining is extracted afterwards, when cementation is carried out. During the setting of the grout, the material has to be topped up from the hole aperture if required. Then, a protection well is installed at the top of the hole, equipped with a carriageable manhole cover.

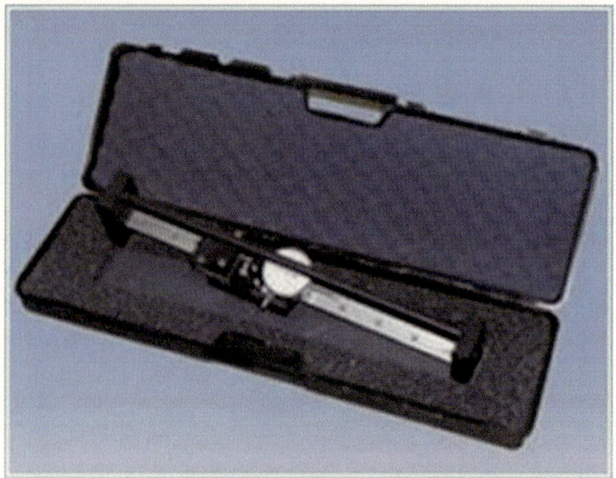

Fig. 8.33 Mechanical crack gauges

Fig. 8.34 Joint meter

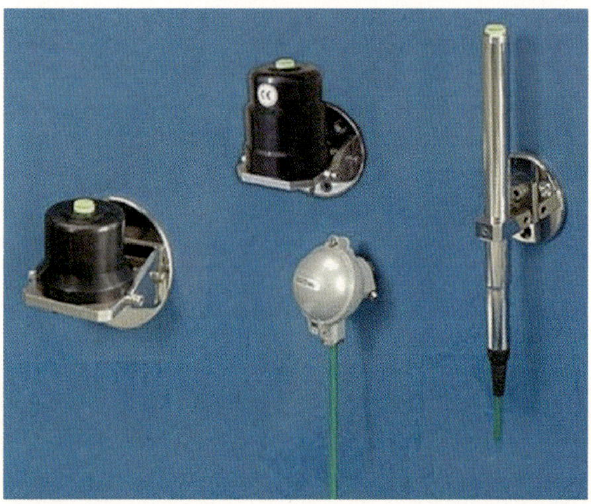

Fig. 8.35 Surface clinometers

Fig. 8.36 Surface clinometer

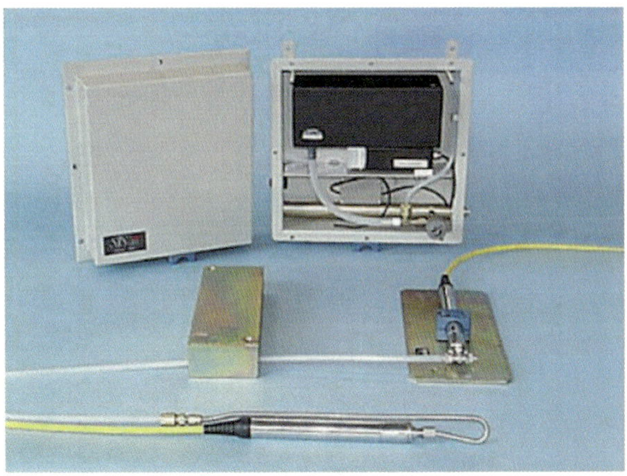

Fig. 8.37 Hydraulic settlement gauge (level metre)

Fig. 8.38 Level metre

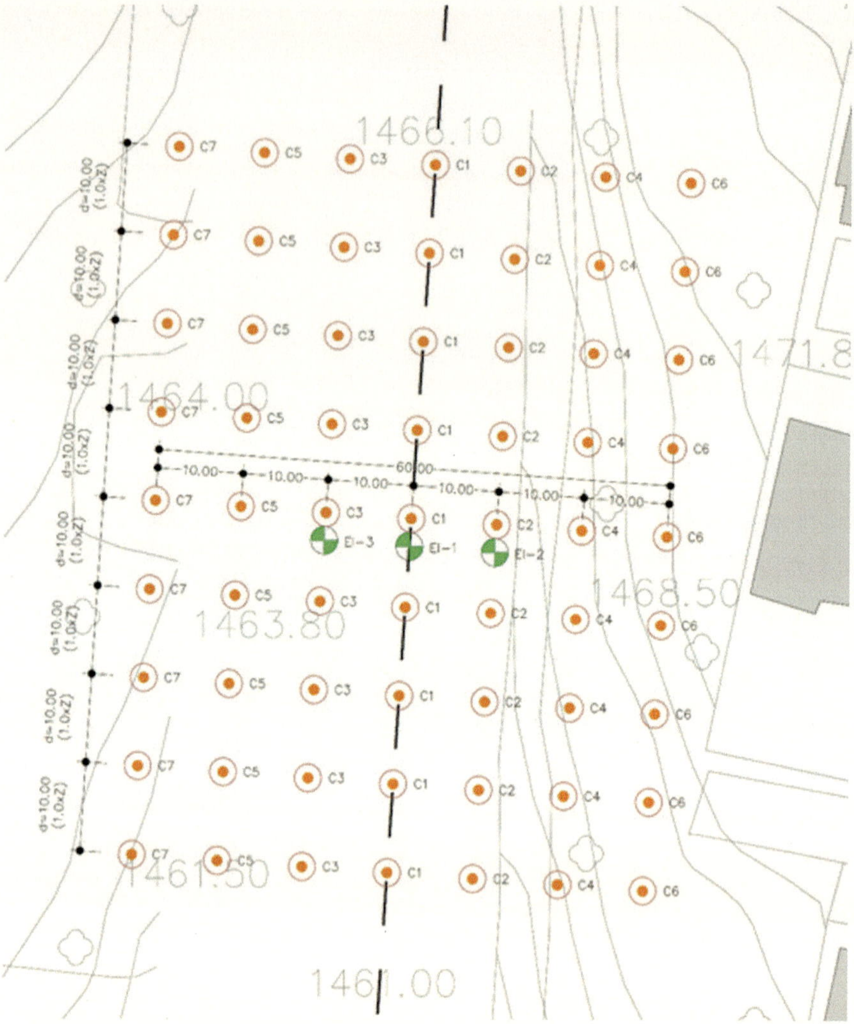

Fig. 8.39 Planimetric layout of benchmarks and strain gauges/clinometers

When the installation is completed, the functionality of the pipe has to be checked and the pipe has to be washed with clean water introduced from the bottom with a suitable pipe.

At each reading affected at a previously set interval, the outer temperature has to be recorded and an adequate thermal stabilization of the equipment in the hole has to be guaranteed. The first zero reading is performed after the cementation grout has set (and anyway at least 4 days after it).

For inclinometric measuring, the zero reading has to be performed proceeding upwards along the four groves. The exercise readings can be performed along the two groves that provided the lower average value of semideviation between opposed

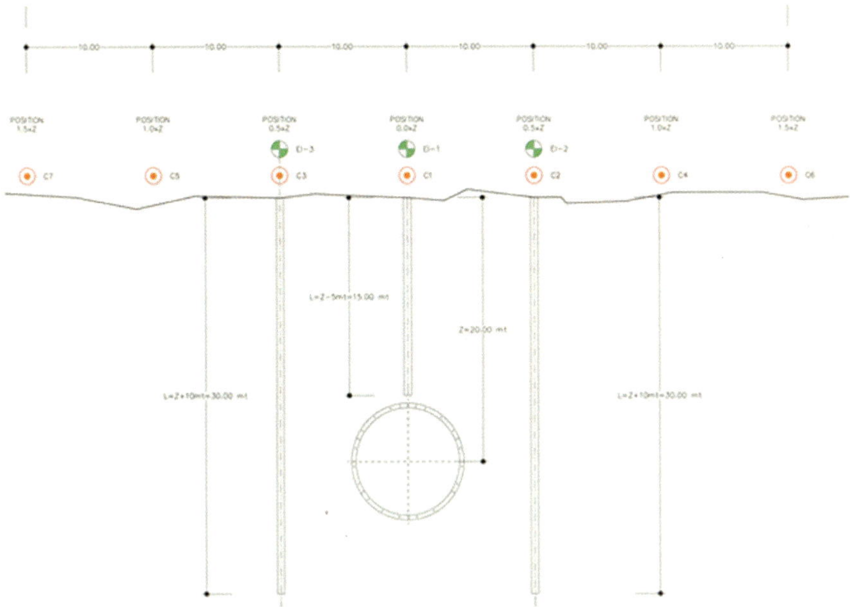

Fig. 8.40 Cross section with benchmarks and strain gauges/clinometers

readings. In case of anomalies or in presence of relevant deformation phenomena, exercise reading can be required on four groves on a single pipe (as for the zero reading).

Measuring data can be provided either in the form of instrumental reading or on paper (as a chart or graphs depth incremental and absolute displacement, Fig. 8.32).

8.9.3 Other Instruments for Buildings and Facilities Monitoring

The buildings and facilities whose foundations are, even only partially, within the subsidence area identified by the tunnel design shall be examined to verify the level of compaction before starting the works or before the potential interference caused by the advance face getting closer. Such verification will make it possible to identify any buildings already subjected to structural problems and therefore classifiable as particularly sensitive to the tunnel underpass. Through a census of deterioration before construction of the work, it will also avoid associating previous anomalies to the construction works for the metro line, which would alter the actual perception of the real behaviour of soils to excavation.

Any previous anomalies shall be documented in special sheets including pictures, a technical description of the compaction level and the kind of instruments installed to monitor any progress in the deformation of the building (mechanical crack gauges, joint metres; Figs. 8.33 and 8.34).

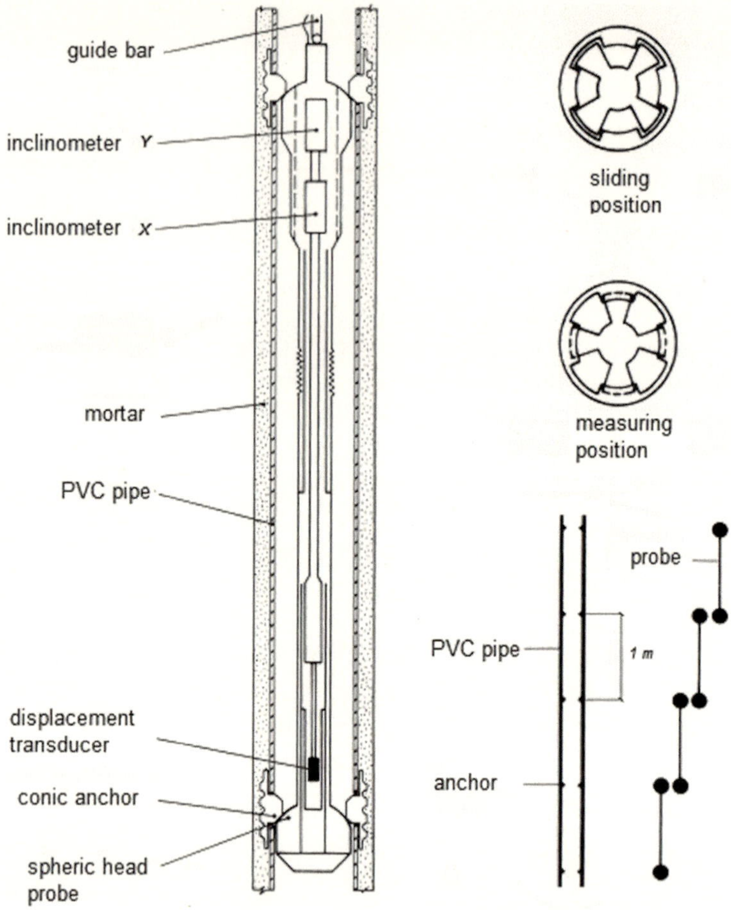

Fig. 8.41 Characteristics of the strain gauge/clinometer. Thanks to a special layout of the pipe, having both groves and anchoring points every 1 m, it is possible to carry out both inclinometric and settlement measures

For all the buildings standing, even only partially, at a distance between the tunnel axis and a distance equal to the tunnel depth (Z) from the axis itself ($0 \div Z$ [m]), displacement monitoring instruments shall be installed. They consist of:

- Clinometer sensors (surface clinometers) (Figs. 8.35 and 8.36)
- Hydraulic settlement gauge (level metre) and/or optical targets for the topographic survey (Figs. 8.37 and 8.38)

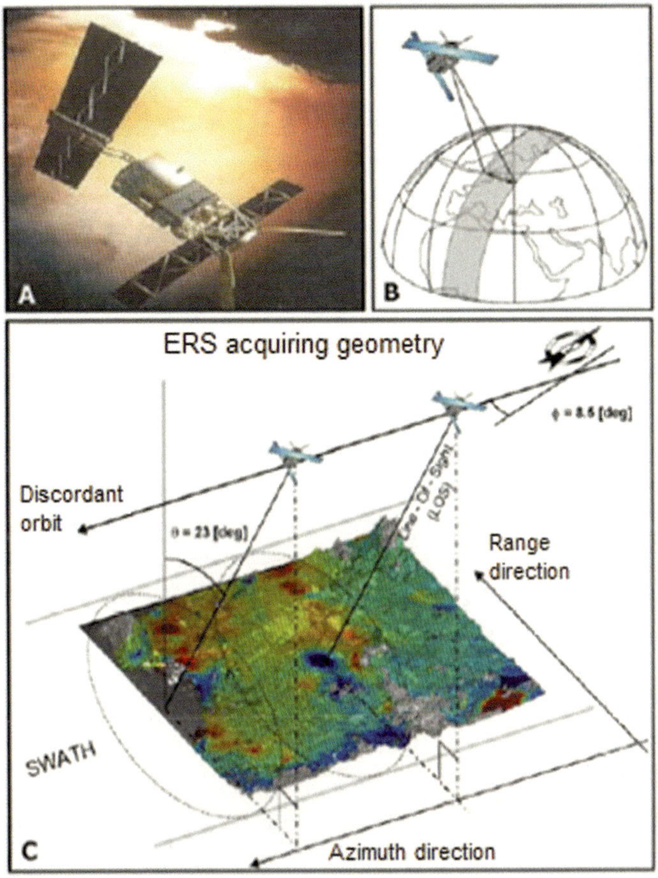

Fig. 8.42 Satellite acquisition geometry

8.9.4 Settlements Monitoring

Different instruments or measurements can be used contemporaneously for monitoring of the subsidence area:

- Topographic measurements of a network of benchmarks arranged on axis with and transversally to the route plan
- Installation of strain gauges/clinometers (Fig. 8.41) arranged on axis with and transversally to the route plan, at given sections of the network of topographic benchmarks

The topographic monitoring of the subsidence area progression based on the excavation face advance should be extended to the areas whose natural surface is—even only partially—free of facilities (roads, residential buildings, constructions, factories etc.).

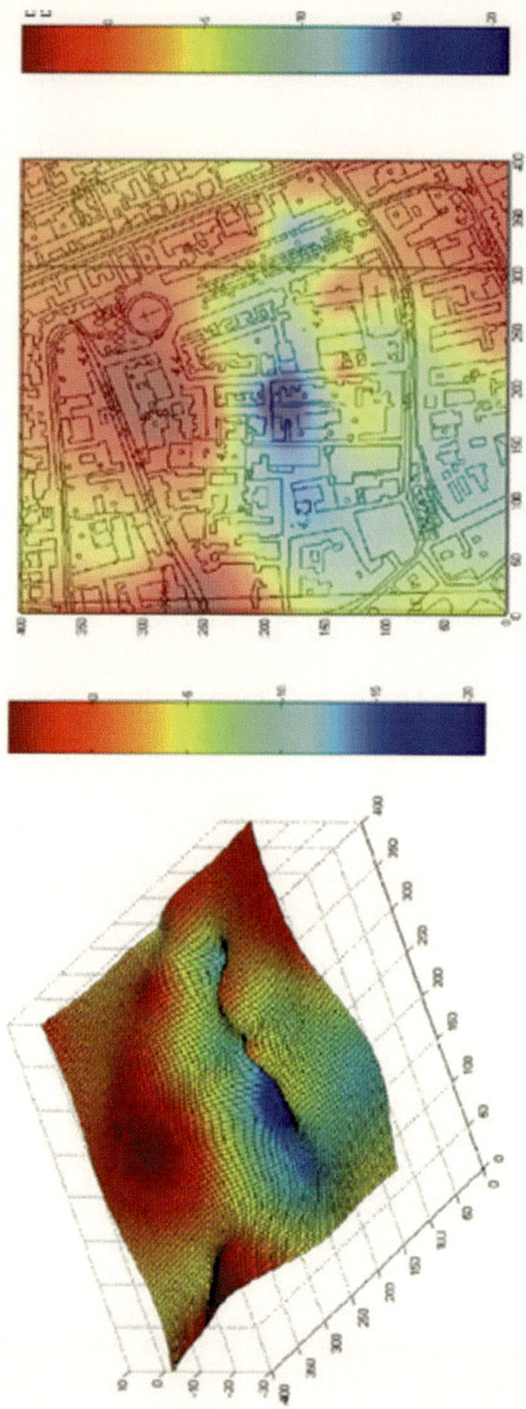

Fig. 8.43 Output of an analysis carried out with the PSInSAR technique

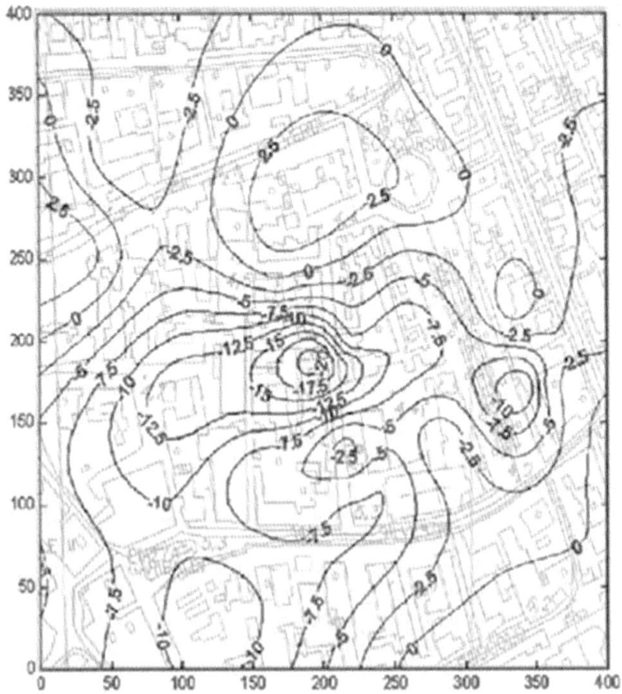

Fig. 8.44 Settlements isolines

Figures 8.39 and 8.40 show an indicative layout of the network of surface benchmarks, including a cross section equipped with three strain gauges/clinometers. Based on the depth of the tunnel centre with respect to the natural surface (Z), the monitoring instruments have to be installed transversally at the following distances from the tunnel centre (Fig. 8.40):

- Topographic benchmarks: 0.0Z, 0.5Z, 1.0Z, 1.5Z
- Strain gauges/clinometers: 0.0Z, 0.5Z

The longitudinal spacing of the cross sections is approximately equal to 0.5Z.

If the tunnel passes under a highly urbanized environment, the subsidence area cannot be monitored continuously as the work proceeds; in addition, in various stretches the network of instruments can be installed only partially. The data not available can be replaced by the data resulting from the displacement measurements of known points (buildings, facilities, trusses, pylons etc.) that can be gathered using the satellite network. Thanks to the combination of the displacement measurements by means of optical and satellite systems, it is possible to fully check the deformation effect of the excavation soils, as well as the covering effect on the surface and the structures involved by the tunnel construction.

The PSlnSAR technique (permanent scatters SAR interferometry technique) for processing the satellite data is an Italian patent (Tele-Rilevamento Europa:

T.R.E., www.treuropa.com), and represents a proven improvement of the traditional techniques of SAR (synthetic aperture radar) satellite interferometry.

The operating principle consists of reading the electromagnetic waves transmitted by a satellite and reflected by the earth surface, which is also partially received by the transmitting (as well as receiving) station (Fig. 8.42).

This technique is particularly effective when checking subsidence in the urban environment, since readings relate to a subgroup of radar targets (PS, permanent scatters) that already exist on site, consisting of cornices, trusses, concrete blocks or rock outcrops, particularly suitable to generate reflecting elements. The lapse of time between transmission and reception makes it possible to evaluate the distance between target and emitter; if there are any differences in the return signal between two consecutive measurements, it should be verified if the target has moved during a given period of time.

The results can be represented on maps (interferograms) highlighting the smallest millimetric differential movements of the area under examination (Fig. 8.43).

In urbanized areas, thanks to the high spatial density of radar targets (PS), the described technique (PSlnSAR) provides displacement measurements with great accuracy (Fig. 8.44).

The images in this chapter show, as an application example, the outputs of an analysis carried out with the PSlnSAR technique in Italy, relating to subsidence in an area where deep excavations are carried out to construct underground car parks in an urban environment.

Index

A
Acoustic emissions, 265, 281, 282
Aggressive waters, 2, 19, 20
Agliardi, F., 38
Agostinelli, G., 3
Anchors, 74, 180-191, 272, 273, 289
Angerer, L., 227
Aquifer, 19, 41, 43, 45, 56, 58, 133, 191, 278
Arch effect, 5, 194, 195
Asbestos, 25, 28, 45, 46

B
Barla, G., 31, 39, 112
Barla, M., 31
Barton equation, 91-93
Barton, N., 75-81, 92, 113
Basic Geotechnical Description, 4
Benussi, G., 108
Bernardos, A.G., 151
Bhasin method, 116
Bhasin, R., 116, 127
Bieniawski classification, 63-67
Bieniawski, Z.T., 64, 66, 76
Block theory, 104-106
Bologna-Firenze (Italy), 38
Bolts, 67, 74, 182, 190, 204, 205, 212
Bonini, M., 22
Bottino, G., 46
Brady, B.H.G., 206
Brittle deformation, 9
Brown, E.T., 11, 81, 87, 107, 120, 206, 248

C
Cables, 190, 273
Cadoppi, P., 38
Caquot method, 131
Caquot, A., 219, 221, 222
Ceiling, 5, 8, 27, 189, 194, 196, 200, 218, 221-223, 225, 228, 229, 240, 273
Characteristic lines, 97, 106-109, 233, 245-249
Civita, M., 40
Cohesion, 4, 16, 65, 79, 85, 91, 93, 102, 107, 121, 127, 128, 172, 191, 194, 221, 222, 233-236
Collapse, 8, 31, 32-34, 103, 104, 114, 147, 149, 150, 153, 155, 158, 169, 176, 246, 257
Collomb, D., 158
Conceptual model, 43, 53, 97, 148
Confinement, 8, 22, 25, 85, 97-99, 103-105, 107, 111, 122, 123, 130, 161, 185, 190, 201, 209, 232, 236-238, 261, 265
Convergence, 98, 99, 104, 107, 109, 111, 112, 117, 128, 129, 138, 141, 198, 206, 235, 245, 246, 261, 266, 267, 273, 275
Curve method, 129
Cutter parameter, 167, 180, 285
Cutter Soil Mix (CSM), 180

D
Decision Aid in Tunneling (DAT), 157
Deep-Seated Gravitational Deformation, 38
Deere, D., 174
Deere, D.U., 83, 84
Deformation, 1, 5, 12, 15, 22, 32-34, 45, 60, 62, 67, 81, 97, 99, 101, 103, 105, 107, 112, 114, 119, 137, 173, 180, 205, 218, 235, 249, 257, 258, 262, 265, 274, 277, 281, 282, 287, 295
Dematteis, A., 42, 43
Dilatancy, 92, 96, 102, 103, 137
Discontinuity, 8, 91, 93, 94, 104-106, 111, 183
Discontinuous medium, 98
Distinct Element Method, 111
Dormieux, L., 123, 124
Drainage, 36, 112, 136, 191, 192, 229

P. Gattinoni et al., *Engineering Geology for Underground Works*,
DOI 10.1007/978-94-007-7850-4,

Draining process, 43, 58, 133, 135
Ductile deformation, 1, 5, 9, 59, 102
Dust, 28, 29, 31, 46, 203

E

Einstein, H.H., 157
Entrance, 288
Environmental impact, 31, 46, 145
Equivalent continuum, 98, 111
Excavation, 32, 36, 47, 97-105, 123, 161-169, 243, 267, 282-285
Extrusion, 112, 272, 273

F

Face
  deformation, 112
  stability, 118, 121, 127-129, 169, 243, 261
  stabilization, 193
Faults, 7, 8, 12, 18, 23, 41, 282
Fechtig, R., 24, 165
Federic, F., 135, 136
Fillibeck, J., 227
Final lining, 19, 198, 200, 206, 207, 213, 219, 235, 261, 265, 266, 275
First stage lining, 198, 201-205, 266
Focaracci, A., 41
Folds, 7, 8, 18
Foliated rocks, 9
Forepoling, 161, 188, 192, 194, 195, 241
Freezing, 130, 178-180
Friction, 8, 85, 87, 91, 93, 106, 123, 127, 205, 221, 244

G

Gas, 25, 27, 28, 60, 179, 180, 280
Gattinoni, P., 34, 43, 134, 145
Geoelectric, 59, 282, 285
Geognostic test, 59
Geological risk, 43, 143, 147, 148
Geological Strength Index, 85
Geomechanical survey, 54, 58, 91, 150, 265-267
Geoseismic, 57
Geothermal gradient, 2, 22, 23
Gesta, P., 102, 103
Gisotti, G., 42
Goel, R.K., 113, 224
Golser, J., 215
Gonzalez de Vallejo, L.I., 54, 59, 62, 70
Goodman, R.E., 93, 105, 134
Grain material, 123
Gran Sasso (Italy), 38, 40
Ground Reaction, 107, 129, 209, 261
Ground-structure interaction
  forepoling, 241
  horizontal loads, 225
  hyperstatic reaction, method of, 216
  inclined loads, 226
  nailing, 231
  Rabcewicz theory, 215
  rock masses, 223
  spiling, 238
  vertical loads, 219
Groundwater, 2, 8, 16, 18, 19, 31, 32, 35, 55, 56, 64, 128, 176, 192, 229, 251, 280
  influence, 132-141
  interaction with surface water, 38-45
Guenot, A., 107, 108, 250, 252, 253
Guo, C., 108

H

Hard rocks, 4, 119, 120
Hazard, 132, 145, 147, 149, 159, 161, 184, 191
Hazardous material, 25-29
Herget, G., 11
High overburden, 55, 60, 98, 104, 117
Hoek, E., 11, 81, 84, 85, 87, 88, 91, 93, 95, 99, 100, 107, 114-117, 119, 120, 127, 248, 250
Hoek-Brown constitutive model, 63, 81
Homogenization method, 233, 234
Horizontal loads, 216, 225, 226
Houska, J., 225-227
Hudson, J.A., 43, 149
Hydrogeological setting, 18, 41, 132
Hydro-Potential, 133

I

In situ cast concrete, 206
In situ stress state, 11-14
Inclined loads, 226, 227
Inclinometer, 55, 266, 288, 289
Inert waste, 45, 46
Inflow, 18-21, 38, 42, 43, 56, 67, 132, 133, 135, 136, 278
Injections, 31, 120, 131, 170-177, 180, 191, 194, 208, 209, 230
Invert, 25, 112, 198, 207, 218, 237, 249, 261, 275
Hyperstatic reactions, 216, 219

J

Jehtwa and Singh Method, 115
Jet-grouting, 161, 175-177, 180, 193, 195, 196
Jethwa, J.L., 224
John, M., 115, 116
Joint, 5, 10, 64, 69, 81, 93, 98, 105, 106, 133, 187, 205, 230, 231, 240, 292, 295

K
Kaliampakos, D.C., 149
Karst, 15, 18, 41, 43, 55
Kerisel, J., 219, 221, 222
Kihlström, B., 48
Kleinensiel/Dedesdorf tunnel (Germany), 46
Kolymbas, D., 134, 228
Kommerell, O., 219
Kovári, K., 24, 165

L
Ladany and Archanbault criterion, 91, 94-96
Lane, K.S., 246
Lanfranchi, P., 47
Langefors, U., 48
Léca, E., 123, 124
Lee, S.-W., 137
Lembo-Fazio, A., 139-141
Limit state, 150, 151
Lining
  first stage
    shotcrete, 28, 196-198, 201-205, 240, 248, 249, 261, 275
    steel ribs, 189, 197, 198, 203-205, 216, 240, 243, 248, 277
  final
    prefabricated, 164, 208-213
    single shell, 200, 213
Load
  horizontal, 216, 225, 226
  hydraulic, 137, 216
  inclined, 226, 227
  surface, 14, 216
  vertical, 219-223, 227, 228, 243
Loew, S., 41
Lombardi, G., 107, 130, 174
Lugeon parameter, 80
Lunardi, P., 41
Lyons, S., 47-49

M
Mt. Piazzo (Italy), 38
Macori, M., 108
Marinos, P., 82, 114, 115, 127
Modelling, 34, 54, 63, 70, 97, 98, 109, 132, 135, 215, 219, 231, 235, 250-252
Monitoring, 27, 56, 98, 146, 180, 215, 262, 265, 266, 272-276, 279-282, 284-289, 295-300
Monte Bianco (Italy), 23
Monte Carlo method, 154
MuirWood, A.M., 120

N
Nailing, 16, 183, 184, 187-189, 193, 195, 205, 231, 232, 234, 235, 237, 238, 249, 261, 262
Nails, 74, 180, 183, 185-190, 195, 233, 235, 236, 238, 275
Nguyen-Minh, D., 108
Nielsen, N.M., 156
Nilsen, B., 122
Noises, 47-49, 281
Numerical method, 109, 132, 250-253

O
Overthrust, 8, 18, 37, 41, 146
Owen, G.N., 26

P
Paleo-river, 20, 41
Palmstrom, A., 70, 84
Palmström, A., 69, 83
Panet method, 117-119
Panet, M., 100, 101, 107, 108, 123, 128, 250, 257
Patton criterion, 91, 92, 95
Pazzagli, G., 42
Peck, R.B., 34
Permeability, 14, 18, 28, 41, 43, 44, 56, 60, 80, 132, 135-137, 139, 140, 167, 171, 172, 279
Picarelli, L., 39
Piezometer, 55, 278-281
Pilot tunnel, 79, 175, 176
Piovano, G., 49
Plasticity, 81, 82, 88, 101, 127
Pollution, 31, 32, 38, 46, 191
Pont Ventoux (Italy), 38
Porosity, 18, 41, 135, 171, 173
Portal, 14, 15, 31, 36, 53-55, 190, 198, 287-289
Precast segment, 167, 208-210, 212, 275
Precutting, 161, 164, 192, 196, 198, 199
Prefabricated lining, 164, 208-213
Pressure
  confinement, 104, 107, 122, 123, 161, 185, 190, 201, 232, 236, 237
  hydrostatic, 137, 167, 230
  interstitial, 140, 191, 192, 279-281
  radial, 99, 100, 107, 108, 130, 131, 137, 216, 229
Pressurized TBM, 164, 167, 208, 209
Pretunnel, 161, 164, 193, 198

Q
Q-system, 77, 80, 81

QTBM classification, 81

R
Rabcewicz, L.V., 215, 216, 219
Rabcewicz's theory, 219
Radial deformation, 273, 274
Radioactivity, 2, 25, 27
Radon, 25, 27, 28
Random fabric, 9
Reinforced Protective Umbrella Methods (RPUM), 192-198
Ribacchi, R., 107, 140, 141
Riccioni, R., 107, 250
Risk
  analysis, 97, 145, 148, 149, 154-157
  assessment, 145, 148, 149, 159
  control, 161
  evaluation, 145, 155, 156
  management, 143, 144, 146, 148
  mitigation, 146, 149, 161, 164, 166
Rock burst, 1, 119, 120, 190, 282
Rock Engineering System (RES), 43, 149
Rock mass, 2, 3, 5, 7, 8, 10, 22, 42, 56, 62, 70, 74, 81, 85, 87, 89, 91, 97, 98, 105,-107, 109-112, 115, 117, 119, 129, 130, 133, 140, 150, 189, 192, 266, 282, 283, 287
Rock Mass Excavability Index (RME), 63, 67, 68
Rock Mass index (RMi), 63, 69
Rock Mass Rating (RMR), 63, 267
Rossi, S., 38
Russenes, B.F., 120
Rybach, L., 42

S
Safety factor, 124, 127, 128, 131, 150-152
Sakurai, S., 129
San Pellegrino Terme (Italy), 39
Scale effect, 10, 11
Scesi, L., 43, 134, 143
Scholl, R.E., 26
Schuck,W., 229, 230
Seepage, 41, 132, 137, 228, 229
Seidenfuß, T., 33
Seismic, 23, 25, 55-59, 80, 216, 253, 257-259, 284
Seismic effects, 23, 216
Selmer-Olsen, R., 120
Sempione (Italy-Switzerland), 23, 40
Settlement gauges, 289
Shallow-overburden, 55
Shear strength, 5, 22, 65, 74, 91-95, 98, 126, 128, 176, 183, 194, 213, 215, 216, 226
Sheory, P.R., 11
Shi, G., 105
Shielded TBM, 164, 166, 167, 285
Shotcrete, 28, 68, 117, 162, 163, 195-198, 200, 203-205, 213, 216, 240, 248, 249, 261, 275, 277
Sibson, R.H., 8
Singh, B., 112, 113
Singh, R., 83, 84
Single-shell lining, 200, 214
Slabbing, 120, 146, 163, 282
Slope instability, 36-38
Soil, 8, 23, 25, 28, 34, 36, 56, 62, 81, 99, 101, 164, 167, 169-171, 173-180, 192, 229, 235, 238
Sorlini, A., 49
Spalling, 120, 146, 223, 282
Speakman, C., 47-49
Spiling, 238-240
Splitting, 120, 282
Spring, 41, 42, 146, 218
Spritz beton see Shotcrete
Squeezing, 111-113, 115, 119, 127, 206, 224
St. Gotthard Base Tunnel (Italy- Switzerland), 47
Steel rib, 38, 189, 195-198, 202-205, 216, 224, 240, 243, 248, 249, 277
Steel sheet, 208, 210, 277
Strain, 25, 114, 115, 232, 265
Strain gauge, 274-277, 294-297, 299
Strength, 10, 63, 91-96, 127, 128, 261
Stress1, 5, 8, 10-14, 74, 76, 118, 129, 137, 140, 180, 229, 257, 258
Strozzi, T., 38
Supper, R., 38
Support, 16, 67, 70, 100, 107, 112, 114, 166,169, 215, 241, 246, 248, 249, 266
Surface Rock Classification (SRC), 63, 70
Surface settlements, 31, 32, 132, 166, 265, 287
Survey
  geological, 8, 28, 54
  geomechanical, 54, 58, 91, 150, 265-267
  geoseisimic, 57
  hydrogeological, 56-58
Swelling, 2, 21, 22, 74, 112, 176

T
Tunnel-boring machine (TBM), 38, 63, 67-69, 81, 164, 166, 167, 169, 208-210, 265, 275, 282, 283, 287

Terzaghi, K., 219, 223-225
Thidemann, A., 122
Tiebacks, 183, 190, 191
Time-dependent behaviour, 111-119

U
Unal, E., 224

V
Vault, 195, 197, 207, 212
Venturini, G., 38
Vertical loads, 219-224
Vibration, 47
Vincenzi, V., 38
Vulnerability, 23, 25, 32, 147

W
Wagner, P., 134
Wannick, H.P., 33
Water table, 19, 41, 123, 127, 132, 135, 136, 139, 169, 178, 191, 224, 228, 279
Waterproofing, 197, 200, 201, 207-211, 213, 230
Weak rocks, 4, 81, 104, 116, 180, 196
Weathering, 2, 3, 14, 21, 22, 59, 64, 74, 85
Wilhelm, J., 42

Printed by Printforce, the Netherlands